MONOGRAPHS ON STATISTICS AND APPLIED PROBABILITY

General Editors

V. Isham, N. Keiding, T. Louis, N. Reid, R. Tibshirani, and H. Tong

Topics in Modelling of Clustered Data

Edited by
Marc Aerts
Helena Geys
Geert Molenberghs
Louise M. Ryan

CHAPMAN & HALL/CRC

A CRC Press Company
Boca Raton   London   New York   Washington, D.C.

Library of Congress Cataloging-in-Publication Data

Topics in modelling of clustered data / edited by Marc Aerts ... [et al.].
    p.   cm. — (Monographs on statistics and applied probability ; 96)
Includes bibliographical references and index.
ISBN 1-58488-185-2 (alk. paper)
1. Cluster analysis. I. Aerts, Marc. II. Series.

QA278 .T665 2002
519.5′3—dc21   2002023758
                                                                        CIP

Visit the CRC Press Web site at www.crcpress.com

To Els, Anouk, and Lauren

To Geert and Sara

To Conny, An, and Jasper

To Nick, Nick, and Sarah

Contents

List of Contributors

Marc Aerts *Limburgs Universitair Centrum* and *transnationale Universiteit Limburg*, Diepenbeek, Belgium

Paul J. Catalano *Harvard School of Public Health*, Boston, Massachusetts

Gerda Claeskens *Texas A & M University*, College Station, Texas

Chris Corcoran *Utah State University*, Logan, Utah

Lieven Declerck *European Organization for Research and Treatment of Cancer*, Brussels, Belgium

Christel Faes *Limburgs Universitair Centrum* and *transnationale Universiteit Limburg*, Diepenbeek, Belgium

Helena Geys *Limburgs Universitair Centrum* and *transnationale Universiteit Limburg*, Diepenbeek, Belgium

Geert Molenberghs *Limburgs Universitair Centrum* and *transnationale Universiteit Limburg*, Diepenbeek, Belgium

Meredith M. Regan *Beth Israel Deaconess Medical Center* and *Harvard Medical School*, Boston, Massachusetts

Didier Renard *Limburgs Universitair Centrum* and *transnationale Universiteit Limburg*, Diepenbeek, Belgium

Louise Ryan *Harvard School of Public Health*, Boston, Massachusetts

Paige L. Williams *Harvard School of Public Health*, Boston, Massachusetts

Editors' Biographical Sketches

Marc Aerts is Professor of Biostatistics at the Limburgs Universitair Centrum in Belgium. He received the PhD in Statistics from Limburgs Universitair Centrum. Dr. Aerts published methodological work on non- and semiparametric smoothing and bootstrap methods, quantitative risk assessment, etc. Dr. Aerts is director of the *Master of Science in Applied Statistics* program. He serves as an associate editor for *Statistical Modelling*.

Helena Geys is postdoctoral researcher at the Limburgs Universitair Centrum in Belgium. Her research is supported by the Institute for the Promotion and Innovation of Science and Technology in Flanders and by DWTC, Belgium. She received MSc and PhD degrees in Biostatistics from the Limburgs Universitair Centrum. Dr. Geys published methodological work on clustered non-normal data with applications in developmental toxicity, risk assessment, and pseudo-likelihood inference. She serves on the editorial panel of *Archives of Public Health* and has worked on several environmental health related projects.

Geert Molenberghs is Professor of Biostatistics at the Limburgs Universitair Centrum in Belgium. He received a PhD in Biostatistics from the Universiteit Antwerpen. Dr. Molenberghs published work on the analysis of longitudinal and incomplete data, on repeated and clustered categorical data, and on the evaluation of surrogate endpoints in clinical trials. He is director of the Center for Statistics. He serves as joint editor of the *Journal of the Royal Statistical Society, Series C* and is associate editor of *Biostatistics*. He is an officer of the International Biometric Society (IBS), of the Quetelet Society, the Belgian Region of the IBS, and of the Belgian Statistical Society. He is member of the Dutch Health Council. He has held visiting positions at the Harvard School of Public Health (Boston, Massachusetts).

Louise Ryan is Professor of Biostatistics at the Harvard School of Public Health in Boston. She obtained MSc and ScD degrees in Biostatistics from Harvard. Dr. Ryan works on statistical methods related to environmental risk assessment for cancer, developmental and reproductive toxicity, and other non-cancer endpoints such as respiratory disease. She also works on cancer clinical trials and epidemiological methods for the study of birth defects and adverse reproductive outcomes, with emphasis on multiple outcomes. Dr. Ryan has a strong interest in community based environmental health research. Dr. Ryan also works with the Eastern Cooperative Oncology Group on the design, implementation, analysis and reporting of clinical trials for a variety of different types of cancer. Dr. Ryan is a fellow of the American Statistical

Association and in the International Statistics Institute. She was co-editor of *Biometrics* and president of the Eastern North American Region of the IBS. She has served on advisory boards for several government agencies, including the National Toxicology Program and the Environmental Protection Agency, as well as several committees for the National Academy of Science.

Preface

This book has grown out of research undertaken at the Department of Biostatistics of the Harvard School of Public Health and the Dana-Farber Cancer Institute in Boston on the one hand and at the Limburgs Universitair Centrum (transnational University Limburg) in Belgium on the other hand, in close collaboration with a number of colleagues located at various institutions. Research interests in the modeling of clustered and repeated categorical data have been brought together with research in the modeling of data from toxicological experiments in general and developmental toxicity studies in particular.

Several local grants have provided the impetus to undertake this work, but an invaluable binding force has been the support from a NATO Collaborative Research Grant, allowing contributors to travel and meet.

While formally an edited volume, this work interpolates between a standard monograph and a collection of stand-alone contributions. The reasons for this choice are manifold.

First and foremost, while in most of the chapters at least one editor was involved, we have found the chosen form of authored chapters to be a fair way to give credit to our colleagues who have contributed in a generous fashion to the genesis of this book.

Second, the field of developmental toxicity is exciting and raises a large range of substantive and methodological questions. Nevertheless, the methodology presented here has much wider ramifications than just this field of application. Therefore, a modular concept seemed most appropriate. In particular, the motivating examples have been collected in a separate chapter, followed by chapters on model building and on particular estimation procedures (generalized estimating equations and pseudo-likelihood). The "modeling chapters" have been written with a general clustered or even correlated data setting in mind and are therefore of use far beyond the developmental toxicology context. In later chapters, specific issues have been tackled. Some of these are rather particular to toxicology and dose-response modeling (e.g., the chapters on quantitative risk assessment and exact dose-response inference) while others are more general in scope (e.g., the chapters on goodness-of-fit, model misspecification, individual level covariates, and combined continuous and discrete outcomes). In this way, both the very focused as well as the more broadly interested reader will have no difficulty selecting the material of interest to her. To underscore the large potential of methods for clustered data, a chapter has been included on the analysis of clustering effects in complex survey data.

Third, we have chosen to "strongly edit" the text to achieve a smooth flow, in spite of the multitude of chapter authors that have contributed. As far as possible, a common set of notation has been used by all authors. The editors have provided ample cross-references between chapters, not only to refer back, in later chapters, to motivating examples and models, but also to point to cross links between chapters. Thus, the book should be suitable to either read a selected number of chapters or the integral text.

In our choice of topics, we have tried to cover a wide variety of choices. Asymptotic inference is contrasted with exact and simulation-based (bootstrap) methods. Classical maximum likelihood is supplemented with such alternatives as generalized estimating equations and pseudo-likelihood methods. Not only fully parametric modeling, but also semi-parametric and non-parametric methods are discussed. Nevertheless, it is impossible to study all available techniques within the scope of a single text. For example, apart from brief mention, no thorough study has been undertaken of Bayesian methodology. We believe that this highly interesting but specialized area would deserve separate treatment.

We hope the book gives pleasure and satisfaction to the more methodologically interested as well as to the substantially motivated reader.

Marc Aerts (transnationale Universiteit Limburg, Diepenbeek-Hasselt)
Helena Geys (transnationale Universiteit Limburg, Diepenbeek-Hasselt)
Geert Molenberghs (transnationale Universiteit Limburg,
Diepenbeek-Hasselt)
Louise Ryan (Harvard School of Public Health, Boston)

Acknowledgments

This book has been accomplished with considerable help from several people. We would like to gratefully acknowledge their support.

First and foremost, we would like to express our gratitude for the hard work and dedication of our contributors: Paul Catalano (Harvard School of Public Health, Boston, MA), Gerda Claeskens (Texas A & M University, College Station, TX), Chris Corcoran (Utah State University, Logan, UT), Lieven Declerck (European Organization for Research and Treatment of Cancer, Brussels, Belgium), Christel Faes (transnationale Universiteit Limburg, Diepenbeek, Belgium), Meredith Regan (Beth Israel Deaconess Medical Center and Harvard Medical School, Boston, MA), Didier Renard (transnationale Universiteit Limburg, Diepenbeek, Belgium), and Paige Williams (Harvard School of Public Health, Boston, MA).

We gratefully acknowledge support from *Fonds voor Wetenschappelijk Onderzoek Vlaanderen*, from *Instituut voor Wetenschap en Technologie*, and from NATO Collaborative Research Grant CRG950648: "Statistical Research for Environmental Risk Assessment".

We are grateful for generous access to various sources of data. In particular, we would like to mention the U.S. Environmental Protection Agency for kindly providing the data of the National Toxicology Program, G. Kimmel for the heat shock data, the Scientific Institute of Public Health–Louis Pasteur in Brussels for access to the Belgian Health Interview Survey 1997, the Project on Pre term and Small for Gestational Age Infant Data, and R. Klein of the University of Wisconsin, Madison, for kindly providing the data on the Wisconsin Diabetic Retinopathy Study NIH grant EY 03083).

We apologize to our partners and children for the time not spent with them during the preparation of this book and we are very grateful for their understanding.

Marc, Helena, Geert, and Louise
Diepenbeek and Boston, January 2002

List of tables

List of figures

CHAPTER 1

Introduction

1.1 Correlated Data Settings

In applied sciences, one is often confronted with the collection of *correlated data*. This generic term embraces a multitude of data structures, such as multivariate observations, clustered data, repeated measurements, longitudinal data, and spatially correlated data.

Historically, multivariate data have received the most attention in the statistical literature (e.g., Seber 1984, Krzanowski 1988, Johnson and Wichern 1992). Techniques devised for this situation include multivariate regression and multivariate analysis of variance. In addition, a suite of specialized tools exists such as principal components analysis, canonical correlation analysis, discriminant analysis, factor analysis, cluster analysis, and so forth.

The generic example of multivariate continuous data is given by Fisher's iris data set (e.g., Johnson and Wichern 1992), where, for each of 150 specimens, petal length, petal width, sepal length, and sepal width are recorded. This is different from a *clustered setting* where, for example, for a number of families, body mass index is recorded for all of their members. A design where, for each subject, blood pressure is recorded under several experimental conditions is often termed a *repeated measures* study. In the case that body mass index is measured repeatedly over time for each subject, we are dealing with *longitudinal data*. Although one could view all of these data structures as special cases of multivariate designs, there clearly are many fundamental differences, thoroughly affecting the mode of analysis. First, certain multivariate techniques, such as principal components, are hardly useful for the other designs. Second, in a truly multivariate set of outcomes, the variance-covariance structure is usually unstructured and hardly of direct scientific interest, in contrast to, for example, clustered or longitudinal data. Therefore, the methodology of the general linear model is too restrictive to perform satisfactory data analyses of these more complex data.

Replacing the time dimension in a longitudinal setting with one or more spatial dimensions leads naturally to spatial data. While ideas in the longitudinal and spatial areas have developed relatively independently, efforts have been spent in bridging the gap between both disciplines. In 1996, a workshop was devoted to this idea: "The Nantucket Conference on Modeling Longitudinal and Spatially Correlated Data: Methods, Applications, and Future Directions" (Gregoire *et al.* 1997).

Among the clustered data settings, longitudinal data perhaps require the most elaborate modeling of the random variability. Diggle, Liang, and Zeger (1994) distinguish among three components of variability. The first one groups traditional random effects (as in a random-effects ANOVA model) and random coefficients (Longford 1993). It stems from interindividual variability (i.e., heterogeneity between individual profiles). The second component, serial association, is present when residuals close to each other in time are more similar than residuals further apart. This notion is well known in the time-series literature (Ripley 1981, Diggle 1983, Cressie 1991). Finally, in addition to the other two components, there is potentially also measurement error. This results from the fact that, for delicate measurements (e.g., laboratory assays), even immediate replication will not be able to avoid a certain level of variation. In longitudinal data, these three components of variability can be distinguished by virtue of both *replication* as well as a clear *distance* concept (time), one of which is lacking in classical spatial and time-series analysis and in clustered data.

These considerations imply that adapting models for longitudinal data to other data structures is in many cases relatively straightforward. For example, clustered data of the type considered in this book can often be analyzed by leaving out all aspects of the model that refer to time. In some cases, a version of serial association can be considered for clustered data with individual-level exposures. We refer to Chapter 4 for an overview of the modeling families that arise in this context.

A very important characteristic of data to be analyzed is the type of outcome. Methods for continuous data form no doubt the best developed and most advanced body of research; the same is true for software implementation. This is natural, since the special status and the elegant properties of the normal distribution simplify model building and ease software development. A number of software tools, such as the SAS procedure MIXED, the SPlus function `lme`, and MLwiN, have been developed in this area. However, also categorical (nominal, ordinal, and binary) and discrete outcomes are very prominent in statistical practice. For example, quality of life outcomes are often scored on ordinal scales. In many surveys, all or part of the information is recorded on a categorical scale.

Two fairly different views can be adopted. The first one, supported by large-sample results, states that normal theory should be applied as much as possible, even to non-normal data such as ordinal scores and counts. A different view is that each type of outcome should be analyzed using instruments that exploit the nature of the data. Extensions of GLIM to the longitudinal case are discussed in Diggle, Liang, and Zeger (1994), where the main emphasis is on generalized estimating equations (Liang and Zeger 1986). Generalized linear mixed models have been proposed by, for example, Breslow and Clayton (1993). Fahrmeir and Tutz (1994) devote an entire book to GLIM for multivariate settings. Subscribing to the second point of view, we will present methodology specific to the case of categorical data. The main emphasis will be on clustered binary data from developmental toxicity studies

(Section 1.2) and from survey data (Section 1.3). However, the modeling and analysis strategies described in this text have a much broader applicability.

In clustered settings, each unit typically has a *vector* Y of responses. This leads to several, generally nonequivalent, extensions of univariate models. In a *marginal model*, marginal distributions are used to describe the outcome vector Y, given a set X of predictor variables. The correlation among the components of Y can then be captured either by adopting a fully parametric approach or by means of working assumptions, such as in the semiparametric approach of Liang and Zeger (1986). Alternatively, in a *random-effects model*, the predictor variables X are supplemented with a vector b of random (or cluster-specific) effects, conditional upon which the components of Y are usually assumed to be independent. This does not preclude that more elaborate models are possible if residual dependence is detected (Longford 1993). Finally, a *conditional model* describes the distribution of the components of Y, conditional on X but also conditional on (a subset of) the other components of Y. Well-known members of this class of models are log-linear models. Several examples are given in Fahrmeir and Tutz (1994).

For normally distributed data, marginal models can easily be fitted, for example, with the SAS procedure MIXED, the SPlus function `lme`, or within the MLwiN package. For such data, integrating a mixed-effects model over the random effects produces a marginal model, in which the regression parameters retain their meaning and the random effects contribute in a simple way to the variance-covariance structure. For example, the marginal model corresponding to a random-intercepts model is a compound symmetry model that can be fitted without explicitly acknowledging the random-intercepts structure. In the same vein, certain types of transition models induce simple marginal covariance structures. For example, some first-order stationary autoregressive models imply an exponential or AR(1) covariance structure. As a consequence, many marginal models derived from random-effects and transition models can be fitted with mixed-models software.

It should be emphasized that the above elegant properties of normal models do not extend to the general GLIM case. For example, opting for a marginal model for clustered binary data precludes the researcher from answering conditional and transitional questions in terms of simple model parameters. This implies that each model family requires its own specific analysis and, consequently, software tools. In many cases, standard maximum likelihood analyses are prohibitive in terms of computational requirements. Therefore, specific methods such as generalized estimating equations (Chapter 5) and pseudo-likelihood (Chapters 6 and 7) have been developed. Both apply to marginal models, whereas pseudo-likelihood methodology can be used in the context of conditional models as well. In case random-effects models are used, the likelihood function involves integration over the random-effects distribution for which generally no closed forms are available. Estimation methods then either employ approximations to the likelihood or score functions, or resort to numerical integration techniques. Some estimation methods have been implemented in standard software. For example, an analysis based on generalized estimating equations can be performed within the GENMOD procedure in SAS. Mixed-effects models for non-Gaussian data can be fitted using the MIXOR program (Hedeker and Gibbons 1994, 1996), MLwiN, or the SAS procedure NLMIXED. In many cases, however, specialized software, either commercially available or user-defined, will be needed.

In this book, we will focus on clustered binary data, arising from developmental toxicity studies, complex surveys, etc. These contexts will be introduced in the remainder of this chapter, whereas actual motivating examples will be introduced in Chapter 2. After discussing specific and general issues in modeling such data (Chapter 3), and reviewing the model families (Chapter 4), specific tools for analysis will be presented and exemplified in subsequent chapters. While the emphasis is on binary data, we also deal with the specifics of continuous outcomes (Chapter 13) and mixtures of binary and continuous outcomes (Chapter 14). Apart from model formulation and parameter estimation, specific attention is devoted to assessing model fit (Chapter 9), quantitative risk assessment (Chapter 10), model misspecification (Chapter 11), exact dose-response inference (Chapter 12), and individual-level covariates (Chapter 13), as opposed to cluster-level covariates.

In the next sections, we will deal with the specifics of developmental toxicity studies and complex surveys.

1.2 Developmental Toxicity Studies

Lately, society has been increasingly concerned about problems related to fertility and pregnancy, birth defects, and developmental abnormalities. Consequently, regulatory agencies such as the U.S. Environmental Protection Agency (EPA) and the Food and Drug Administration (FDA) have given increased priority to protection against drugs, harmful chemicals, and other environmental hazards. As epidemiological evidence of adverse effects on fetal development may not be available for specific chemicals present in the environment, laboratory experiments in small mammalian species provide an alternative source of evidence essential for identifying potential developmental toxicants. For ethical reasons, animal studies afford a greater level of control than epidemiological studies. Moreover, they can be conducted in advance of human exposure. Unfortunately, there have been cases in which animal studies have not been run properly. The thalidomide tragedy is a prominent example (Salsburg 1996). Thalidomide was present in at least 46 countries under many different brand names. In Belgium it is best known as "Softenon". The drug was described as being "safe" because it was not possible to develop toxic lesions in animal trials. Unfortunately, this was not the case. An estimated 10,000 children were born throughout the world as deformed, some with fin-like hands grown directly on the shoulders, with stunted or missing limbs, deformed eyes and ears, ingrown genitals, absence of a lung, a great many of them stillborn or dying shortly after birth, etc. The animal tests performed

Figure 1.1 *Time line for a typical Segment II study.*

by the inventor of the drug were very superficial and incomplete. They did not carry out animal tests specifically to demonstrate teratogenetic effects. This runs contrary to the basic ideas behind such studies. According to Paracelsus all compounds are potential poisons: "Only the dose makes a thing not a poison". Malformations, like cancer, could occur when practically any substance, including sugar and salt, is given in excessive doses. A proper animal study should therefore always include a dose at which a toxic lesion happens.

As a consequence of the thalidomide tragedy, there has been a marked upsurge in the number of animals used in testing of new drugs. Also, drugs are now specifically tested on pregnant animals to safeguard against possible teratogenic effects on the human foetus. However, methods for extrapolating the results to humans are still being developed and refined. Differences in the physiological structure, function, and biochemistry of the placenta that exist between species make reliable predictions difficult.

Since laboratory studies further involve considerable amounts of time and money, as well as huge numbers of animals, it is essential that the most appropriate and efficient statistical models are used (Williams and Ryan 1996). Three standard procedures (Segments I, II, and III) have been established to assess specific types of effects.

- Segment I or fertility studies are designed to assess male and female fertility and general reproductive ability. Such studies are typically conducted in one species of animals and involve exposing males for 60 days and females for 14 days prior to mating.

- Segment II studies are also referred to as "teratology studies", since historically the primary goal was to study malformations (the origin of the word "teratology" lies in the Greek word "tera", meaning monster). In Section 1.2.1, we will describe standard teratology studies in greater detail. The time line for a typical Segment II study is depicted in Figure 1.1.

- Segment III tests are focused on effects later in gestation and involve exposing pregnant animals from the 15th day of gestation through lactation.

In addition, we will describe alternative animal test systems, such as the so-called "heatshock studies" in Section 1.2.2. The methodology described in this work will be applied primarily to standard Segment II designs and to heatshock studies.

Figure 1.2 *Dissected mouse with removed uterus.*

1.2.1 *The Segment II Study: a Standard Experimental Design*

A Segment II experiment involves exposing timed-pregnant animals (rats, mice, and occasionally rabbits) during major organogenesis (days 6 to 15 for mice and rats) and structural development. A graphical representation is given in Figure 1.1. Administration of the exposure is generally by the clinical or environmental routes most relevant for human exposure. Dose levels consist of a control group and 3 or 4 dose groups, each with 20 to 30 pregnant dams. The dams are sacrificed just prior to normal delivery, at which time the uterus is removed and thoroughly examined (Figures 1.2 and 1.3).

An interesting aspect of Segment II designs is the hierarchical structure of the developmental outcomes. Figure 1.4 illustrates the data structure. An implant may be resorbed at different stages during gestation. If the implant survives being resorbed, the developing foetus is at risk of fetal death. Adding the number of resorptions and fetal deaths yields the number of non viable foetuses. If the foetus survives the entire gestation period, growth reduction such

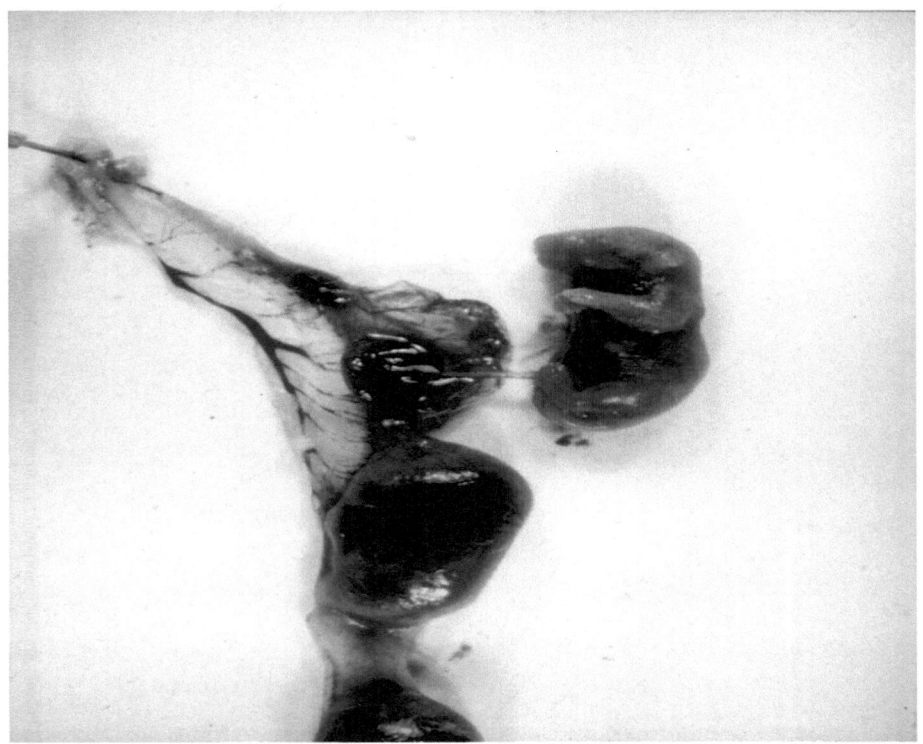

Figure 1.3 *Uterus with removed foetus.*

as low birth weight may occur. The foetus may also exhibit one or more types of malformation. These are commonly classified into three broad categories:

- external malformations are those visible by naked eye, for instance missing limbs;
- skeletal malformations might include missing or malformed bones;
- visceral malformations affect internal organs such as the heart, the brain, the lungs, etc.

Each specific malformation is typically recorded as a dichotomous variable (present or absent). Adding the number of resorptions, the number of fetal deaths, and the number of viable foetuses yields the total number of implantations. Since exposure to the test agent takes place after implantation, the number of implants, a random variable, is not expected to be dose-related.

The analysis of developmental toxicity data as described above, combining hierarchical, multivariate, and clustered data issues, raises a number of challenges (Molenberghs *et al.* 1998, Zhu and Fung 1996). These will be described in detail in Chapter 3.

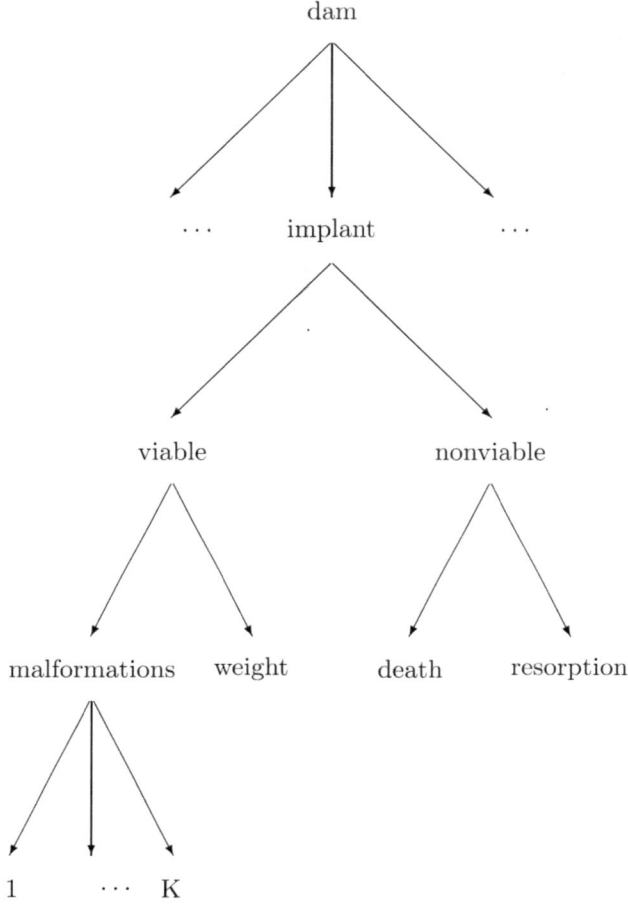

Figure 1.4 *Data structure of developmental toxicity studies.*

1.2.2 Heatshock Studies

A unique type of developmental toxicity study was originally developed by Brown and Fabro (1981) to assess the impact of heat stress on embryonic development. Subsequent adaptations by Kimmel *et al.* (1994) allows the investigation of effects, related to both temperature and duration of exposure. These heatshock experiments are described in Section 2.2. The embryos are explanted from the uterus of the maternal dam and cultured in vitro. Next, each embryo is exposed to a short period of heat stress by placing the culture vial into a warm water bath, involving an increase over body temperature of 4 to 5°C for a duration of 5 to 60 minutes. The embryos are examined 24 hours later for impaired or accelerated development. This type of developmental test system has several advantages over the standard Segment II design. First of all, the exposure is directly administered to the embryo, so that controversial issues regarding the unknown relationship between the exposure level to the maternal dam and that which is actually received by the embryo need not be taken into account. Second, the exposure pattern can be easily controlled, since target temperature levels in the warm water baths can be achieved within 2 minutes. Further, information regarding the effects of exposure are quickly obtained, in contrast to the Segment II study which requires 8 to 12 days after exposure to assess impact. And finally, this animal test system provides a convenient mechanism for examining the joint effects of both duration of exposure and exposure levels.

1.3 Complex Surveys

1.3.1 Introduction

Complex surveys can be seen as another, very important, area of motivation for and application of the methods presented in this book. The reasons for this are manifold.

First, survey applications are very common in many areas of statistics and related fields. Classically associated with sociological and psychometric research, they are very prominent also in the biomedical sciences and in epidemiology. Examples include quality-of-life assessment in clinical studies (Mesbah, Cole, and Ting Lee 2001), health interview surveys (see Section 2.3), needs assessment, etc. There is a well established literature on survey literature (Kish 1965, Cochran 1977, Foreman 1991, Scheaffer, Mendenhall, and Ott 1990, and Lemeshow and Levy 1999). To the casual user of survey methodology, it may appear as if the survey literature is completely segregated from other branches of statistics, such as, for example, biostatistics. This is not true, and brings us to the next point.

Second, survey sampling goes in almost all cases far beyond simple random sampling. To cope with stratification, unequal selection probabilities, and multistage sampling, the survey researcher has developed a specific language, littered with *weights*, *Horvitz-Thompson estimators*, and *design effects*. We will zoom in on the designs in Section 1.3.2, but at this point it is important to realize that the design effect is closely related to the clustering as it occurs in developmental toxicity studies and, to some extent, to the concept of overdispersion, which is so familiar to the user of generalized linear models (Hinde and Demétrio 1998).

On the one hand, it is to be applauded that standard software packages such as SUDAAN, STATA, and SAS are devoting more and more attention to the correct analysis of survey data. In the past, the lack of software, and also a lack of knowledge, has often produced incorrect analyses. Indeed, often survey data, arising from complicated designs, have been analyzed as if they comprise a simple random sample. There is also a danger, however, arising from these software tools, since they reinforce the view that survey data analysis is completely divorced from other types of data analysis. This may lead to an entirely different treatment of the same phenomenon, such as clustering, dependent on the context. While this can be defended in some situations, it is not always the case. Essentially, the data analyst has to choose between either advanced data analysis tools such as linear and nonlinear mixed models, thereby ignoring the design aspects, or correctly accounting for the design, but then restricting the estimands to such simple quantities as means, totals, and proportions (the main themes in the standard survey texts). The book by Skinner, Holt, and Smith (1989) tries to bridge this gap and considers explicitly the case of complex designs.

1.3.2 Sampling Designs

As mentioned in the previous section, a key characteristic of many survey samples is their relatively complex design. We refer to the standard texts mentioned in the previous section for a thorough treatment, and only give a brief overview at this point.

Most surveys aim to be representative for a certain population (e.g., the Belgian population in 2001), unlike, for example, a clinical study, which borrows its authority from randomization (Piantadosi 1997). To reach the population, a sampling frame is selected (e.g., the phone directory, the National Register) in order to strike a balance between avoiding over- and underrepresentation on the one hand and keeping the field work manageable on the other hand. For example, using a phone survey will lead to companies as well as individuals (a benign problem) but will fail to reach those without a phone (a more serious problem since those respondents without a phone may form a specific subgroup in the socio-economic sense, the public health sense, etc.). At the same time a phone survey will be much cheaper than a face-to-face survey.

To obtain a representative sample of a population, one often resorts to strata to ensure that not only overall, but also within certain subgroups, the number of respondents is under control. Typical stratification variables are age, sex, and geographical location. In federal countries, states or provinces may be given a share of the sample which is proportional or disproportional to the population size. In the latter case, the selection probabilities of individuals from different states will be unequal, a feature that needs to be accounted for in the analysis, just as the stratification itself. Further, in order to reach respondents (target units), one often resorts to a multi-stage sampling scheme. For example, one first selects towns (primary sampling units), then a number of households within towns (secondary sampling units), and finally a number of household members within a household (target or tertiary sampling units).

A consequence of such a sampling scheme is that a number of respondents stem from the same household and the same town. One then cannot ignore the possibility of individuals within families being more alike than between families, with the same to a lesser extent holding for towns. In the way described above, clustering arises as a by-product of the chosen multi-stage sampling design. In some cases, clusters can be selected without the detour via multi-stage sampling. In any case, one can distinguish between, broadly, three ways of dealing with such clustering:

- Clustering is ignored. While this typically leaves the consistency of point estimation intact, the same is not true for measures of precision. In case of a "positive" clustering effect (i.e., units within a cluster are more alike than between clusters), then ignoring this aspect of the data, just as ignoring overdispersion, overestimates precision and hence underestimates standard errors and lengths of confidence intervals.

- In answer to the previous strategy, one often *accounts for* clustering. This means the existence of clustering is recognized but considered a nuisance characteristic. A crude way of correcting for clustering is by means of computing a so-called *design effect*. Roughly, the design effect is a factor comparing the precision under simple random sampling with the precision of the actual design. Standard errors, computed as if the design had been simple random sampling, can then be inflated using the design effect. This is similar to a heterogeneity factor for overdispersion.

- In contrast to the previous viewpoint, one can have a genuine scientific interest in the clustering itself. When the design encompasses families, familial association can be of direct interest. In genetic studies, pedigrees are sampled with the explicit goal to study the association in general and the genetic component thereof in particular.

While the second viewpoint is closely connected to the available methodology in the survey sampling context, obviously the third one is much broader and analysis strategies consistent with an interest in the intra-cluster dependence provided in this book can be applied in this context as well.

1.4 Other Relevant Settings

It will be clear from the above that developmental toxicity studies and complex surveys, as exemplified in the next chapter and analyzed throughout the book, form very versatile areas of application. Needless to say that there are many others, a few of which are given in Sections 2.3–2.6. In texts such as Fahrmeir and Tutz (1994) and Pendergast *et al.* (1996) a multitude of other relevant settings can be found.

1.5 Reading Guide

The remaining chapters are grouped so as to facilitate access for readers with different backgrounds and interests.

Case studies and issues. These are presented in Chapter 2. While some issues coming from them are indicated in the chapter itself, a formal but nontechnical overview of the issues arising from clustered binary data is given in Chapter 3. Both chapters provide ample references to later chapters where the data are used and/or issues are addressed.

Modeling chapter. Chapter 3 gives a thorough and general overview of model families and relevant members of each family. In Chapter 8, flexible polynomial methods are introduced which, in combination with the model families, yield a broad framework to model clustered binary data.

Inference chapters. Inferential procedures are developed in Chapters 5 (where the focus is on generalized estimating equations), 6 and 7 (pseudo-likelihood), and 12 (exact inference).

Selected topics are given in the remaining chapters. Some are concerned with the quality of a model (Chapter 9 on goodness-of-fit and Chapter 11 on model misspecification), while others are rather specific to risk assessment (Chapter 10 on quantitative risk assessment). Further, Chapter 13 deals with individual-level covariates and Chapter 14 is dedicated to a combination of continuous and discrete outcomes. These two topics are illustrated using developmental toxicity studies, but their relevance reaches well beyond this area. Finally, Chapter 15 illustrates both the analysis of complex clustered survey data, where in addition weights are incorporated, as well as the use of the multilevel model paradigm.

The more methodologically interested reader can start from the modeling chapter and explore the inferential chapters and selected topics from there. The reader who is primarily substantively motivated can go directly to an appropriately selected topic, and use the cross-references to earlier chapters to fill in holes. While still acknowledging the edited nature of the book, we hope the chapters are sufficiently structured as to cater a wide audience.

CHAPTER 2

Motivating Examples

This chapter introduces the sets of data which will be used throughout the book. The National Toxicology Program data are presented in Section 2.1. The heatshock studies are introduced in Section 2.2. Section 2.3 describes the Belgian Health Interview Survey data. The POPS data, the low-iron rat teratology data and data from the Wisconsin diabetes study are presented in Sections 2.4–2.6. In Sections 2.7 and 2.8, two examples are introduced that will illustrate the developments related to exact inference.

2.1 National Toxicology Program Data

The developmental toxicity studies introduced in this section are conducted at the Research Triangle Institute, which is under contract to the National Toxicology Program of the U.S. (NTP data). These studies investigate the effects in mice of five chemicals: ethylene glycol (Price *et al.* 1985), triethylene glycol dimethyl ether (George *et al.* 1987), diethylene glycol dimethyl ether (Price *et al.* 1987), di(2-ethylhexyl)phthalate (Tyl *et al.* 1988) and theophylline (Lindström *et al.* 1990).

2.1.1 Ethylene Glycol

Ethylene glycol (EG) is also called 1,2-ethanediol and can be represented by the chemical formula $HOCH_2CH_2OH$. It is a high-volume industrial chemical with many applications. EG is used as an antifreeze in cooling and heating systems, as one of the components of hydraulic brake fluids, as an ingredient of electrolytic condensers and as a solvent in the paint and plastics industries. Furthermore, EG is employed in the formulation of several types of inks, as a softening agent for cellophane and as a stabilizer for soybean foam used to extinguish oil and gasoline fires. Also, one uses EG in the synthesis of various chemical products, such as plasticizers, synthetic fibers and waxes (Windholz 1983).

EG may represent little hazard to human health in normal industrial handling, except possibly when used as an aerosol or at elevated temperatures. EG at ambient temperatures has a low vapor pressure and is not very irritating to the eyes or skin. However, accidental or intentional ingestion of antifreeze products, of which approximately 95% is EG, is toxic and may result in death (Rowe 1963, Price *et al.* 1985).

EG Study in Mice

Price *et al.* (1985) describe a study in which timed-pregnant CD-1 mice were dosed by gavage with EG in distilled water. Dosing occurred during the period of organogenesis and structural development of the foetuses (gestational days 8 through 15). The doses selected for the study were 0, 750, 1500 or 3000 mg/kg/day. Table 2.1 shows, for each dose group and for all five NTP toxic agents, the number of dams containing at least one implant, the number of dams having at least one viable fetus, the number of live foetuses, the mean litter size and the percentage of malformation for three different classes: external malformations, visceral malformations and skeletal malformations. While for EG, skeletal malformations are substantial in the highest dose group, external and visceral malformations show only slight dose effects. The distribution of the number of implants is given in Table 2.2 for each of these five chemicals. It is shown that clusters consisting of 10–15 implants occur frequently.

Figure 2.1 represents some of the data of this study. For each dose group, cumulative relative frequencies of the number of clusters are plotted for the number of implants in a cluster, the number of viable foetuses, the number of dead foetuses, the number of abnormals (i.e., dead or malformed foetuses), the number of external, skeletal and visceral malformations and the number of foetuses with at least one type of malformation.

Figures 2.2–2.4 show for each of these studies and for each dose group the observed and averaged malformation rates in mice.

EG Study in Rats

Price *et al.* (1985) also describe a developmental toxicity experiment, investigating the effect of EG in rats. The doses selected for the present teratology study were 0, 1.25, 2.50 and 5.0 g/kg/day. A total of 1368 live rat foetuses were examined for low birth weight (continuous) or defects (binary). This joint occurrence of continuous and binary outcomes will provide additional challenges in model development. Table 2.3 summarizes the malformation and fetal weight data from this experiment. The data show clear dose-related trends for both outcomes. The rate of malformation increases with dose, ranging from 1.3% in the control group to 68.6% in the highest dose group. The mean fetal weight decreases monotonically with increasing dose, ranging from 3.40 g to 2.48 g in control and highest dose group, respectively. The fetal weight variances, however, do not change monotonically with dose. In the lower dose groups, the variances remain approximately constant. However, in the highest dose group, the fetal weight variance is elevated. Further, it can be observed that simple Pearson correlation coefficients (ρ) between weight and malformation tend to strengthen with increasing doses. As doses increase, the correlation becomes more negative, because the probability of malformation is increasing and fetal weight is decreasing. This is illustrated in Figure 2.5, which shows the observed malformation rates for all clusters, the averaged

Table 2.1 *Summary Data by NTP studies in mice. The dose is in mg/kg/day.*

Exposure	Dose	# dams, ≥ 1 impl.	viab.	Live	Litter Size (mean)	Malformations Ext.	Visc.	Skel.
EG	0	25	25	297	11.9	0.0	0.0	0.3
	750	24	24	276	11.5	1.1	0.0	8.7
	1500	23	22	229	10.4	1.7	0.9	36.7
	3000	23	23	226	9.8	7.1	4.0	55.8
DEHP	0	30	30	330	13.2	0.0	1.5	1.2
	44	26	26	288	11.1	1.0	0.4	0.4
	91	26	26	277	10.7	5.4	7.2	4.3
	191	24	17	137	8.1	17.5	15.3	18.3
	292	25	9	50	5.6	54.0	50.0	48.0
DYME	0	21	21	282	13.4	0.0	0.0	0.0
	62.5	20	20	225	11.3	0.0	0.0	0.0
	125	24	24	290	12.1	1.0	0.0	1.0
	250	23	23	261	11.3	2.7	0.1	20.0
	500	22	22	141	6.1	66.0	19.9	79.4
TGDM	0	27	26	319	12.3	0.003	0.000	0.000
	250	26	26	275	10.6	0.000	0.000	0.000
	500	26	24	262	10.9	0.004	0.000	0.004
	1000	28	26	286	11.0	0.042	0.003	0.073
THEO	0	26	25	296	11.8	0.003	0.000	0.000
	282	26	25	278	11.1	0.007	0.000	0.000
	372	33	29	300	10.3	0.017	0.003	0.003
	396	23	17	197	11.6	0.020	0.005	0.000

malformation rates for each dose group, the average weight outcomes for all clusters and the average weight outcomes for each dose group.

2.1.2 Di(2-ethylhexyl)Phthalate

Di(2-ethylhexyl)phthalate (DEHP) is also called octoil, dioctyl phthalate or 1,2-benzenedicarboxylic acid bis(2-ethylhexyl) ester. It can be represented by $C_{24}H_{38}O_4$. DEHP is used in vacuum pumps (Windholz 1983). Furthermore, this ester as well as other phthalic acid esters are used extensively as plas-

Table 2.2 *NTP Data in Mice. Frequency distribution of the number of implants.*

Number of implants	EG	TGDM	DYME	DEHP	THEO
1	0	1	0	1	2
2	0	0	0	1	2
3	1	1	1	0	1
4	0	3	1	2	1
5	1	1	0	0	0
6	0	1	0	2	3
7	2	2	2	0	0
8	1	0	2	4	0
9	8	2	2	5	6
10	4	7	7	7	4
11	8	21	10	18	14
12	19	26	15	21	17
13	16	19	27	26	21
14	11	10	19	21	19
15	16	8	9	10	12
16	6	4	10	8	3
17	1	1	5	2	2
18	0	0	0	2	1
19	1	0	0	1	0
	95	107	110	131	108

ticizers for numerous plastic devices made of polyvinyl chloride. DEHP provides the finished plastic products with desirable flexibility and clarity (Shiota, Chou and Nishimura 1980).

It has been well documented that small quantities of phthalic acid esters may leak out of polyvinyl chloride plastic containers in the presence of food, milk, blood or various solvents. Due to their ubiquitous distribution and presence in human and animal tissues, considerable concern has developed as to the possible toxic effects of the phthalic acid esters (e.g., Autian 1973).

In particular, the developmental toxicity study described by Tyl *et al.* (1988) has attracted much interest in the toxicity of DEHP. The doses selected for the study were 0, 0.025, 0.05, 0.1 and 0.15%, corresponding to a DEHP consumption of 0, 44, 91, 191 and 292 mg/kg/day respectively. Females were observed daily during treatment, but no maternal deaths or distinctive clinical signs were observed. The dams were sacrificed, slightly prior to normal

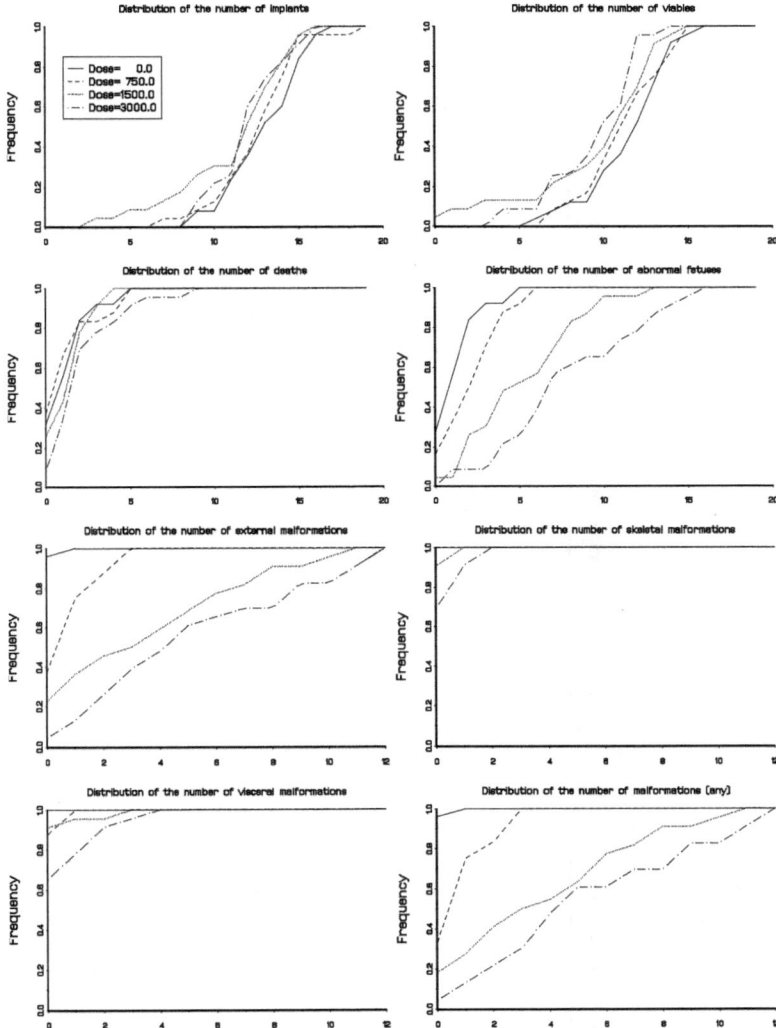

Figure 2.1 *EG Study in Mice. Cumulative relative frequencies of the number of clusters representing some of the data.*

delivery, and the status of uterine implantation sites recorded. A total of 1082 live foetuses were dissected from the uterus, anaesthetized and examined for external, visceral and skeletal malformations.

Some of the data of this study are shown in Figure 2.6. Table 2.1 suggests clear dose-related trends in the malformation rates. The average litter size (number of viable animals) decreases with increased levels of exposure to DEHP, a finding that is attributable to the dose-related increase in fetal deaths.

18

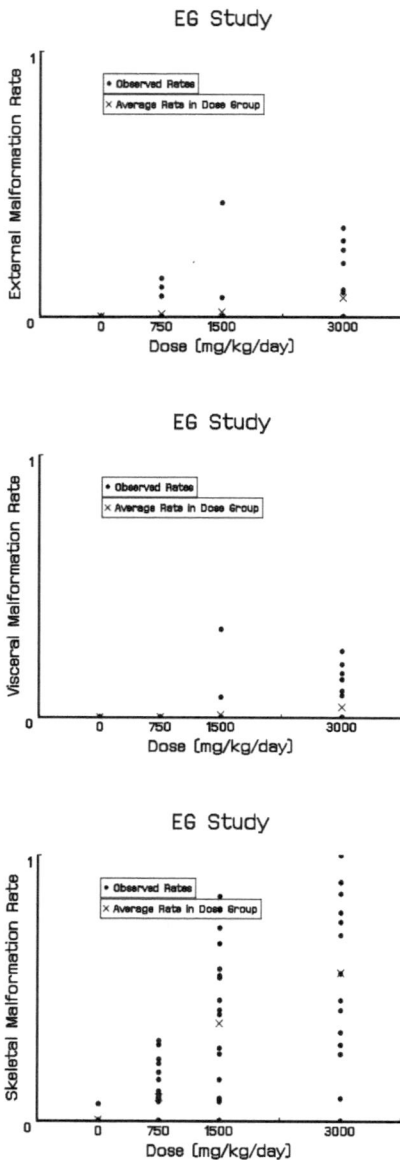

Figure 2.2 *EG Study in Mice. Observed and averaged malformation rates.*

2.1.3 Diethylene Glycol Dimethyl Ether

Other names for diethylene glycol dimethyl ether (DYME) are diglyme and bis(2-methoxyethyl) ether. DYME has as its chemical formula

$$CH_3O(CH_2)_2O(CH_2)_2OCH_3$$

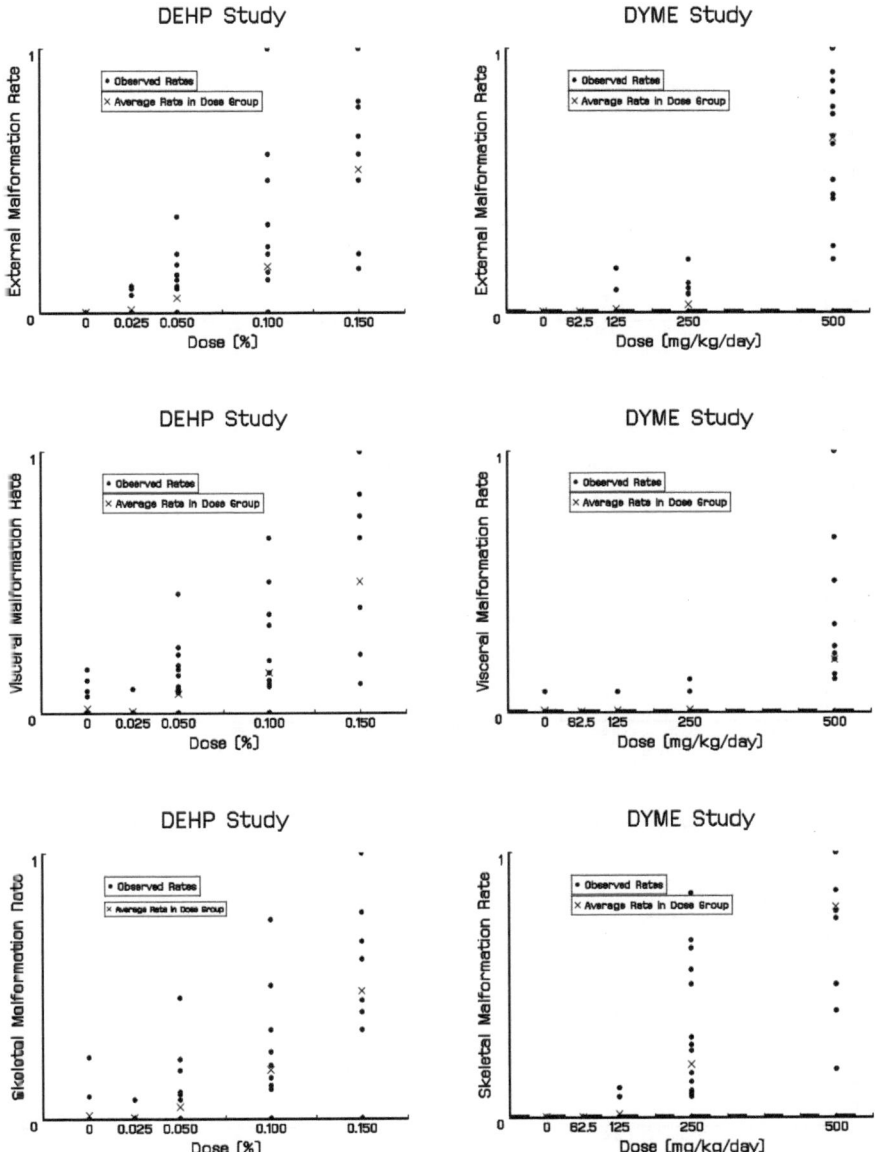

Figure 2.3 *DEHP and DYME Studies. Observed and averaged malformation rates.*

(Windholz 1983). It is a component of industrial solvents. These are widely used in the manufacture of protective coatings such as lacquers, metal coatings, baking enamels, etc. (NIOSH 1983). Although to date, several attempts have proven inadequate to evaluate the potential of glycol ethers to produce

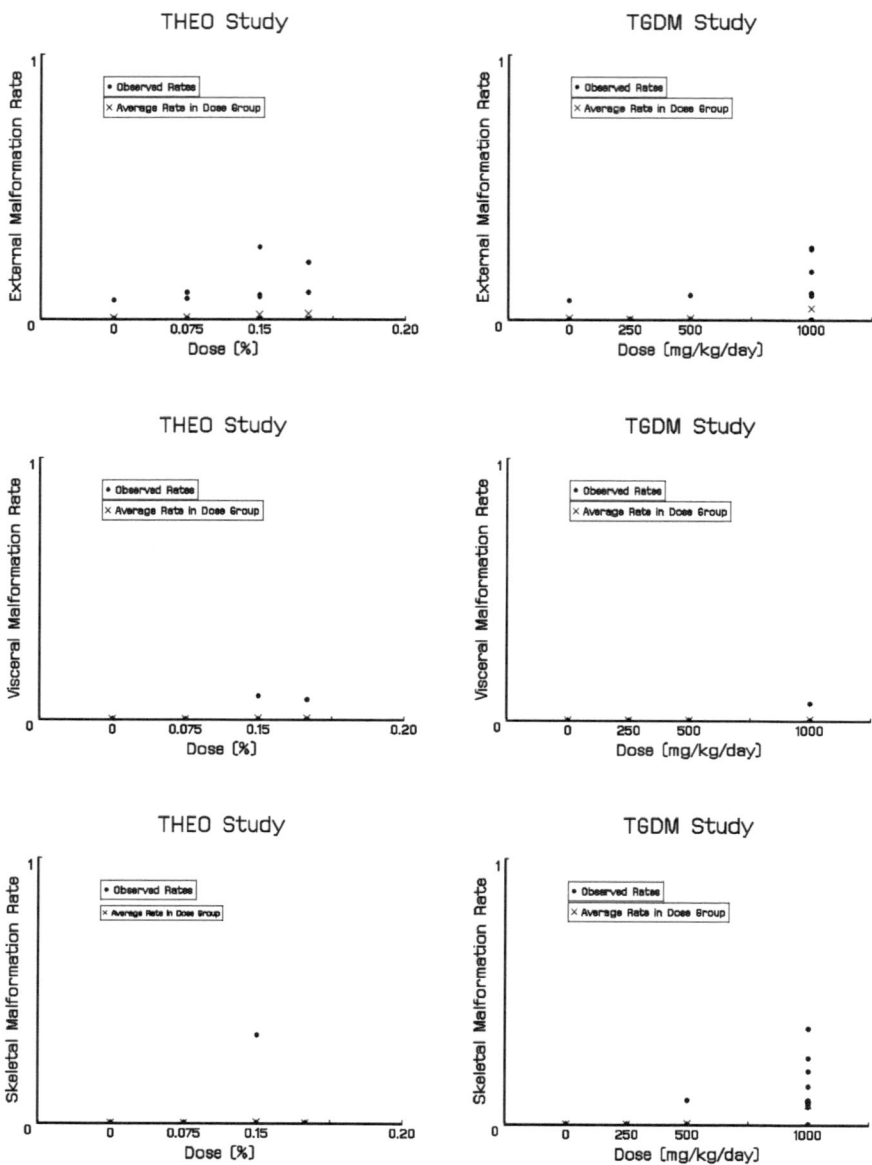

Figure 2.4 *THEO and TGDM Studies. Observed and averaged malformation rates.*

human reproductive toxicity, structurally related compounds have been identified as reproductive toxicants in several mammalian species, producing (1)
testicular toxicity and (2) embryotoxicity.

Price *et al.* (1987) describe a study in which timed-pregnant mice were dosed

Table 2.3 *EG Study in Rats. Summary data.*

Dose (1)	Dams (2)	Live	Size (3)	Malf. Nr.	Malf. %	Weight Mean	Weight SD	Pearson Corr. (ρ)
0	28	379	13.50	5	1.3	3.40	0.38	0.07
125	28	357	12.75	21	5.8	3.30	0.37	0.00
250	29	345	11.89	86	24.9	2.90	0.36	-0.29
500	26	287	11.04	197	68.6	2.48	0.46	-0.37

(1) Dose is in mg/kg/day.

(2) The number of dams with at least one implant is given.

(3) Mean litter size.

with DYME throughout major organogenesis (gestational days 8 through 15). The doses selected for the study were 0, 62.5, 125, 250 and 500 mg/kg/day. Table 2.1 summarizes the data and a representation of the data in the DYME study is given in Figure 2.7.

2.1.4 Triethylene Glycol Dimethyl Ether

Similar to DEHP, triethylene glycol dimethyl ether (TGDM), also referred to as triglyme or tetraoxadodecane, is an industrial solvent with diverse applications. The solvent's chemical formula can be written as

$$CH_3O(CH_2)_2O(CH_2)_2O(CH_2)_2OCH_3$$

(Windholz 1983). TGDM is a member of the glycol ether class of industrial solvents. These solvents are widely used in the manufacture of protective coatings (NIOSH 1983).

Although field studies have not adequately evaluated the potential of glycol ethers to produce human reproductive toxicity, some glycol ethers have been identified as reproductive toxicants in several mammalian species (Clapp, Zaebst and Herrick 1984, George *et al.* 1987).

The pregnant dams of the TGDM study are exposed to 0, 250, 500 or 1000 mg/kg/day (George *et al.* 1987). Table 2.1 summarizes the data from their study. Clearly, visceral malformations are very infrequent with TGDM (only one malformation observed).

In Figure 2.8, some of the data of the TGDM study are shown.

2.1.5 Theophylline

Theophylline (THEO) has many other names, among others 1,3-dimethyl-xanthine, theocin and 3,7-dihydro-1,3-dimethyl-1H-purine-2,6-dione. One can

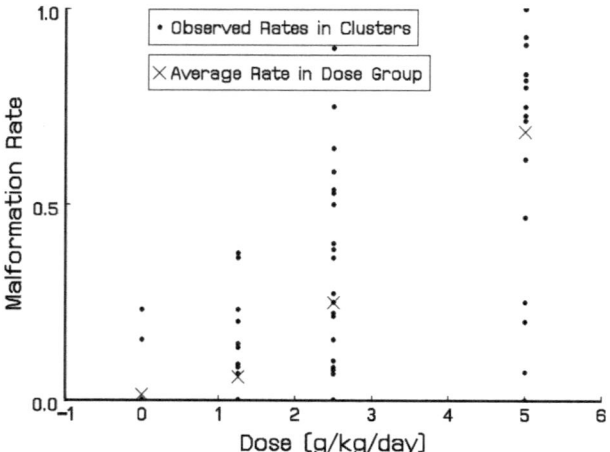

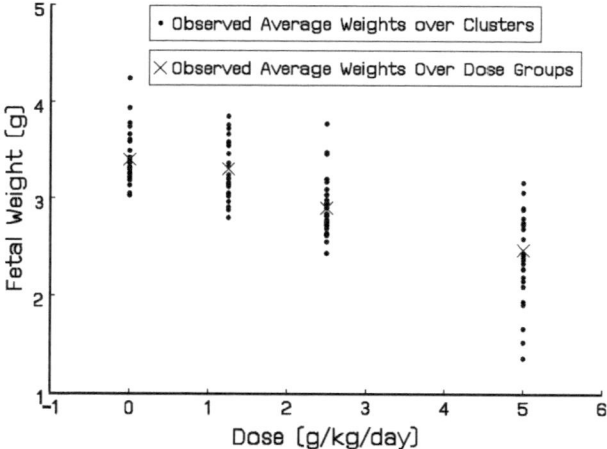

Figure 2.5 *EG Study in Rats. Observed malformation rates and average weights for all clusters.*

represent THEO by $C_7H_8N_4O_2$ (Windholz 1983). The developmental toxicity of orally administered theopylline (THEO) in mice has been described by Lindström *et al.* (1990). Theophylline belongs to the class of compounds used in the treatment of respiratory diseases, as an anti-asthmatic, diuretic, etc. Theophylline has been shown to cross the human placenta and is secreted in breast milk. Therefore, there has been an increased interest in the teratogenetic potential of theophylline in rodents.

Table 2.1 summarizes the data from a developmental toxicity study, investigating the effect of theophylline in mice. The doses selected for the study were

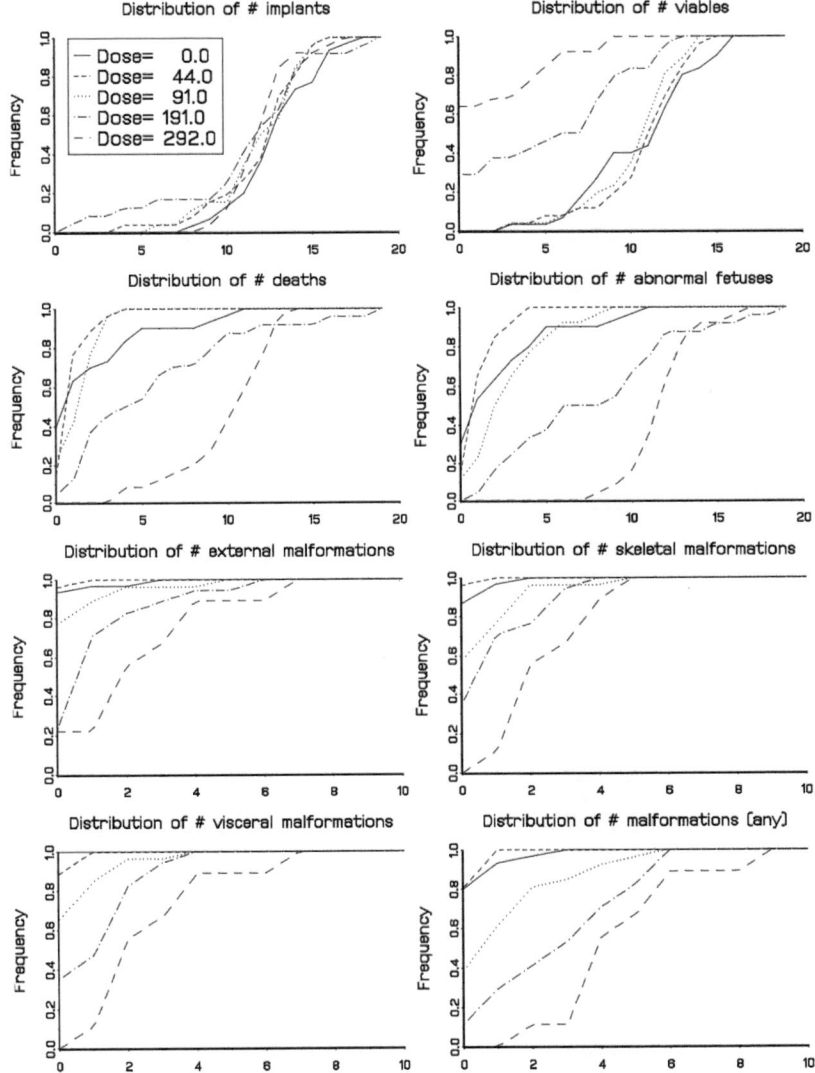

Figure 2.6 *DEHP Study in Mice. Cumulative relative frequencies of the number of clusters representing some of the data.*

0, 0.075, 0.15 or 0.20 % THEO, which correspond to a consumption of 0, 282, 372 and 396 mg/kg/day respectively. The table suggests small dose-related trends in the malformation rates. Figure 2.9 represents some of the data of this experiment.

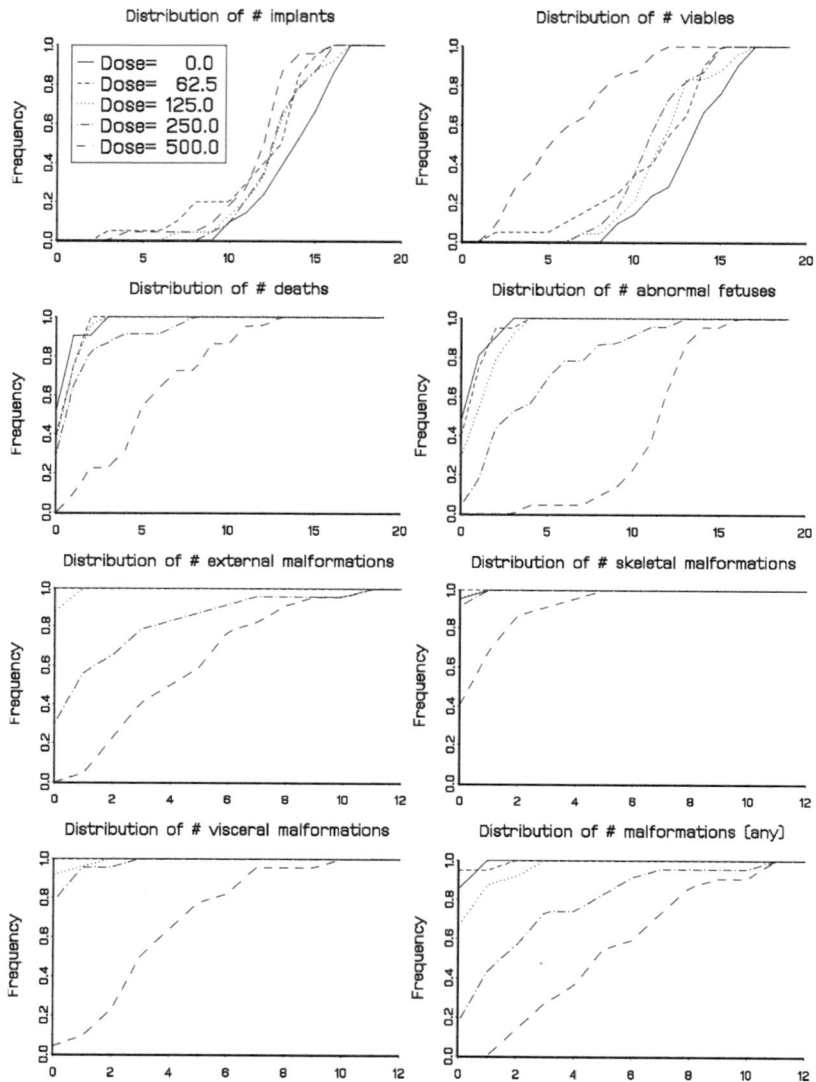

Figure 2.7 *DYME Study in Mice. Cumulative relative frequencies of the number of clusters representing some of the data.*

2.2 Heatshock Studies

Heatshock studies have been described by Brown and Fabro (1981) and Kimmel *et al.* (1994). In these experiments, embryos are explanted from the uterus of a maternal dam (rats, mice or rabbits) during the gestation period and cultured *in vitro*. Each subject is subjected to a short period of heat stress by placing the culture vial into a water bath, usually involving an increase

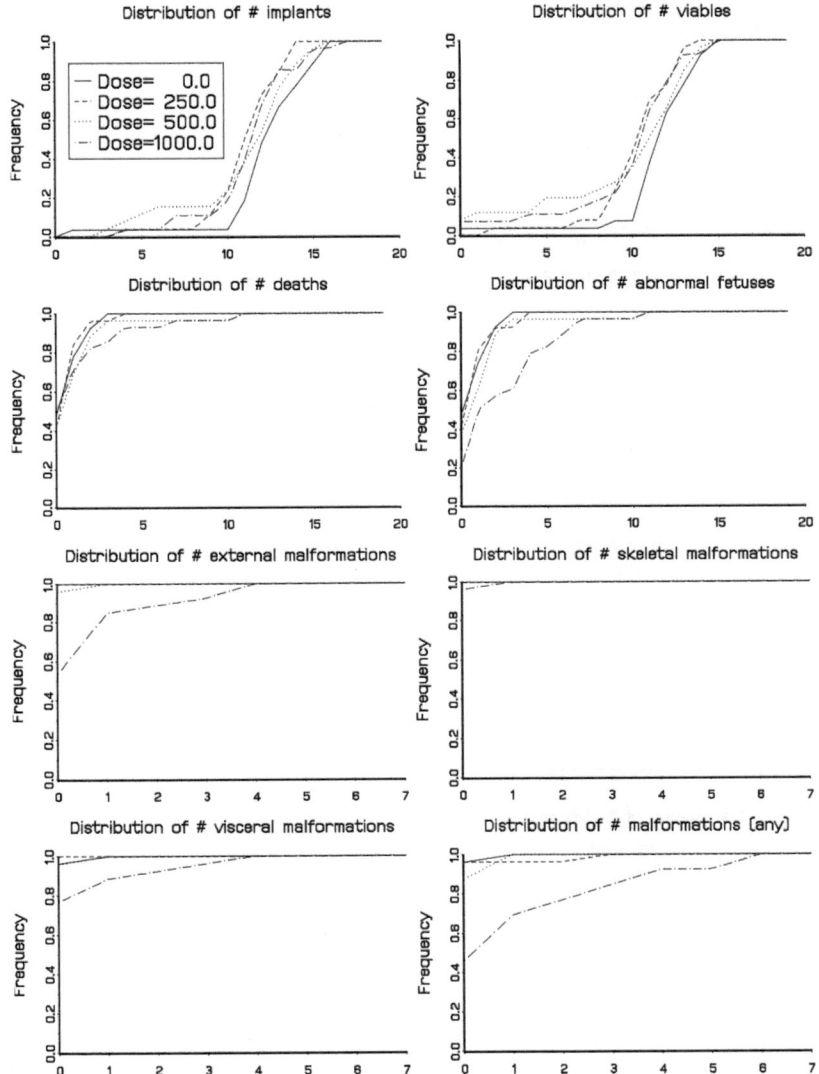

Figure 2.8 *TGDM Study. Cumulative relative frequencies of the number of clusters representing some of the data.*

over body temperature of 4 to 5°C for a duration of 5 to 60 minutes. The embryos are examined 24 hours later for impaired and/or accelerated development. The studies collect measurements on 13 morphological variables. Three of these are: olfactory system (OLF), optic system (OPT) and midbrain (MBN). We can assess the effects of both duration and level of exposure on each morphological endpoint, coded as affected (1) versus normal (0).

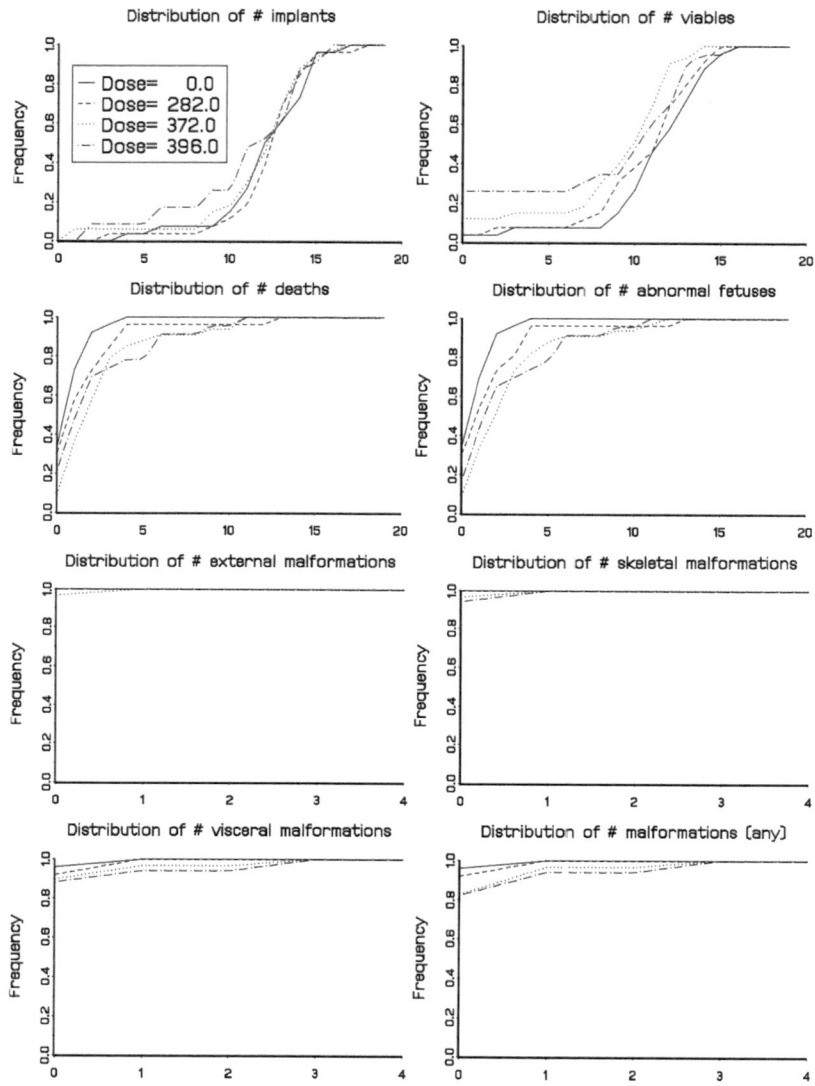

Figure 2.9 *THEO Study in Mice. Cumulative relative frequencies of the number of clusters representing some of the data.*

In addition to the discrete outcomes, there are several continuous outcomes recorded in the heatshock study, such as size measures on crown rump, yolk sac, and head. These will be studied in Chapter 13.

While the heatshock studies do not represent a standard developmental toxicity test system (Tyl *et al.* 1988), they have several advantages. These include direct exposure to the embryo rather than the dam, easily controlled

Table 2.4 *Heatshock Studies. Number of (surviving) embryos exposed to each combination of duration and temperature.*

Temperature	Duration of Exposure							Total
	5	10	15	20	30	45	60	
37.0	11	11	12	13	12	18	11	88
40.0	11	9	9	8	11	10	11	69
40.5	9	8	10	9	11	10	7	64
41.0	10	9	10	11	9	6	0	55
41.5	9	8	9	10	10	7	0	53
42.0	10	8	10	5	7	6	0	46
Total	60	53	60	56	60	57	29	375

Table 2.5 *Heatshock Studies: Distribution of cluster sizes.*

cluster size n_i	1	2	3	4	5	6	7	8	9	10	11
# clusters of size n_i	6	3	6	12	13	11	8	5	2	3	2

exposures, quick results, and a mechanism for exploring dose-rate effects. The study design for the set of experiments conducted by Kimmel *et al.* (1994) is shown in Table 2.4, which indicates the number of embryos cultured in each temperature-duration combination. A total of 375 embryos, arising from 71 initial dams, survived the heat exposure. These were further examined for any affections and used for analysis.

The distribution of cluster sizes, ranging between 1 and 11, is given in Table 2.5. The mean cluster size is about 5. Since only surviving foetuses were included, cluster sizes are smaller than those observed in most other developmental toxicity studies and do not reflect the true original litter size.

Figure 2.10 shows the actual percentages of affected embryos for each experimental temperature-duration combination. Historically, the strategy for comparing responses among exposures of different durations to a variety of environmental agents (e.g., radiation, inhalation, chemical compounds) has relied on a conjecture called Haber's Law, which states that adverse response levels should be the same for any equivalent level of dose times duration (Haber 1924). In other words, a 15-minute exposure to an increase of 3 degrees should produce the same response as a 45-minute exposure to an increase of 1 degree. Clearly, the appropriateness of applying Haber's Law depends on the phar-

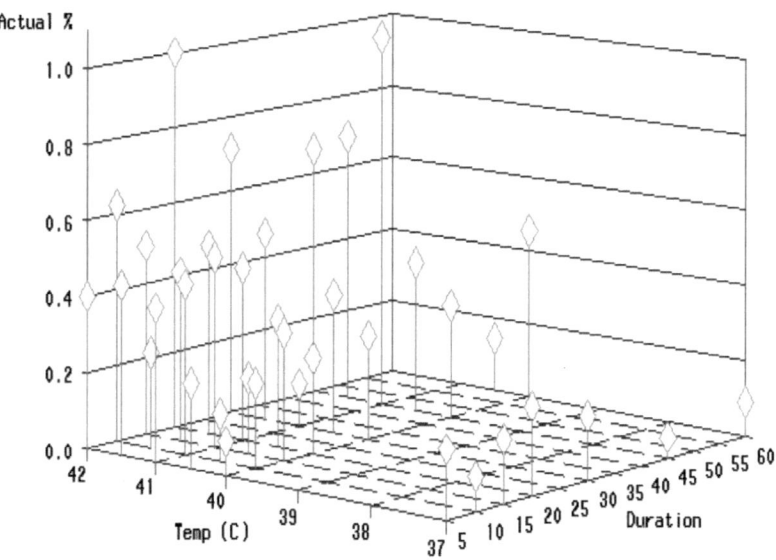

Figure 2.10 *Heatshock Studies. Actual percentage of affected embryos (experimental data points only).*

macokinetics of the particular agent, the route of administration, the target organ and the dose/duration patterns under consideration. Although much attention has been focused on documenting exceptions to this rule, it is often used as a simplifying assumption in view of limited testing resources and the multitude of exposure scenarios. However, given the current desire to develop regulatory standards for a range of exposure durations, models flexible enough to describe the response patterns over varying levels of both exposure concentration and duration are greatly needed.

For the heatshock studies, the vector of exposure covariates must incorporate both exposure level (also referred to as temperature or dose), d_{ij}, and duration (time), t_{ij}, for the jth embryo within the ith cluster. Furthermore, models must be formulated in such a way that departures from Haber's premise of the same adverse response levels for any equivalent multiple of dose times duration can easily be assessed. The exposure metrics in these models are the cumulative heat exposure, $(dt)_{ij} = d_{ij}t_{ij}$, referred to as *durtemp* and the effect of duration of exposure at positive increases in temperature (the

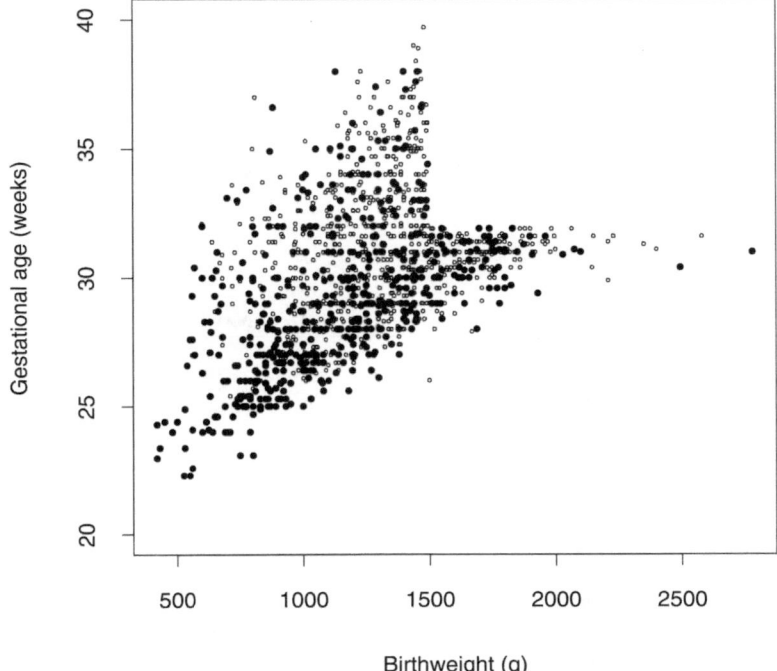

Figure 2.11 *POPS Data. The open circles correspond to zero outcomes.*

increase in temperature over the normal body temperature of $37°C$):

$$(pd)_{ij} = t_{ij}I(d_{ij} > 37).$$

We refer to the latter as *posdur*. We will return to this subject in Chapter 13.

2.3 Belgian Health Interview Survey

In 1997, the second Belgian Health Interview Survey took place. The HIS1997 was conducted to evaluate the usefulness of a periodic health-related survey, with the idea to collect information on the subjective health of the Belgian population, as well as on important predictor variables.

The main goal of the HIS is to give a description of the health status of the overall population in Belgium as well as of the three regional subpopulations (Flemish, Walloon and Brussels region), and in addition of the German community. The idea is to obtain a reflection of how specific groups of people experience their health, to what extent they utilize health care facilities, and how they look after their own health by adopting a certain life-style or by relying on preventive and other health services. Precisely, the focus is on: (1) identification of health problems, (2) description of the health status and

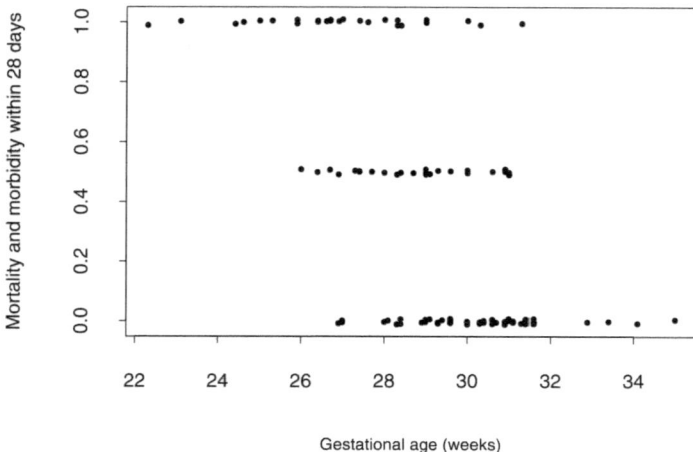

Figure 2.12 *Twins data.*

health needs of the population, (3) estimation of prevalence and distribution of health indicators, (4) analysis of social (in)equality in health and access to the health services and (5) study of health consumption and its determinants.

The target population is defined as all people residing in Belgium at a particular point in time. Due to the selection of a sample frame and practical considerations, not all persons belonging to this target population will or can be considered for the survey. Because the National Register is being used as the sampling frame, only people listed in this register can participate in the survey. This implies that no information about the health status of, for example, the homeless can easily be collected.

The total number of successful interviews for the sample was set to 10,000. The sampling of the households and respondents is a combination of several sampling techniques: stratification, multistage sampling and clustering, and differential selection probabilities. The sampling of respondents took place in the following steps: (1) stratification by region and province, (2) selection of the municipalities within each stratum, (3) selection of a cluster of households within each municipality and (4) selection of respondents within a household.

The use of the National Register and interviewers' travel are two important concerns, which need to be balanced carefully against coverage of the population. Even when a relatively good list is available (such as the National Register), a direct selection from this list would be too expensive, because the spread would be too wide. Cost savings may allow the investigators to use a larger sample size than they could use for a simple random sample of the same cost. Therefore, a multi-stage design with municipalities as primary selection units is a feasible solution.

The main advantage of stratification is that it typically produces a smaller bound on the error of estimation than would be produced by a simple random sample of the same size. This result is enforced if strata are largely homogeneous. In the HIS, there are two stratification levels (at the regional and provincial levels). Within a region, a proportional representation per province in the base sample of 10,000 is sought. A simple random sample of municipalities within a region would ascertain this condition from the sampling framework point of view. Resulting differences are regarded as purely random. However, stratifying proportionally over provinces further controlled this random variation.

Second- and third-stage selection units are households within municipalities and individuals within households, respectively. Municipalities are established administrative units, they are stable (in general they do not change during the time the survey is conducted) and they are easy to use in comparison with other specialized sources of data related to the survey. Municipalities are preferred above regions or provinces, because the latter are too large and too few. The large variation in the size of the municipalities is controlled for by systematically sampling within a province with a selection probability proportional to their size.

Within each municipality, a sample of households is drawn such that blocks (also referred to as groups) of 50 individuals in total can be interviewed. Whereas the stratification effects and the systematic sampling according to municipalities have the effect of increasing the precision, the clustering effect (selecting blocks of 50) might slightly reduce precision, since units will resemble each other more than in a simple random sample. However, since stratification is based on unequal probabilities (to guarantee a meaningful sample size per stratum) a slight decrease in overall efficiency is to be expected.

As a result, data are clustered at the levels of municipality and household. The study of this is undertaken in Chapter 15. In Chapter 8, these data were used to illustrate the use of flexible polynomial models.

2.4 POPS Data

The Project On Preterm and Small-for-gestational age infants (POPS) collected information on 1338 infants born in the Netherlands in 1983 and having gestational age less than 32 weeks and/or birthweight less than 1500 g (Verloove *et al.* 1988). The outcome of interest here concerns the situation after two years. The binary variable is 1 if an infant has died within two years after birth or survived with a major handicap, and 0 otherwise. Some of the recorded observations are from twins or triplets. So, one might have to account for the association between siblings of the same twin (or triplet, ...). Another interesting aspect is that there are observations on both cluster and individual level. For example, for a twin, the mother's age and the gestational age is the same for both siblings, while birthweight is subject specific.

The POPS data are shown graphically in Figure 2.11, and Figure 2.12 is a jittered scatter plot of the individual neonatal mortality and morbidity outcomes of the 107 twins as a function of gestational age. The data are used in

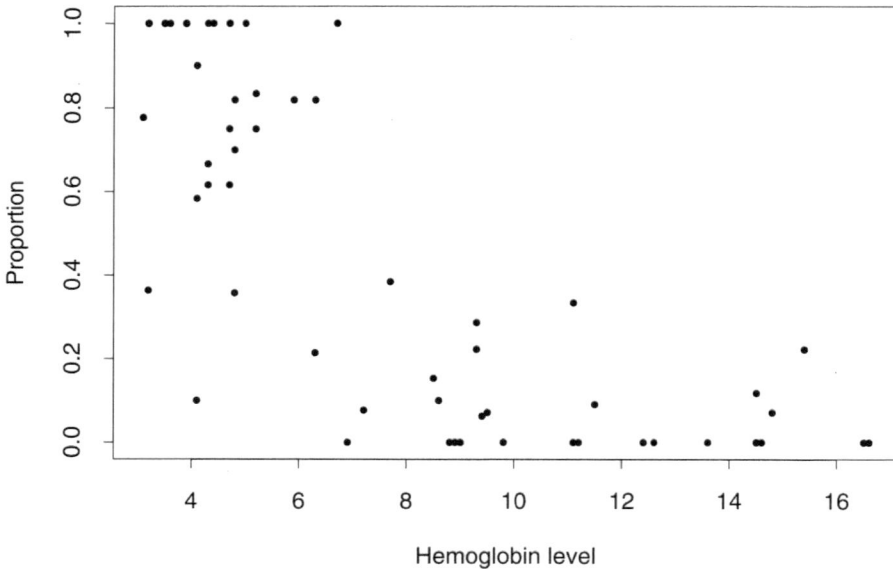

Figure 2.13 *Low-iron rat teratology data.*

Chapters 8 and 9 to illustrate how classical regression smoothers and omnibus lack-of-fit tests can be generalized to a multiparameter likelihood framework.

2.5 Low-iron Rat Teratology Data

This example uses data from the experimental setup from Shepard, Mackler and Finch (1980). A total number of 58 female rats was given different amounts of iron-supplement (ranging from normal to zero level). The rats were made pregnant and sacrificed 3 weeks later. For each female rat the hemoglobin levels were recorded, as well as its total number of foetuses (ranging from 1 to 17) and the number of dead foetuses. The proportion of dead foetuses as a function of the mother animal's hemoglobin level is shown in Figure 2.13. This plot does not give any information about the correlation structure within a litter. For this kind of clustered binary data it would be interesting to have an estimation method resulting in two smooth curves: the estimated probability of death *and* the estimated correlation within a cluster, both as a function of the covariate. Such a method will be discussed in Chapter 8, together with a bootstrap method to construct simultaneous confidence intervals for both curves.

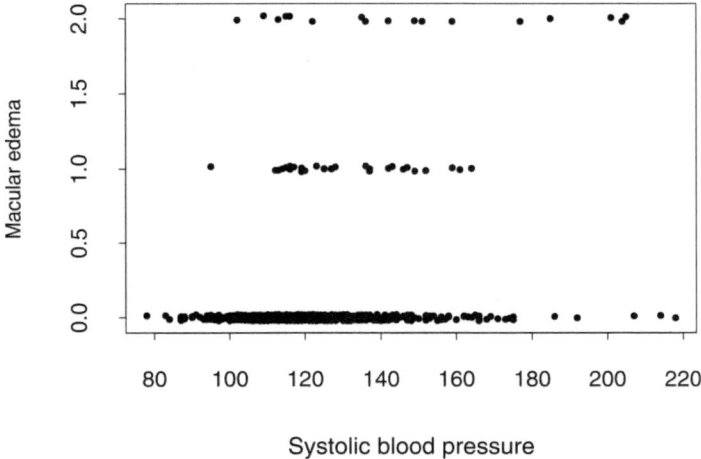

Figure 2.14 *Wisconsin diabetes study.*

2.6 The Wisconsin Diabetes Study

In this data set there are records from 720 younger onset diabetic persons. Both eyes of each person are examined for the presence of macular edema. See Klein, Klein, Moss, Davis and DeMets (1984) for more details. In the study there were 29 individuals where macular edema is present at only one eye, for 17 of them it is observed at both eyes, and for the remaining 674 persons, it was completely absent. We will study the probability of macular edema as a function of the patient's systolic blood pressure, hereby taking the clustered nature of the data into account, as indeed the response values of both eyes are likely to be correlated. A graphical representation of these data, as in Figure 2.14, can easily yield some idea of how the proportion of macular edema-infected eyes varies with the person's systolic blood pressure.

Such a graph, however, does not give any information about the correlation between the outcomes of both eyes. Most often, this correlation is just assumed to be some constant, which can be estimated from the data. Chapters 8 and 9 present some graphical diagnostic tools and a formal lack-of-fit test to examine the validity of such an assumption.

2.7 Congenital Ophthalmic Defects

Table 2.6 contains the proportions (z_i/n_i), within two age groups, of rejected corneal grafts (z_i) out of grafts received (n_i) for 9 children diagnosed with an ophthalmic dysfunction congenital hereditary endothelial dystrophy (CHED). A study of these children was made by Schaumberg *et al.* (1999), in part to assess the impact of potential risk factors on the success of corneal implants to correct the loss of visual acuity resulting from CHED. Seven of the children

Table 2.6 *Congenital Ophthalmic Defects. Proportion (z_i/n_i) of rejected corneal grafts (z_i) out of number of grafts given (n_i) for nine children affected by CHED.*

Age at Diagnosis (years)	
≤ 3	≥ 3
0/2	0/2
0/2	1/2
0/2	1/2
0/2	1/1
	1/1

received implants in both eyes, and two received implants in only a single eye. All of the children were diagnosed before the age of 6 years, and all received implants before the age of 12 years. Early intervention in cases of CHED may be critical to prevent amblyopia, commonly known as "lazy eye", due to opacification of the corneas. However, the necessary surgery carries greater risk when performed on young children as opposed to older children and adults. The investigators were therefore interested in comparing the results between two age groups: those younger than 3 years versus those aged at least 3 years at the time of surgery.

The small sample sizes reported in this and other studies of CHED reflect the extreme rarity of the condition. The investigators were unfortunately unable to apply inferential procedures to their data because of this. The outcome of interest in Table 2.6 is whether or not an eye of a given child rejected the initial implant. Of the eight eyes among the younger children, there were no rejections, whereas four of the eight implants were rejected among the older children. We wish to know whether this implies a higher average rejection rate among children over 3 years of age compared to those who are younger. Likelihood or quasi-likelihood procedures that require estimation of an age effect clearly fail for these data, since all 4 rejections occurred in older children.

2.8 A Developmental Toxicology Study

The data of Table 2.7 are reprinted from Bradstreet and Liss (1995). One hundred female mice were randomized to either control or one of three dose levels (8, 80 or 800 mg/kg) of a potentially harmful drug, then mated with untreated male mice. While 25 animals were assigned to each treatment, not all were impregnated (one each in the control and 800 mg/kg groups, and two in the 80 mg/kg group). The drug was administered orally from day 6 through day 15 of the gestational period. At day 17 of gestation, the animals were sacrificed and their offspring observed for external, visceral and skeletal

Table 2.7 *Oral Teratogenic Data. Proportion (z_i/n_i) of foetuses with ≥ 1 malformation (z_i) out of total foetuses (n_i) born to exposed dams.*

	Dose (mg/kg)			
	0	8	80	800
	0/7	0/10	0/11	0/11
	0/8	3/10	0/11	0/11
	0/9	0/11	0/11	0/11
	0/10	0/11	0/11[†]	2/11[†]
	0/10	0/12	0/12	0/12
	0/11	0/12	0/12	1/12
	0/11	0/12	0/12[†]	0/13
	0/11	0/12	1/12	0/13
	1/11	0/12	2/12	0/13
	0/12	0/12	0/13	0/13
	0/12	0/13	0/13	1/13
	1/12	0/13	0/13	2/13
	0/13	0/13	1/13	0/14
	0/13	1/13	0/14	0/14
	0/13	0/14	0/14	0/14
	0/13	0/14	0/14	1/14
	0/13	0/14	1/14	1/14
	1/13	0/14	0/15	0/15
	0/14	1/14	1/15	1/15[†]
	0/14	1/14[†]	2/15	0/16
	0/14	0/15	0/16	0/16
	0/14[†]	0/15	0/16	1/16
	1/14	0/15[†]	0/16	0/17
	0/16	0/15		0/17
		0/17		
Total	4/288	6/327	8/305	10/328
(% Malformed)	(1.4%)	(1.8%)	(2.6%)	(3.0%)

[†]One additional foetus born dead but not examined for malformations.

malformations. The table records the proportion (z_i/n_i) of foetuses (z_i) per litter (n_i) who were observed to have ≥ 1 malformation.

We wish to know whether the increasing marginal empirical probability of response across dose groups provides evidence of a dose effect, after adjusting appropriately for the clustering. While the number of clusters (96) and total sample size (1248) are relatively large, the number of affected foetuses (28) seems relatively small. The investigator may feel uncomfortable using asymptotic approximations to assess a dose-response relationship.

CHAPTER 3

Issues in Modeling Clustered Data

Louise M. Ryan

Harvard School of Public Health, Boston, MA

In Figure 1.4, the complexity of developmental toxicity studies is clearly illustrated. This, of course, implies there are a number of non-trivial challenges for model development (Molenberghs *et al.* 1998, Zhu and Fung 1996). Let us list the most important ones.

- Because of genetic similarity and the same treatment conditions, offspring of the same mother behave more alike than those of another mother. This has been termed *litter effect*. As a result, responses on different foetuses within a cluster are likely to be correlated, inducing extra variation in the data relative to those associated with the common binomial or multinomial distribution. This extra variation must be taken into account in statistical analyses (Chen and Kodell 1989; Kupper *et al.* 1986).

- Since deleterious events can occur at several points in development, an interesting aspect lies in the staging or *hierarchy* of possible adverse fetal outcomes (Williams and Ryan 1996). Ultimately, a model should take into account this hierarchical structure in the data: (i) a toxic insult early in gestation may result in a resorbed fetus; (ii) thereafter an implant is at risk of fetal death; (iii) foetuses that survived the entire gestation period are threatened by low birth weight and/or several types of malformation.

- While some attempts have been made for the joint analysis of prenatal death and malformation (Chen *et al.* 1991, Ryan 1992), the analysis of developmental toxicity data has usually been conducted on the number of viable foetuses alone. An appropriate statistical model should then account for possible *correlations among the different fetal endpoints*.

- As the number of viable foetuses can sometimes affect the chance of an adverse effect (in a large litter a larger number of animals have to compete for the same maternal resources and therefore the probability of malformation may be larger), a model should also be flexible enough to allow *litter size* to affect response probabilities.

- Finally, one may have to deal with outcomes of a *mixed* continuous (e.g., low birth weight) *versus* discrete (e.g., malformation indicator) nature.

In subsequent sections, we will discuss several of these issues in turn. A detailed treatment will then be given in following chapters.

3.1 Choosing a Model Family

As a result of the research activity over the past 10 to 15 years, there are presently several different schools of thought regarding the best approach to the analysis of correlated binary data. See also Chapter 4. Unlike in the normal setting, marginal, conditional, and random-effects approaches tend to give dissimilar results, as do likelihood, quasi-likelihood, and generalized estimating equations (GEE) based inferential methods. Prentice (1988), Fitzmaurice, Laird and Rotnitzky (1993), Diggle, Liang and Zeger (1994), and Pendergast *et al.* (1996) present excellent reviews.

Several likelihood based methods have been proposed. Fitzmaurice and Laird (1993) incorporate marginal parameters for the main effects in this model and quantify the degree of association by means of conditional odds ratios. Fully marginal models are presented by Bahadur (1961) and Cox (1972), using marginal correlations, and by Ashford and Sowden (1970), using a multiply dichotomized version of a multivariate normal to analyze multivariate binary data. Alternatively, marginal odds ratios can be used, as shown by Dale (1986) and Molenberghs and Lesaffre (1994, 1999). Cox (1972) also describes a model with parameters that can be interpreted in terms of conditional probabilities. Similar models were proposed by Rosner (1984) and Liang and Zeger (1989). Random-effects approaches have been studied by Stiratelli, Laird and Ware (1984), Zeger, Liang and Albert (1988), Breslow and Clayton (1993), and Wolfinger and O'Connell (1993). Generalized estimating equations were developed by Liang and Zeger (1986).

The debate continues about the relative merits of the different approaches. For several years it seemed that marginal models, particularly GEEs, were the most popular, perhaps due to their relative computational ease and the availability of good software. It is noteworthy that the recent renewed interest in random-effects models is partly provoked by the availability of the GLIM-MIX macro in SAS and most recently by the NLMIXED procedure in SAS. There are merits and disadvantages to all three model families and generally no simple transformations between the three families exist. Arguably, model choice has to depend not only on the application of interest but also on the specific analysis goals.

Because of the need to account for litter effects, all these issues of modeling strategy arise with developmental toxicity data. Several additional issues complicate the analysis. For example, cluster sizes vary and can affect response rates, perhaps due to competition between littermates or underlying health of the mother (Rai and Van Ryzin 1985). Also, it is often important to account for the multivariate nature of the outcomes measured on each littermate. Random-effects models (beta-binomial, Williams 1975) were among the first proposed for developmental toxicity data (see also Chen and Kodell 1989).

However, they do not extend naturally to multivariate outcomes. Lefkopoulou, Moore and Ryan (1989) apply the generalized estimating equations ideas to model multiple binary outcomes measured on clusters of individuals. While their approach is simple to apply, and leads to easily interpreted tests, a disadvantage is lack of a likelihood basis. Furthermore, there are some regions of the parameter space where the method can be quite inefficient (Lefkopoulou and Ryan 1993). Also, therefore, the approach does not lend itself well to quantitative risk assessment.

Due to the popularity of marginal (especially GEE) and random-effects models for correlated binary data, conditional models have received relatively little attention, especially in the context of multivariate clustered data. A noticeable exception is Liang and Zeger (1989), although this approach was criticized (Diggle, Liang and Zeger, 1994, p. 147) because the interpretation of the dose effect on the risk of one outcome is conditional on the responses of other outcomes for the same individual, outcomes of other individuals, and the litter size. Nevertheless, there are some advantages to conditional models and with appropriate care the disadvantages can be overcome. See also Section 3.4.

3.2 Joint Continuous and Discrete Outcomes

Developmental toxicity studies may seek to determine the effects of dose on fetal weight (continuous) and malformation incidence (binary) simultaneously, as both have been found to be indicative of a toxic effect. This motivates the formulation of a joint distribution with mixed continuous and discrete outcomes. However, this is not standard.

Catalano and Ryan (1992) note that latent variable models provide a useful and intuitive way to motivate the distribution of the discrete outcome. Such models presuppose the existence of an unobservable, normally distributed random variable, underlying the binary outcome. The binary event is then assumed to occur if the latent variable exceeds some threshold value. They further note that this notion of latent variables has much appeal to toxicologists, because it provides a natural and intuitive framework for the biological mechanism leading to adverse events such as malformation.

A flexible latent variable approach to model an arbitrary number of continuous and discrete outcomes, each of which follows an exponential family distribution, is proposed by Sammel, Ryan and Legler (1997). They introduce a modified EM algorithm for parameter estimation with either a simple Monte Carlo expectation or a numerical integration technique based on, e.g., Gauss-Hermite quadrature to approximate the E-step which is not necessarily available in closed form. The method allows for arbitrary covariate effects and estimates of the latent variable are produced as a by-product of the analysis. However, their approach does not extend to correlated (i.e., clustered) data.

In the context of developmental toxicity studies, the dose-response model is often characterized in each of the two outcomes (weight and malformation) separately, using appropriate methods to account for correlation induced by the clustering of foetuses within litters, or the well-known "litter-effect". The more sensitive of the two outcomes is determined based on the dose-response patterns and used for risk assessment purposes. However, because these outcomes are correlated (Ryan *et al.* 1991), jointly modeling the outcomes and using the bivariate outcome as a basis for risk assessment may be more appropriate (Regan and Catalano 1999a). A standard approach is to apply a conditioning argument that allows the joint distribution to be factorized in a marginal component and a conditional component, where the conditioning can be done on either the discrete or continuous outcome (Catalano and Ryan 1992, Cox and Wermuth 1992, Cox and Wermuth 1994, Fitzmaurice and Laird 1995, Olkin and Tate 1961). Cox and Wermuth (1992, 1994) consider various factorization methods and tests for independence. Let us discuss some factorization methods.

Catalano and Ryan (1992) apply the latent variable concept to derive the joint distribution of a continuous and a discrete outcome and then extend the model, using GEE ideas, to incorporate clustering. They parameterize the model in a way that allows to write the joint distribution as the product of the marginal distribution of the continuous response, and the conditional distribution of the binary response given the continuous one. The marginal distribution of the continuous response is related to covariates, using a linear link function, while for the conditional distribution they use a probit link. Due to the non-linearity of the link function relating the conditional mean of the binary response to the covariates, the regression parameters in the probit model of Catalano and Ryan (1992) have no direct marginal interpretation. Furthermore, if the model for the mean has been correctly specified, but the model for the association between the binary and continuous outcomes is misspecified, the regression parameters in the probit model are not consistent. The lack of marginal interpretation and lack of robustness may be considered unattractive features of this approach. An important advantage, however, is that it can be readily extended to allow for clustering. Fitzmaurice and Laird (1995) circumvent the difficulties in the approach of Catalano and Ryan (1992) by factorizing the joint distribution as the product of a marginal Bernoulli distribution for the discrete response, and a conditional Gaussian distribution for the continuous response given the discrete one. Under independence, their method yields maximum likelihood estimates of the marginal means that are robust to misspecification of the association between the binary and continuous response. They also consider an extension of their model that allows for clustering. By using GEE methodology, they avoid the computational complexity of maximum likelihood in this more elaborate setting. A conceptual difficulty with this model is the interpretation of the parameters, which depends on cluster size.

A drawback of mixed outcome models based on factorization (as above) is that they may be difficult to apply for quantitative risk assessment (Geys *et al.* 2001, Regan and Catalano 1999a). While taking into account the dependence between weight and malformation, the intrafoetus correlation itself cannot be directly estimated. Thus, an expression for the joint probability that a fetus is affected (i.e., malformed and/or of low birth weight) is difficult to specify. Catalano et al. (1993) used a factorization model for quantitative risk assessment, in which direct estimation of the bivariate correlation is approximated using a conditioning argument. To overcome this problem, one needs joint models that incorporate the correlation between outcomes directly. Thus, a desirable model should have three properties:

- it allows separate dose-response functions for each component of the bivariate outcome,

- it accounts for the correlations due to clustering within litters,

- it estimates the bivariate intrafoetus association.

In Chapter 14, we will propose models that satisfy these properties (see also Geys et al. 2001 and Regan and Catalano 1999a).

3.3 Likelihood Misspecification and Alternative Methods

Likelihood methods enjoy many desirable properties, such as efficiency under appropriate regularity and the ability to calculate functions of interest based on the posited parametric model and the maximum likelihood estimates of the parameters (Edwards 1972, Welsh 1997). These properties come at a price.

First, not only the specification of a likelihood function but also the estimation of the parameters thereof can be computationally intensive. That is why several chapters of this book are devoted to alternatives, such as generalized estimating equations (Chapter 5) or pseudo-likelihood (Chapters 6 and 7). Broadly, generalized estimating equations are a replacement for the often complicated true first derivatives of the log-likelihood function (i.e., the score equations). The theory of Liang and Zeger (1986) establishes the conditions under which solving such equations yields correct inference. Pseudo-likelihood is based on directly replacing the complicated log-likelihood function with a simpler pseudo-likelihood function. Again, it has been established that valid inference can be obtained from such models (Arnold and Strauss 1991, Geys, Molenberghs and Ryan 1999).

Second, the advantage of fully specifying the joint probability model for a vector (e.g., a cluster) of outcomes also carries a danger: there is an increased risk for misspecification. This is well known throughout the statistical literature when a choice has to be made between classical or robust methods (e.g., a mean *versus* a median, a linear *versus* a quadratic function in discriminant analysis, etc.). In addition, it is well known that the impact on misspecification is different depending on the goal of inference. In linear mixed models theory (Verbeke and Molenberghs 2000), it is known that the fixed-effects structure is less sensitive to misspecification than the variance-covariance structure. When modeling dichotomous outcomes, it has been established that the choice between classical link functions is almost immaterial *for the central area of the unit probability interval*. In other words, for probabilities between 30–70%, or even 20–80%, one can choose between a logit or a probit link on other grounds than merely model fit.

However, in some cases, *as is the case here*, one wants to explicitly address probabilities at the extreme of the scale. Indeed, the ultimate goal of developmental toxicity studies is to perform risk assessment, i.e., to set safe limits for human exposure, based on the fitted model (Crump 1984). Risk assessment is developed in detail in Chapter 10; see also Section 3.4. To this end, models should fit the data well. This has implications for both the model family chosen, as well as for the form of the linear predictors. Since classical polynomial predictors are often of poor quality, especially when low dose extrapolation is envisaged, there is a clear need for alternative specifications of the predictors describing main effects and associations. Non-linear predictors pose non-trivial challenges. We apply the fractional polynomial approach of Royston and Altman (1994), which provides more flexibly shaped curves than conventional polynomials. They argue that conventional low order polynomials offer only a limited family of shapes and that high order polynomials may fit poorly at the extreme values of the covariates. Moreover, polynomials do not have finite asymptotes and cannot fit the data where limiting behavior is expected. The method was also applied by Royston and Wright (1998) for the construction of age-specific reference intervals and by Sauerbrei and Royston (1999) for building prognostic and diagnostic indices for multivariate models. An attractive feature is that conventional polynomials are included as a subset of this extended family. Fractional polynomial models are flexible but nevertheless parametric and user-defined. More non-parametrically inspired approaches, fully data driven, can be very useful. In an explorative way, a parametric model can be graphically compared with its non-parametric alternative. In the one-parameter case, several authors have examined strategies to implement nonparametric estimation procedures in likelihood based regression models. The local polynomial fitting, which has become the standard in kernel smoothing, produces smoothers that have several advantages in comparison with other linear smoothers, such as the behavior at the boundary. Chapter 8 provides a detailed treatment of both fractional polynomial methods and local polynomial smoothers.

The construction of more flexible predictors is a major step forward. At the same time, it is imperative to investigate the fit of the models considered.

The lack-of-fit of a regression model is investigated by testing the hypothesis that a function has a prescribed parametric form. The function of interest can be one of the parameters in a regression model; typically the mean of the response, but it might also be its variance, or the correlation between different outcomes. In other cases, it might be a complete density function of which we want to investigate the goodness-of-fit. Parametric testing methods are designed to detect very specific types of departures from the hypothesized model. For example, likelihood ratio, Wald, or score tests are employed to contrast a linear and a quadratic dose-response curve. While very powerful for this particular class of alternative models, these tests quickly lose power when the truth is more complicated. In Chapter 9 the omnibus nonparametric methods of this chapter are appealing in that they are consistent against virtually any departure from the hypothesized parametric model. An adaption of the Hosmer-Lemeshow (1989) approach, for application to clustered binary data, is constructed. Further, order selection tests based on orthogonal series estimators are discussed.

Chapter 11 is devoted to the specific implications of model misspecification on the dose-effect assessment on the one hand, and on safe-dose determination on the other hand. When the data do not come from the assumed parametric model, the usual asymptotic chi-squared distribution under the null hypothesis remains valid for "robustified" Wald and score test statistics. For full likelihood models, robust Wald and score tests have been described in the literature (Kent 1982, Viraswami and Reid 1996). The modified tests again have an asymptotic chi-squared distribution, even when the assumed model is not correct. Robust test statistics are also used in the context of generalized estimating equations (Liang and Zeger 1986, Rotnitzky and Jewell 1990) and of pseudo-likelihood (Geys, Molenberghs and Ryan 1999). Aerts and Claeskens (2001) compare the performance of such a chi-squared approximation to that of a semiparametric bootstrap method. In the context of likelihood-based estimation, Williams and Ryan (1996) have indicated it is preferable to define the BD using the likelihood ratio statistic. As indicated earlier, a full likelihood technique will perform best when the likelihood is correctly specified, but one might expect problems in case of misspecification (Aerts, Declerck and Molenberghs 1997). Therefore it is important to look at robust estimation methods such as quasi-likelihood, GEE, and pseudo-likelihood. Further, the likelihood method is unavailable in quasi-likelihood settings, and hence also in GEE, since there is no analogue to the likelihood ratio statistic. Precisely, the use of a profile score approach has been proposed by Claeskens, Aerts, Molenberghs and Ryan (2002).

The justification of inferences usually rests upon the approximate normality of the statistics of interest. Such a distributional assumption may be untenable when samples are small or sparse. If a normal approximation is not accurate, the result might be tests that do not preserve the *a priori* testing level established by the investigator. Likewise, actual coverage probabilities for confidence intervals may be much lower or higher than the nominal confidence level. Moreover, where likelihood or quasi-likelihood methods are applied, inference can be further complicated when parameter estimates lie at or near the boundary of the parameter space. Exact inference then provides a sensible alternative. Strategies developed to this end are studied in Chapter 12.

3.4 Risk Assessment

In this section we zoom in on risk assessment, which is further studied in detail in Chapter 10.

Risk assessment can be defined as (Roberts and Abernathy 1996): "the use of available information to evaluate and estimate exposure to a substance and its consequent adverse health effects." An important goal in the risk assessment process is to determine a safe level of exposure. Traditionally, quantitative risk assessment in developmental toxicology has been based on the NOAEL, or No Observable Adverse Effect Level, which is the dose immediately below that deemed statistically or biologically significant when compared with controls. The NOAEL, however, has been criticized for its poor statistical properties (see, for example, Williams and Ryan 1996). Therefore, interest in developing techniques for dose-response modeling of developmental toxicity data has increased, and new regulatory guidelines (U.S. EPA 1991) emphasize the need of quantitative methods for risk assessment. The standard approach requires the specification of an adverse event, along with $r(d)$ representing the probability that this event occurs at dose level d. For developmental toxicity studies where offspring are clustered within litters, there are several ways to define the concept of an adverse effect. First, one can state that an adverse effect has occurred if a particular offspring is abnormal (fetus based). Alternatively, one might conclude that an adverse effect has occurred if at least one offspring from the litter is affected (litter based). Based on this probability, a common measure for the excess risk over background is defined as

$$r^*(d) = r(d) - r(0)$$

or as

$$r^*(d) = \frac{r(d) - r(0)}{1 - r(0)}, \tag{3.1}$$

where definition (3.1) puts greater weight on outcomes with large background risks. The benchmark dose (BMD$_q$), sometimes also called the effective dose (ED$_q$), is then defined as the dose satisfying $r^*(d) = q$, where q corresponds to the pre-specified level of increased response and is typically specified as $0.01, 1, 5$, or 10% (Crump 1984).

In practice, calculation of the BMD follows several steps. After choosing and fitting an appropriate dose-response model, the excess risk function is solved for the dose, d, that yields $r^*(d) = q$. Since the dose-response curve is estimated from data and has inherent variability, the BMD is itself only an estimate of the true dose that would result in this level of excess risk. The final step therefore consists of acknowledging this sampling uncertainty for the model on which the BMD$_q$ is based, by replacing the BMD$_q$ by its lower confidence limit (Williams and Ryan 1996). Several approaches have been proposed.

Using the delta method, a Wald based method can be used:

$$\widehat{BMDL}_q = \widehat{BMD}_q - 1.645\sqrt{\widehat{\text{Var}}(\widehat{BMD}_q)}.$$

Assume that $\boldsymbol{\beta}$ is the vector of parameters included in the dose-response model; then the BMD$_q$ variance can be obtained from the variance matrix of $\boldsymbol{\beta}$. Several authors have indicated that this method suffers from drawbacks, especially with low dose extrapolation (Aerts, Declerck and Molenberghs 1997,

Crump 1984, Crump and Howe 1985, Krewski and Van Ryzin 1981) in which case the method may yield negative lower limits. Furthermore, Catalano, Ryan and Scharfstein (1994) have empirically found that this method can yield unstable estimates.

Alternatively, an upper limit for the risk function can be computed, and thus the dose that corresponds to a $q\%$ increased response above background is determined from this upper limit curve by solving:

$$\hat{r}^*(d) + 1.645\sqrt{\widehat{\text{Var}}(\hat{r}^*(d))} = q,$$

where the variance of the estimated increased risk function $\hat{r}^*(d)$ is estimated as:

$$\widehat{\text{Var}}(\hat{r}^*(d)) = \left(\frac{\partial r^*(d)}{\partial \boldsymbol{\beta}}\right)^T \widehat{\text{Cov}}(\hat{\boldsymbol{\beta}}) \left(\frac{\partial r^*(d)}{\partial \boldsymbol{\beta}}\right)\Bigg|_{\boldsymbol{\beta}=\hat{\boldsymbol{\beta}}}$$

and where $\widehat{\text{Cov}}(\hat{\boldsymbol{\beta}})$ is the estimated covariance matrix of $\hat{\boldsymbol{\beta}}$. The resulting dose level is referred to as the lower effective dose (LED_q) (Kimmel and Gaylor 1988).

Crump and Howe (1985) recommend using the asymptotic distribution of the likelihood ratio (if available). According to this method, an approximate $100(1-\alpha)\%$ lower limit for the BMD, denoted by BMD(1), corresponding to an excess risk of q is defined as

$$\min\{d(\boldsymbol{\beta}) : r(d;\boldsymbol{\beta}) = q \text{ over all } \boldsymbol{\beta} \text{ such that } 2(\ell(\hat{\boldsymbol{\beta}}) - \ell(\boldsymbol{\beta})) \leq \chi_p^2(1-\alpha)\},$$

where ℓ denotes the log-likelihood and p is the number of model parameters. A second approach, denoted BMD(2), is based on the profile likelihood method (Morgan 1992). First, construct a profile likelihood based confidence interval for the dose effect parameter β_d. Second, transform this interval into an interval for d and check that the transformation is monotonic. Aerts, Declerck and Molenberghs (1997) compare the different lower limits for the BMD and show that, in general, BMD(1) yields lower results than BMD(2). Furthermore, they note that for conditionally specified models, the transformation is not monotonic, and hence the BMD(2) should not be applied to such models. A variation on this theme, suggested by many authors (Chen and Kodell 1989, Ryan 1992, Gaylor 1989), first determines a lower confidence limit, e.g., corresponding to an excess risk of 1%, and then linearly extrapolates it to a BMD. The main advantage quoted for this procedure is that the determination of a BMD is less model dependent.

CHAPTER 4

Model Families

Geert Molenberghs

transnationale Universiteit Limburg, Diepenbeek–Hasselt, Belgium

In most developmental toxicity studies, exposure is administered to the dam, rather than directly to the developing foetuses. Because of genetic similarity and the same treatment conditions, offspring of the same mother behave more similar than those of another mother. This has been termed "litter effect" and is one important form of clustering. There are several ways to handle clustering. While dose-response modeling is relatively straightforward in uncorrelated settings, it is less so in the clustered context. Of course, one can ignore the clustering altogether by treating the littermates as if they were independent. However, this will in general be too strong an assumption. Also, the litter effect issue can be avoided by modeling the probability of an affected cluster via, e.g., a logistic regression model. Such models are generally too simplistic but there is a multitude of models which do consider clustering.

Indeed, failure to account for the clustering in the data can lead to serious underestimation of the variances of dose effect parameters and, hence, inflated test statistics. The need for methods that appropriately account for the heterogeneity among litters, especially with regard to binary outcomes, has long been recognized. When the response is continuous and assumed to be approximately Gaussian, there is a general class of linear models that is suitable for analyses (see Section 4.3.2). However, when the response variable is categorical, fewer techniques are available. This is partly due to the lack of a discrete analogue to the multivariate normal distribution. The use of binomial or Poisson models in toxicological testing has frequently been criticized on the grounds that they generally poorly fit actual experimental data. This is caused by extra-binomial variation, i.e., more variability among litters than would be expected based on binomial or Poisson models. In an attempt to explain this variation, a number of generalized linear models have been proposed. Williams (1975) assumes that foetuses in the same litter provide a set of independent Bernoulli responses conditional on the litter-specific success probability, and that the variation in this probability from litter to litter follows a beta-distribution. Haseman and Kupper (1979) provide an early survey of likelihood generalizations of standard distributions to account for clustering. Later, Pendergast *et al.* (1996) gave an overview of methods for clustered binary data. The texts by Fahrmeir and Tutz (1994) and Diggle, Liang and Zeger (1994) are very useful in this respect.

As indicated in Chapter 1, models for correlated data can be grouped into the following different classes:

- conditionally specified models,
- marginal models,
- cluster-specific models.

Note that the term *cluster-specific models* is similar to the more commonly used *random-effects models*. If a cluster is understood in the broad sense (i.e., representing one of the following: litters, families, repeated measures, longitudinal measures, etc.) then the term is more general than the random-effects terminology. Indeed, the presence of cluster-specific effects in a model can be handled in several ways. Broadly, one can treat them as either fixed effects or as random effects. A third alternative consists in conditioning upon the cluster-specific effects, a principle well known in the area of matched case-control studies, where conditional logistic regression is frequently used (Breslow and Day 1987).

The answer to the question of which model family is to be preferred depends principally on the research question(s) to be answered. In conditionally specified models the probability of a positive response for one member of the cluster is modeled conditionally upon other outcomes for the same cluster, while marginal models relate the covariates directly to the marginal probabilities. Cluster-specific models differ from the two previous models by the inclusion of parameters that are specific to the cluster. What *method* is used to fit the model should not only depend on the assumptions the investigator is willing to make, but also (to some extent) on the availability of computational algorithms. If one is willing to fully specify the joint probabilities, maximum likelihood methods can be adopted. Yet, if only a partial description in terms of marginal or conditional probabilities is given, one has to rely on non-likelihood methods such as: generalized estimating equations (Chapter 5) or pseudo-likelihood methods (Chapters 6 and 7).

4.1 Marginal Models

In marginal models, the parameters characterize the marginal probabilities of a subset of the outcomes, without conditioning on the other outcomes. Advantages and disadvantages of conditional and marginal modeling have been discussed in Diggle, Liang and Zeger (1994), and Fahrmeir and Tutz (1994).

Bahadur (1961) proposed a marginal model, accounting for the association via marginal correlations. This model has also been studied by Cox (1972), Kupper and Haseman (1978) and Altham (1978). While the general form of the Bahadur model requires the specification of a number of parameters, exponential in the cluster size, considerable simplification is possible when assuming exchangeability, in the sense that each foetus within a litter has the same malformation probability, and in addition setting all the three- and higher-way correlations equal to zero (Eq. 4.2). A drawback of the approach is the existence of severe constraints on the correlation parameter when higher-order correlations are removed. Even in the unconstrained case, the parameter space has a peculiar shape. While non-rectangular parameter spaces is a typical feature of marginal models (cf. the parameter space of covariance matrices for the normal distribution), it often poses unsurmountable problems in the case of the Bahadur model. A general study is given in Declerck, Aerts and Molenberghs (1998).

Molenberghs and Lesaffre (1994) and Lang and Agresti (1994) have proposed models which parameterize the association in terms of marginal odds ratios. Dale (1986) defined the bivariate global odds ratio model, based on a bivariate Plackett distribution (Plackett 1965). Molenberghs and Lesaffre (1994, 1999) extended this model to multivariate ordinal outcomes. They generalize the bivariate Plackett distribution in order to establish the multivariate cell probabilities. Their 1994 method involves solving polynomials of high degree and computing the derivatives thereof, while in 1997 generalized linear models theory is exploited, together with the use of an adaption of the iterative proportional fitting algorithm. Lang and Agresti (1994) exploit the equivalence between direct modeling and imposing restrictions on the multinomial probabilities, using undetermined Lagrange multipliers. Alternatively, the cell probabilities can be fitted using a Newton iteration scheme, as suggested by Glonek and McCullagh (1995).

However, even though a variety of flexible models exist, maximum likelihood can be unattractive due to excessive computational requirements, especially when high dimensional vectors of correlated data arise. As a consequence, alternative methods have been in demand. Liang and Zeger (1986) proposed so-called *generalized estimating equations* (GEE) which require only the correct specification of the univariate marginal distributions provided one is willing to adopt "working" assumptions about the association structure (see Chapter 5). le Cessie and van Houwelingen (1994) suggested to approximate the true likelihood by means of a pseudo-likelihood (PL) function that is easier to evaluate and to maximize (Chapters 6 and 7). Both GEE and PL yield consistent and asymptotically normal estimators, provided an empirically corrected variance estimator, often referred to as the sandwich estimator, is used. However, GEE is typically geared towards marginal models, whereas PL can be used with both marginal (Le Cessie and Van Houwelingen 1994, Geys, Molenberghs and Lipsitz 1998) and conditional models (Geys, Molenberghs and Ryan 1997, 1999).

Alternative marginal models include the correlated binomial models of Altham (1978) and the double binomial model of Efron (1986).

4.1.1 The Bahadur model

Let the binary response Y_{ij} indicate if foetus j of cluster i has the adverse event under investigation. To be more specific, some type of malformation is considered here. The marginal distribution of Y_{ij} is Bernoulli with $E(Y_{ij}) = P(Y_{ij} = 1) \equiv \pi_{ij}$, i.e., the probability that the foetus is affected according to the specified malformation type.

Next, to describe the association between binary outcomes, the pairwise probability $P(Y_{ij} = 1, Y_{ik} = 1) = E(Y_{ij}Y_{ik}) \equiv \pi_{ijk}$ has to be characterized. This "success probability" of two foetuses of the same dam can be modeled in terms of the two marginal probabilities π_{ij} and π_{ik}, as well as an association parameter.

Dealing with binary responses, common choices for the association parameter are the marginal odds ratio, the marginal correlation and the kappa coefficient (Agresti 1990). The marginal odds ratio will be treated in Section 4.1.3.

The marginal correlation coefficient assumes the form

$$\text{Corr}(Y_{ij}, Y_{ik}) \equiv \rho_{ijk} = \frac{\pi_{ijk} - \pi_{ij}\pi_{ik}}{[\pi_{ij}(1 - \pi_{ij})\pi_{ik}(1 - \pi_{ik})]^{1/2}}.$$

In terms of this association parameter, the joint probability π_{ijk} can then be written as

$$\pi_{ijk} = \pi_{ij}\pi_{ik} + \rho_{ijk}[\pi_{ij}(1 - \pi_{ij})\pi_{ik}(1 - \pi_{ik})]^{1/2}.$$

Hence, given the marginal correlation coefficient ρ_{ijk} and the univariate probabilities π_{ij} and π_{ik}, the pairwise probability π_{ijk} can easily be calculated. Other expressions for the associations and the pairwise probabilities can be found in Cox (1972). Bahadur (1961) and Cox (1972) consider the marginal correlation ρ_{ijk} to measure the association.

The first and second moments of the distribution have been specified. However, a likelihood-based approach requires the complete representation of the joint probabilities of the vector of binary responses in each litter. The full joint distribution $f(\boldsymbol{y})$ of $\boldsymbol{Y}_i = (Y_{i1}, \ldots, Y_{in_i})^t$ is multinomial with a 2^{n_i} probability vector. Different models put different restrictions on the 2^{n_i} joint probabilities of $\boldsymbol{Y}_i$. We will first introduce the Bahadur model. The model has been used by several authors in the context of toxicological experiments (Altham 1978, Kupper and Haseman 1978). As a consequence, it is treated here as a representative of the marginal family. The Bahadur model gives a closed form expression for the joint distribution $f(\boldsymbol{y})$. The association between binary responses is expressed in terms of marginal malformation probabilities and correlation coefficients of second, third, ... order.

Let

$$\varepsilon_{ij} = \frac{Y_{ij} - \pi_{ij}}{\sqrt{\pi_{ij}(1 - \pi_{ij})}} \quad \text{and} \quad e_{ij} = \frac{y_{ij} - \pi_{ij}}{\sqrt{\pi_{ij}(1 - \pi_{ij})}},$$

where y_{ij} is an actual value of the binary response variable Y_{ij}. Further, let $\rho_{ijk} = E(\varepsilon_{ij}\varepsilon_{ik})$, $\rho_{ijkl} = E(\varepsilon_{ij}\varepsilon_{ik}\varepsilon_{il})$, $\ldots$, $\rho_{i12\ldots n_i} = E(\varepsilon_{i1}\varepsilon_{i2}\ldots\varepsilon_{in_i})$.

Then, the general Bahadur model can be represented by the expression $f(\boldsymbol{y}_i) = f_1(\boldsymbol{y}_i)c(\boldsymbol{y}_i)$, where

$$f_1(\boldsymbol{y}_i) = \prod_{j=1}^{n_i} \pi_{ij}^{y_{ij}}(1 - \pi_{ij})^{1-y_{ij}}$$

and

$$c(\boldsymbol{y}_i) = 1 + \sum_{j<k} \rho_{ijk}e_{ij}e_{ik} + \sum_{j<k<\ell} \rho_{ijk\ell}e_{ij}e_{ik}e_{i\ell} + \ldots + \rho_{i12\ldots n_i}e_{i1}e_{i2}\cdots e_{in_i}.$$

Thus, the probability mass function is the product of the independence model $f_1(\boldsymbol{y}_i)$ and the correction factor $c(\boldsymbol{y}_i)$. The factor $c(\boldsymbol{y}_i)$ can be viewed as a model for overdispersion.

When the focus is on the special case of exchangeable littermates, this implies on the one hand that each foetus within a litter has the same malformation probability, i.e., $\pi_{ij} = \pi_i$ for littermates $j = 1, \ldots, n_i$ and litters $i = 1, \ldots, N$. On the other hand, it implies that within a litter, the associations of a particular order are constant, i.e., $\rho_{ijk} = \rho_{i(2)}$ for $j < k$, $\rho_{ijkl} = \rho_{i(3)}$ for $j < k < l$, $\ldots$, $\rho_{i12\ldots n_i} = \rho_{i(n_i)}$, with $i = 1, \ldots, N$. Under exchangeability, the Bahadur model reduces to

$$f_1(\boldsymbol{y}_i) = \pi_i^{z_i}(1 - \pi_i)^{n_i - z_i}$$

and

$$c(\boldsymbol{y}_i) = 1 + \sum_{r=2}^{n_i} \rho_{i(r)} \sum_{s=0}^{r} \binom{z_i}{s}\binom{n_i - z_i}{r - s}(-1)^{s+r}\lambda_i^{r-2s}, \qquad (4.1)$$

with $\lambda_i = \sqrt{\pi_i/(1 - \pi_i)}$. The probability mass function of Z_i, the number of malformations in cluster i, is given by

$$f(z_i) = \binom{n_i}{z_i}f(\boldsymbol{y}_i).$$

In addition, setting all three- and higher-way correlations equal to zero, the probability mass function of Z_i simplifies further to:

$$f(z_i) \equiv f(z_i|\pi_i, \rho_{i(2)}, n_i) = \binom{n_i}{z_i}\pi_i^{z_i}(1 - \pi_i)^{n_i - z_i}$$

$$\times \left[1 + \rho_{i(2)}\left\{\binom{n_i - z_i}{2}\frac{\pi_i}{1 - \pi_i} - z_i(n_i - z_i)\right.\right.$$

$$\left.\left.+ \binom{z_i}{2}\frac{1 - \pi_i}{\pi_i}\right\}\right]. \qquad (4.2)$$

This very tractable expression of the Bahadur probability mass function is advantageous over other representations, such as an odds ratio representation for which no closed form solution for the joint distribution is possible. However, a drawback is the fact that the correlation between two responses is highly constrained when the higher order correlations are removed. Even when higher order parameters are included, the parameter space of marginal parameters and correlations is known to be of a very peculiar shape. Bahadur (1961) discusses restrictions on the correlation parameters. The second order approximation in (4.2) is only useful if it is a probability mass function. Bahadur indicates that the sum of the probabilities of all possible outcomes is one. However, depending on the values of π_i and $\rho_{i(2)}$, expression (4.2) may fail to be non-negative for some outcomes. The latter results in restrictions on the parameter space which, in case of the second order approximation, are described by Bahadur (1961). From these, it can be deduced that the lower bound for $\rho_{i(2)}$ approaches zero as the cluster size increases. However, it is important to notice that also the upper bound for this correlation parameter is constrained. Indeed, even though it is one for clusters of size two, the upper bound varies between $1/(n_i - 1)$ and $2/(n_i - 1)$ for larger clusters. Taking a (realistic) litter of size 12, the upper bound is in the range $(0.09; 0.18)$. Kupper and Haseman (1978) present numerical values for the constraints on $\rho_{i(2)}$ for choices of π_i and n_i. Restrictions for a specific version where a third order association parameter is included as well are studied by Prentice (1988), while a more general situation is discussed in Appendix A.

The marginal parameters π_i and $\rho_{i(2)}$ can be modeled using a composite link function. Since Y_{ij} is binary, the logistic link function for π_i is a natural choice. In principle, any link function, such as the probit link, the log-log link or the complementary log-log link, could be chosen. A convenient transformation of $\rho_{i(2)}$ is Fisher's z-transform. This leads to the following generalized linear regression relations

$$
\left(\begin{array}{c} \ln\left(\frac{\pi_i}{1-\pi_i}\right) \\ \ln\left(\frac{1+\rho_{i(2)}}{1-\rho_{i(2)}}\right) \end{array} \right) \equiv \boldsymbol{\eta}_i = X_i\boldsymbol{\beta},
\tag{4.3}
$$

where X_i is a design matrix and $\boldsymbol{\beta}$ is a vector of unknown parameters. For example, a linear marginal logit model and a constant association $\rho_{i(2)} = \rho_{(2)}$ implies:

$$
X_i = \left(\begin{array}{ccc} 1 & d_i & 0 \\ 0 & 0 & 1 \end{array} \right) \quad \text{and} \quad \boldsymbol{\beta} = \left(\begin{array}{c} \beta_0 \\ \beta_d \\ \beta_2 \end{array} \right).
\tag{4.4}
$$

Obviously, this model can be extended by changing the design matrix and the vector of regression parameters, such that the logit of π_i depends on dose via, e.g., a quadratic or a higher order polynomial function. Also, the association parameter $\rho_{i(2)}$ can be modeled as some function of dose.

Denote the log-likelihood contribution of the ith cluster by

$$
\ell_i = \ln f(z_i | \pi_i, \rho_{(2)}, n_i).
$$

The maximum likelihood estimator $\hat{\boldsymbol{\beta}}$ for $\boldsymbol{\beta}$ is defined as the solution to the score equations $U(\beta) = 0$. The score function $U(\beta)$ can be written as

$$U(\beta) = \sum_{i=1}^{N} X_i^t (T_i^t)^{-1} L_i \tag{4.5}$$

where N is the number of clusters in the dataset,

$$T_i = \frac{\partial \eta_i}{\partial \Theta_i} = \begin{pmatrix} \frac{\partial \eta_{i1}}{\partial \pi_i} & \frac{\partial \eta_{i2}}{\partial \pi_i} \\ \frac{\partial \eta_{i1}}{\partial \rho_{(2)}} & \frac{\partial \eta_{i2}}{\partial \rho_{(2)}} \end{pmatrix} = \begin{pmatrix} \frac{1}{\pi_i(1-\pi_i)} & 0 \\ 0 & \frac{2}{(1-\rho_{(2)})(1+\rho_{(2)})} \end{pmatrix},$$

$$L_i = \frac{\partial \ell_i}{\partial \Theta_i} = \begin{pmatrix} \frac{\partial \ell_i}{\partial \pi_i} \\ \frac{\partial \ell_i}{\partial \rho_{(2)}} \end{pmatrix} \quad \text{and}$$

$\Theta_i = (\pi_i, \rho_{(2)})^t$, the set of natural parameters.

A Newton-Raphson algorithm can be used to obtain the maximum likelihood estimates $\hat{\beta}$ and an estimate of the asymptotic covariance matrix of $\hat{\beta}$ can be obtained from the observed information matrix at maximum.

When including higher order correlations, implementing the score equations and the observed information matrices becomes increasingly cumbersome. While the functional form (4.5) does not change, the components T_i and L_i become fairly complicated. Fisher's z transform can be applied to all correlation parameters $\rho_{i(r)}$. The design matrix X_i is extended in a straightforward fashion. Unfortunately, fitting a higher order Bahadur model, whether through numerical or analytical maximization, is not straightforward, due to increasingly complex restrictions on the parameter space.

Observing that, in many studies considered, interest will be restricted to the marginal mean function and the pairwise association parameter, one can replace a full likelihood approach by estimating equations where only the first two moments are modeled and working assumptions are adopted about third and fourth order moments. A thorough treatment is found in Liang, Zeger and Qaqish (1992). See also Chapter 5. Obviously, an important special form for these working assumptions is given by setting the higher order parameters equal to zero, thereby avoiding the need for moment-based estimation of nuisance parameters. Consistent point estimates are supplemented with *robust* standard errors (following from the sandwich estimator), rather than with purely model-based (or naive) standard errors. Often, point estimates differ only slightly from their likelihood counterparts, while test statistics may change considerably. This point will be illustrated in Chapter 11.

4.1.2 The George-Bowman model

George and Bowman (1995) propose a model for the analysis of exchangeable binary data. The probability mass function for the number of malformations

Z_i in litter i consisting of n_i viable foetuses is presented as:

$$f(z_i|\lambda_{i,z_i}, \lambda_{i,z_i+1}, \dots, \lambda_{i,n_i}, n_i) =$$

$$\binom{n_i}{z_i} \sum_{\ell=0}^{n_i-z_i} (-1)^\ell \binom{n_i - z_i}{\ell} \lambda_{i,z_i+\ell}, \qquad (4.6)$$

in which

$$\lambda_{i,k} = \begin{cases} P(Y_{i1} = 1, Y_{i2} = 1, \dots, Y_{ik} = 1) & \text{if } k = 1, \dots, n_i, \\ 1 & \text{if } k = 0. \end{cases}$$

As a consequence, the parameter $\lambda_{i,k}$ can be interpreted as the probability that in litter i, all foetuses in a set of k exhibit the adverse event under consideration. The mean of the number of malformed foetuses and the second order correlation between two responses of the same litter can be expressed in terms of $\lambda_{i,k}$ parameters:

$$E(Z_i) = \sum_{j=1}^{n_i} E(Y_{ij}) = n_i P(Y_{ij} = 1) = n_i \lambda_{i,1}$$

and

$$\begin{aligned} \text{Corr}(Y_{ij}, Y_{ik}) &= \frac{E(Y_{ij}Y_{ik}) - E(Y_{ij})E(Y_{ik})}{E(Y_{ij}^2) - (E(Y_{ij}))^2} \\ &= \frac{P(Y_{ij} = 1, Y_{ik} = 1) - P(Y_{ij} = 1)P(Y_{ik} = 1)}{P(Y_{ij} = 1) - P(Y_{ij} = 1)^2} \\ &= \frac{\lambda_{i,2} - \lambda_{i,1}^2}{\lambda_{i,1}(1 - \lambda_{i,1})}. \end{aligned}$$

George and Bowman (1995) also give expressions for higher order moments of Z_i and for higher order correlations.

Under independence of the n_i responses of litter i,

$$\lambda_{i,k} = P(Y_{i1} = 1) \dots P(Y_{ik} = 1) = \lambda_{i,1}^k$$

and (4.6) can be written as:

$$\begin{aligned} f(z_i|\lambda_{i,z_i}, \lambda_{i,z_i+1}, \dots, \lambda_{i,n_i}, n_i) &= \binom{n_i}{z_i} \sum_{\ell=0}^{n_i-z_i} (-1)^\ell \binom{n_i - z_i}{\ell} \lambda_{i,1}^{z_i} \lambda_{i,1}^\ell \\ &= \binom{n_i}{z_i} \lambda_{i,1}^{z_i} \sum_{\ell=0}^{n_i-z_i} \binom{n_i - z_i}{\ell} (-\lambda_{i,1})^\ell \\ &= \binom{n_i}{z_i} \lambda_{i,1}^{z_i} (1 - \lambda_{i,1})^{n_i-z_i}. \end{aligned}$$

Hence, under independence, the George-Bowman model of which the parameters are $\lambda_{i,z_i}, \lambda_{i,z_i+1}, \dots, \lambda_{i,n_i}$ and n_i reduces to a binomial model with parameters n_i and $\lambda_{i,1}$.

George and Bowman focus attention on the so-called *folded logistic* parameterization:

$$\lambda_{i,z_i+\ell}(\boldsymbol{\beta}) = \frac{2}{1 + \exp\left[-X_i\boldsymbol{\beta}\ln(z_i + \ell + 1)\right]} \tag{4.7}$$

where $X_i = (1, d_i)$ and $\boldsymbol{\beta} = (\beta_0, \beta_d)^t$. Hence, (4.7) can be rewritten as:

$$\lambda_{i,z_i+\ell}(\boldsymbol{\beta}) = \frac{2}{1 + (z_i + \ell + 1)^{-\beta_0 - \beta_d d_i}}.$$

However, it turns out that the "specific" George-Bowman model with the
folded logistic parameterization does not simplify to the binomial model in this
case. In addition, the model is not coding invariant, i.e., if the 0/1 coding for
successes and failures is swapped, the model changes and so do the maximum
likelihood estimates. This should be seen as an undesirable feature of the
model.

The maximum likelihood estimates of the George-Bowman model with
this specific parameterization are found by the Newton-Raphson algorithm.
George and Bowman prove that $X_i\boldsymbol{\beta} < 0$ is necessary and sufficient in order
to have a valid probability mass function.

4.1.3 The Dale and Probit Models

The probit and Dale models have been proposed for multivariate and repeated
ordered categorical outcomes, of which binary outcomes are a special case. In
the case of the probit model, the ordinal outcome vector is assumed to arise
from discretizing an underlying multivariate normal, whereas in the case of
the Dale model an underlying Plackett distribution is assumed. In the first
case, the association is captured by means of correlation coefficients, whereas
in the second case global odds ratios are used to model the association.

The outcome for cluster i is a series of measurements y_{ij} ($j = 1, \ldots, n_i$).
Assume that y_{ij} can take on c_j distinct ordered values $k_j = 1, \ldots, c_j$. It is
convenient to define so-called cumulative multi-indicator functions:

$$z_i(\boldsymbol{k}) = z_i(k_1, \ldots, k_{n_i}) = I(\boldsymbol{y}_i \leq \boldsymbol{k}).$$

The corresponding probability is denoted by $\mu_i(\boldsymbol{k})$. The choice to use cumulative indicators is in agreement with the ordinal nature of the outcomes.
Setting one or more of the indices k_j equal to their maximal value c_j has
the effect of marginalizing over the corresponding outcome. Doing this for all
but one index results in the univariate indicators $z_{ijk} = I(y_{ij} \leq k)$ and their
corresponding marginal probability μ_{ijk}.

The ordering needed to stack the multi-indexed counts and probabilities into
a vector will be assumed fixed. Several orderings of both $\boldsymbol{z}_i$ and $\boldsymbol{\mu}_i$ are possible.
A natural choice is the lexicographic ordering, but this has the disadvantage
of dispersing the univariate marginal counts and means over the entire vector.
Therefore, we will group the elements first by dimensionality.

We can now complete the model by choosing appropriate link functions.

For the vector of links $\boldsymbol{\eta}_i$ we consider a function, mapping the C_i-vector $\boldsymbol{\mu}_i$ $(C_i = c_1 \cdot c_2 \cdots \cdot c_{T_i})$ to

$$\boldsymbol{\eta}_i = \boldsymbol{\eta}_i(\boldsymbol{\mu}_i), \tag{4.8}$$

a C_i'-vector. Often, $C_i = C_i'$, and $\boldsymbol{\eta}_i$ and $\boldsymbol{\mu}_i$ have the same ordering. A counterexample is provided by the probit model, where the number of link functions is smaller than the number of mean components, as soon as $n_i > 2$.

We consider particular choices of link functions. The univariate logit link becomes $\eta_{ijk} = \ln(\mu_{ijk}) - \ln(1 - \mu_{ijk}) = \text{logit}(\mu_{ijk})$. The probit link is $\eta_{ijk} = \Phi_1^{-1}(\mu_{ijk})$, with Φ_1 the univariate standard normal distribution.

However, univariate links alone do not fully specify $\boldsymbol{\eta}_i$, and hence leave the joint distribution partly undetermined. Full specification of the association requires addressing the form of pairwise and higher-order probabilities. First, we will consider the pairwise associations. Let us denote the bivariate probabilities, pertaining to the j_1th and j_2th outcomes, by

$$\mu_{i,j_1 j_2, k_1 k_2} = \mu_i(c_1, \ldots, c_{j_1-1}, k_1, c_{j_1+1}, \ldots, c_{j_2-1}, k_2, c_{j_2+1}, \ldots, c_{n_i}).$$

The marginal correlation coefficient is defined as

$$\rho_{i,j_1 j_2, k_1 k_2} = \frac{\mu_{i,j_1 j_2, k_1 k_2} - \mu_{ij_1 k_1} \mu_{ij_2 k_2}}{\sqrt{\mu_{ij_1 k_1}(1 - \mu_{ij_1 k_1})\mu_{ij_2 k_2}(1 - \mu_{ij_2 k_2})}},$$

which is the basis of the Bahadur model discussed in Section 4.1.1.

The Dale model is based on the marginal global odds ratio defined by

$$\psi_{i,j_1 j_2, k_1 k_2} = \frac{(\mu_{i,j_1 j_2, k_1 k_2})(1 - \mu_{ij_1 k_1} - \mu_{ij_2 k_2} + \mu_{i,j_1 j_2, k_1 k_2})}{(\mu_{ij_2 k_2} - \mu_{i,j_1 j_2, k_1 k_2})(\mu_{ij_1 k_1} - \mu_{i,j_1 j_2, k_1 k_2})} \tag{4.9}$$

and usefully modeled on the log scale. Higher order global odds ratios are easily introduced using ratios of conditional odds (ratios). Let

$$\mu_{ij_1 | j_2}(w_{j_2}) = P(z_{ij_1 k_1} = 1 | z_{ij_2 k_2} = w_{j_2}, X_i, \boldsymbol{\theta}) \tag{4.10}$$

be the conditional probability of observing a success at occasion j_1, given the value w_{j_2} is observed at occasion j_2, and write the corresponding conditional odds as

$$\psi_{ij_1 | j_2}(w_{j_2}) = \mu_{ij_1 | j_2}(w_{j_2})/[1 - \mu_{ij_1 | j_2}(w_{j_2})].$$

The pairwise marginal odds ratio, for occasions j_1 and j_2, is defined as

$$\psi_{ij_1 j_2} = \frac{\{P(z_{ij_1 k_1} = 1, z_{ij_2 k_2} = 1)\}\{P(z_{ij_1 k_1} = 0, z_{ij_2 k_2} = 0)\}}{\{P(z_{ij_1 k_1} = 0, z_{ij_2 k_2} = 1)\}\{P(z_{ij_1 k_1} = 1, z_{ij_2 k_2} = 0)\}} = \frac{\psi_{ij_1 | j_2}(1)}{\psi_{ij_1 | j_2}(0)},$$

in accordance with (4.9). This formulation can be exploited to define the higher order marginal odds ratios in a recursive fashion:

$$\psi_{ij_1 \ldots j_m t_{m+1}} = \frac{\psi_{ij_1 \ldots j_m | j_{m+1}}(1)}{\psi_{ij_1 \ldots j_m | j_{m+1}}(0)}, \tag{4.11}$$

where $\psi_{ij_1\ldots j_m|j_{m+1}}(w_{m+1})$ is defined by conditioning all probabilities occurring in the expression for $\psi_{ij_1\ldots j_m}$ on $Z_{ij_{m+1}k_{m+1}} = w_{j_{m+1}}$. The choice of the conditioning variable is immaterial.

The multivariate probit model also fits within the class defined by (4.8). For three categorical outcome variables, the inverse link is specified by

$$\mu_{ijk} = \Phi_1(\eta_{ijk}), \tag{4.12}$$

$$\mu_{i,j_1j_2,k_1k_2} = \Phi_2(\eta_{ij_1k_1}, \eta_{ij_1k_2}, \eta_{i,j_1j_2,k_1k_2}), \tag{4.13}$$

$$\mu_{i,123,k_1k_2k_3} = \Phi_3(\eta_{i1k_1}, \eta_{i2k_3}, \eta_{i3k_3}, \eta_{i,12,k_1k_2}, \eta_{i,13,k_1k_3}, \eta_{i,23,k_2k_3}), \tag{4.14}$$

where the notation for the three-way probabilities is obvious. The association links $\eta_{i,ts,k\ell}$ represent any transform (e.g., Fisher's z-transform) of the polychoric correlation coefficient. It is common practice to keep each correlation constant throughout a table, rather than having it depend on the categories: $\eta_{i,j_1j_2,k_1k_2} \equiv \eta_{i,j_1j_2}$. Relaxing this requirement may still give a valid set of probabilities, but the correspondence between the categorical variables and a latent multivariate normal variable is lost. Finally, observe that univariate links and bivariate links (representing correlations) fully determine the joint distribution. This implies that the mean vector and the link vector will have a different length, except in the univariate and bivariate cases.

Model formulation is completed by specifying appropriate design matrices. Parameter estimation then proceeds by means of, for example, maximum likelihood. Especially for longer sequences, computational requirements are non-trivial and it may be necessary to use alternative estimation procedures such as pseudo-likelihood.

4.2 Conditional Models

In a conditional model the parameters describe a feature (probability, odds, logit, ...) of (a set of) outcomes, given values for the other outcomes (Cox 1972). The best known example is undoubtedly the log-linear model. Rosner (1984) described a conditional logistic model. Due to the popularity of marginal (especially generalized estimating equations) and random-effects models for correlated binary data, conditional models have received relatively little attention, especially in the context of multivariate clustered data. Diggle, Liang and Zeger (1994, pp. 147–148) criticized the conditional approach because the interpretation of the dose effect on the risk of one outcome is conditional on the responses of other outcomes for the same individual, outcomes of other individuals and the litter size. Molenberghs, Declerck and Aerts (1998) and Aerts, Declerck and Molenberghs (1997) have compared marginal, conditional and random-effects models for univariate clustered data. Their results are encouraging for the conditional model, since they are competitive for the dose effect testing and for benchmark dose estimation, and because they are computationally fast and stable. Molenberghs and Ryan (1999) discuss, in the specific context of exchangeable binary data, the advantages of conditional models and show how, with appropriate care, the disadvantages can be overcome. They constructed the joint distribution for clustered multivariate binary outcomes, based on a multivariate exponential family model. A slightly different approach, also based on the exponential family, is presented in Fitzmaurice, Laird, and Tosteson (1996). An advantage of such a likelihood-based approach is that, under correct model specification, efficiency can be gained over other procedures such as generalized estimating equations (GEE) methods. Furthermore, the model provides a natural framework for quantitative risk assessment (Chapter 10). Present approaches estimate benchmark doses (Crump 1984) based on the marginal probability of a single offspring being affected (Chen and Kodell 1989). From a biological perspective, one might argue that it is important to take into account the health of the entire litter when modeling risk as a function of dose. The likelihood basis of the Molenberghs and Ryan (1999) model allows calculation of quantities such as the probability that at least one littermate is affected (probability of an affected litter). In contrast, GEE based models do not provide a way to derive such quantities since they do not specify the joint probability between outcomes but only marginal probabilities and a working correlation matrix. While they could be calculated from a fully specified marginal model, fitting these models is hampered by lengthy computations and/or parameter restrictions (Molenberghs, Declerck and Aerts 1998 and Aerts, Declerck and Molenberghs 1997).

The flexibility of the Molenberghs and Ryan (1999) model partly relies on the exponential family framework. However, maximum likelihood estimation can be unattractive, due to excessive computational requirements. For example, with multivariate exponential family models, the normalizing constant can have a cumbersome expression, rendering it hard to evaluate (Arnold and Strauss 1991). Several suggestions have been made to overcome this problem, such as Monte Carlo integration (Tanner 1991). For example, Geyer and Thompson (1992) use Markov Chain Monte Carlo simulations to construct a Monte Carlo approximation to the analytically intractable likelihood. Arnold and Strauss (1991) and Arnold, Castillo and Sarabia (1992) propose the use of a so-called *pseudo-likelihood* (PL). Pseudo-likelihood (or pseudo-maximum-likelihood) methods are alternatives to maximum likelihood estimation that retain the methodology and properties while trying to eliminate some of the difficulties such as strong distributional assumptions or intensive computations. The idea is that a parametric family of models is specified, to which likelihood methodology is applied; the method is denoted "pseudo", as there is no assumption that this family is the true distribution generating the data. Geys, Molenberghs and Ryan (1997, 1999) implemented a pseudo-likelihood method for the Molenberghs and Ryan (1999) model that replaces the joint distribution of the responses, a multivariate exponential-family model, by a product of conditional densities that do not necessarily multiply to the joint distribution (see also Chapters 6 and 7). In this approach, the normalizing constant cancels, thus greatly simplifying computations, especially when litter sizes are large and variable (since the normalizing constant depends on litter size). In following chapters we will show that pseudo-likelihood estimation is an attractive alternative for maximum likelihood estimation in the context of clustered binary data. Moreover, since the pseudo-likelihood still reflects the underlying likelihood it can be useful for dose-response modeling (e.g., to determine a benchmark dose). Pseudo-likelihood estimation turned out to be also extremely useful in the context of spatial statistics (Cressie 1991). Besag (1975) used pseudo-likelihood estimation in the context of a general Markov random field and established consistency of the estimators. A selection of other applications of this technique can be found in Connolly and Liang (1988), Liang and Zeger (1989) and le Cessie and van Houwelingen (1994).

As before, we consider an experiment involving N clusters, the ith of which contains n_i individuals, each of whom are examined for the presence or absence of M different responses. Suppose for the moment that $Y_{ijk} = 1$ when the kth individual in cluster i exhibits the jth response and 0 otherwise. Let Y_i represent the vector of outcomes for the ith cluster, and x_i an associated vector of cluster level covariates.

4.2.1 No Clustering

Let us first suppose there is no clustering ($n_i = 1; i = 1, \ldots, N$). Because $k \equiv 1$ in this setting, we drop this index temporarily from our notation. The observable outcome is thus $Y_i = (Y_{i1}, \ldots, Y_{iM})^T$. Next, consider the following probability mass function proposed by Cox (1972):

$$f_{Y_i}(y_i; \Theta_i) = \exp \left\{ \sum_{j=1}^{M} \theta_{ij} y_{ij} + \sum_{j<j'} \omega_{ijj'} y_{ij} y_{ij'} + \ldots \right. \tag{4.15}$$

$$\left. + \omega_{i1\ldots M} y_{i1} \cdots y_{iM} - A(\Theta_i) \right\}.$$

The θ parameters can be thought of as "main effects", whereas the ω parameters are association parameters or interactions. Models that do not include all interactions are derived by replacing W_i, the vector of the ω parameters, by one of its subvectors. A useful special case is found by setting all three and higher order parameters equal to zero, which is a member of the quadratic exponential family discussed by Zhao and Prentice (1990). Thélot (1985) studied the case where $M = 2$. If $M = 1$, the model reduces to ordinary logistic regression.

We will briefly outline standard procedures for likelihood based parameter estimation in this setting. Modeling in terms of a parsimonious parameter vector of interest can be achieved using a linear model of the form $\Theta_i = X_i \beta$, where Θ_i is a vector of natural parameters, X_i is a $q \times p$ design matrix and β a $p \times 1$ vector of unknown regression coefficients. Let the mean parameter be π_i. Then it is a basic property of exponential families (e.g., Brown 1986, p. 36) that π_i is related to the natural parameter Θ_i by $\pi_i = \partial A(\Theta_i)/\partial \Theta_i$.

Here, $A(\boldsymbol{\Theta}_i)$ is a normalizing constant. Next, the log-likelihood can be written as

$$\ell = \sum_{i=1}^{N} \ln f(\boldsymbol{y}_i; \boldsymbol{\Theta}_i) = \sum_{i=1}^{N} \left\{ \boldsymbol{\beta}^T X_i^T \boldsymbol{w}_i - A(X_i \boldsymbol{\beta}) \right\},$$

and the score function is

$$\boldsymbol{U}(\boldsymbol{\beta}) = \sum_{i=1}^{N} X_i^T (\boldsymbol{w}_i - \boldsymbol{\pi}_i).$$

The maximum likelihood estimator for $\boldsymbol{\beta}$ is defined as the solution to $\boldsymbol{U}(\boldsymbol{\beta}) = \boldsymbol{0}$. It is usually found by applying a Newton-Raphson procedure, which coincides with a Fisher scoring algorithm for exponential family models with canonical link functions.

4.2.2 Single Clustered Outcome

Let us now consider a single clustered outcome. Because the index j always equals 1, we drop it temporarily from our notation. We re-introduce however the subscript k to indicate an individual within a cluster.

Similarly to the Thélot model (a bivariate conditional model, Thélot 1985), Molenberghs and Ryan (1999) derived the joint distribution of the clustered binary data $\boldsymbol{Y}_i$ as:

$$f_{\boldsymbol{Y}}(\boldsymbol{y}_i; \boldsymbol{\Theta}_i^*, n_i) = \exp \left\{ \sum_{k=1}^{n_i} \theta_i^* y_{ik} + \sum_{k<k'} \delta_i^* y_{ik} y_{ik'} - A(\boldsymbol{\Theta}_i^*) \right\}, \tag{4.16}$$

with δ_i^* describing the association between pairs of individuals within the ith cluster.

They code $Y_{ijk} = 1$ when the kth individual in cluster i exhibits the jth response and -1 otherwise. They use this coding rather than 1 and 0 since it provides a parameterization that more naturally leads to desirable properties when the roles of success and failure are reversed (see Cox and Wermuth 1994). Defining the number of individuals from cluster i with positive response to be z_i, (4.16) then becomes

$$
\begin{aligned}
f_{\boldsymbol{Y}}(\boldsymbol{y}_i; \boldsymbol{\Theta}_i^*, n_i) &= \exp \left\{ \theta_i^* z_i - \theta_i^* (n_i - z_i) \right. \\
&\quad \left. + \delta_i^* \left[\binom{z_i}{2} + \binom{n_i - z_i}{2} - z_i(n_i - z_i) \right] - A(\boldsymbol{\Theta}_i^*) \right\} \\
&= \exp \left\{ \theta_i^* (2z_i - n_i) + \delta_i^* \left[\binom{n_i}{2} - 2z_i n_i + 2z_i^2 \right] \right. \\
&\quad \left. - A(\boldsymbol{\Theta}_i^*) \right\}.
\end{aligned}
\tag{4.17}
$$

Upon absorbing constant terms into the normalizing constant and using the reparametrization $\theta_i = 2\theta_i^*$ and $\delta_i = 2\delta_i^*$ this becomes

$$f_Y(y_i; \Theta_i, n_i) = \exp\left\{\theta_i z_i^{(1)} + \delta_i z_i^{(2)} - A(\Theta_i)\right\}, \tag{4.18}$$

with $z_i^{(1)} = z_i$ and $z_i^{(2)} = -z_i(n_i - z_i)$. For this model, independence corresponds to $\delta_i = 0$. A positive δ_i corresponds to classical clustering or overdispersion, whereas a negative parameter value occurs in the underdispersed case. It is worthwhile to note that even for underdispersion, no restrictions are required on the parameter space. Molenberghs and Ryan (1999) show that model (4.18) has several additional desirable properties. First, the model is clearly invariant to interchanging the codes of successes and failures, whence both estimation and testing will be invariant for this change as well. Second, the conditional probability of observing a positive response in a cluster of size n_i, given that the remaining littermates yield $z_i - 1$ successes, is given by:

$$P(y_{ik} = 1 | z_i - 1, n_i) = \frac{\exp[\theta_i - \delta_i(n_i - 2z_i + 1)]}{1 + \exp[\theta_i - \delta_i(n_i - 2z_i + 1)]}, \tag{4.19}$$

which decreases to zero when n_i increases and z_i is bounded, and approaches unity for increasing n_i and bounded $n_i - z_i$, whenever there is a positive association between outcomes. From (4.19) it is clear that the conditional logit of an additional success, given $z_i - 1$ successes, equals $\theta_i - \delta_i(n_i - 2z_i + 1)$. Thus, upon noting that the second term vanishes if $z_i - 1 = (n_i - 1)/2$, θ_i is seen to be the conditional logit for an additional success when about half of the littermates exhibit a success already. Similarly, the log odds ratio for the responses between two littermates is equal to $2\delta_i$, confirming the association parameter interpretation of the δ-parameter. Finally, the marginal success probability in a cluster of size n_i is clearly a (non-linear) function of n_i:

$$E\left(\frac{Z_i}{n_i}\right) = \frac{\sum_{z=0}^{n_i} z\binom{n_i}{z} \exp\{\theta_i z - \delta_i z(n_i - z)\}}{\sum_{z=0}^{n_i} n_i\binom{n_i}{z} \exp\{\theta_i z - \delta_i z(n_i - z)\}}.$$

Because this model is conditional in nature, this marginal quantity does not simplify in general. Nevertheless, this expectation can be easily calculated and plotted to explore the relationship between cluster size and response probability.

4.2.3 Clustered Multivariate Outcomes

Suppose again that $y_{ijk} = 1$ when the kth individual in cluster i exhibits response j and -1 otherwise. It is convenient to group the outcomes for the ith cluster in an Mn_i vector $Y_i = (Y_{i11}, \ldots, Y_{i1n_i}, \ldots, Y_{iMn_i})$. Molenberghs and Ryan (1999) proposed the following model for the joint distribution of

62

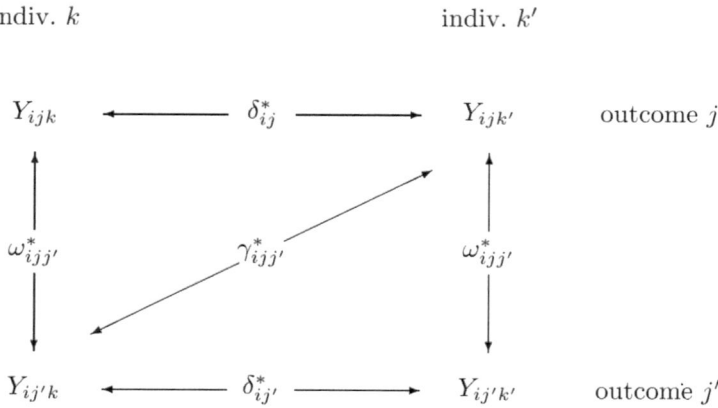

indiv. k indiv. k'

Y_{ijk} ←———— δ^*_{ij} ————→ $Y_{ijk'}$ outcome j

$\omega^*_{ijj'}$ $\gamma^*_{ijj'}$ $\omega^*_{ijj'}$

$Y_{ij'k}$ ←———— $\delta^*_{ij'}$ ————→ $Y_{ij'k'}$ outcome j'

Figure 4.1 *Association structure for outcomes j and j' on individuals k and k' in cluster i.*

clustered multivariate binary data:

$$
f_{\boldsymbol{Y}_i}(\boldsymbol{y}_i; \boldsymbol{\Theta}^*_i) = \exp\left\{ \sum_{j=1}^{M}\sum_{k=1}^{n_i} \theta^*_{ij} y_{ijk} + \sum_{j=1}^{M}\sum_{k<k'} \delta^*_{ij} y_{ijk} y_{ijk'} \right.
$$
$$
+ \sum_{j<j'}\sum_{k=1}^{n_i} \omega^*_{ijj'} y_{ijk} y_{ij'k}
$$
$$
\left. + \sum_{j<j'}\sum_{k\neq k'} \gamma^*_{ijj'} y_{ijk} y_{ij'k'} - A(\boldsymbol{\Theta}^*_i) \right\}, \qquad (4.20)
$$

where $A(\boldsymbol{\Theta}^*_i)$ is the normalizing constant, resulting from summing (4.20) over all 2^{Mn_i} possible outcomes. The building blocks of this model are clearly the "main effects" (θ^*) and three types of association parameters, reflecting three different types of association. For example, δ^*_{ij} refers to the association between two different individuals from the same cluster on the same outcome j, $\omega^*_{ijj'}$ refers to the association between outcomes j and j' for a single individual within cluster i and $\gamma^*_{ijj'}$ gives the association between outcomes j and j' for two different individuals in the same cluster. The three different types of associations captured in the model are depicted in Figure 4.1.

The absence of individual-specific subscripts reflects the implicit exchangeability assumption between any two individuals within the same cluster. This assumption will now be used to simplify the model. Defining z_{ij} as the number of individuals from cluster i positive on outcome j and $z_{ijj'}$ as the number of individuals in cluster i, positive on both outcomes j and j', Molenberghs and Ryan (1999) derived (after reparameterization):

$$f_{\boldsymbol{Y}_i}(\boldsymbol{y}_i; \boldsymbol{\Theta}_i) = \exp\left\{ \sum_{j=1}^{M} \theta_{ij} z_{ij}^{(1)} + \sum_{j=1}^{M} \delta_{ij} z_{ij}^{(2)} \right.$$

$$\left. + \sum_{j<j'} \omega_{ijj'} z_{ijj'}^{(3)} + \sum_{j<j'} \gamma_{ijj'} z_{ijj'}^{(4)} - A(\boldsymbol{\Theta}_i) \right\}, \quad (4.21)$$

where

$$\begin{aligned}
z_{ij}^{(1)} &= z_{ij} \\
z_{ij}^{(2)} &= -z_{ij}(n_i - z_{ij}) \\
z_{ijj'}^{(3)} &= 2z_{ijj'} - z_{ij} - z_{ij'} \\
z_{ijj'}^{(4)} &= -z_{ij}(n_i - z_{ij'}) - z_{ij'}(n_i - z_{ij}) - z_{ijj'}^{(3)}.
\end{aligned} \quad (4.22)$$

Advantages of this model are the flexibility with which both main effects and associations can be modeled, and the absence of constraints on the parameter space, which eases interpretability. Both foetus-based and litter-based malformation probabilities (Chapter 10) have appealing and simple expressions. This aspect is important when using the model in a dose-response setting. Further, the fact that the probability model depends explicitly (see (4.22)) and implicitly on the cluster size is an advantage since it is in line with the observation that litter size itself may depend on the level of exposure. Note that model (4.21) is conditional in nature, since it describes a feature of (a set of) outcomes conditional on the other outcomes. More precisely, it implies conditional odds and conditional odds ratios that are log-linear in the natural parameters. Molenberghs and Ryan (1999) construct the conditional logit associated with the presence and absence of outcome j for an individual k in cluster i, given all other outcomes in the same cluster, and show that this function depends on cluster size and on the observed pattern of the remaining outcomes. Let $\kappa_{ijk} = 1$ if the kth individual exhibits a success on the jth variable and 0 otherwise. Then

$$\log \frac{\text{pr}(Y_{ijk} = 1 | y_{ij'k'}, j' \neq j \text{ or } k' \neq k)}{\text{pr}(Y_{ijk} = -1 | y_{ij'k'}, j' \neq j \text{ or } k' \neq k)} = \theta_{ij} + \delta_{ij}(2z_{ij} - n_i - 1)$$

$$+ \sum_{j' \neq j} \omega_{ijj'}(2\kappa_{ij'k} - 1) + \sum_{j' \neq j} \gamma_{ijj'}(2z_{ij'} - n_i - 2\kappa_{ij'k} + 1). \quad (4.23)$$

Marginal quantities are fairly complicated functions of the parameters and are best represented graphically.

4.2.4 Implications for Risk Assessment

The model described in the previous section can be used in various ways as a basis of risk assessment. First, we can focus on the number of dead or resorbed foetuses out of the number of implant versus the number of malformed foetuses out of the number of viable foetuses. Second, an adverse event can be defined in terms of the foetus or the litter.

Representing the number of dead foetuses in cluster i by R_i out of m_i implants (similar to z_i malformations out of n_i viable foetuses), expression (4.17) can be used for this situation as well:

$$f(\boldsymbol{y}_i|\psi_i^*, \phi_i^*, m_i) = \exp\{\psi_i^*(2r_i - m_i)$$

$$+\phi_i^* \left[\binom{m_i}{2} - 2r_i m_i + 2r_i^2 \right] - A_i^* \right\}. \quad (4.24)$$

After absorbing the constant terms into the normalizing constant and after a simple reparameterization ($\psi_i = 2\psi_i^*$ and $\phi_i = 2\phi_i^*$), one obtains from (4.24):

$$f(\boldsymbol{y}_i|\psi_i, \phi_i, m_i) = \exp\{\psi_i r_i - \phi_i r_i(m_i - r_i) - A_i\}. \quad (4.25)$$

From formula (4.25), the probability mass function of the number of fetal deaths follows:

$$f(r_i|\psi_i, \phi_i, m_i) = \binom{m_i}{r_i} \exp\{\psi_i r_i - \phi_i r_i(m_i - r_i) - A_i\}. \quad (4.26)$$

Hence, the normalizing constant A_i can be written as:

$$A_i = \ln \left\{ \sum_{r_i=0}^{m_i} \binom{m_i}{r_i} \exp\{\psi_i r_i - \phi_i r_i(m_i - r_i)\} \right\} \equiv A(\psi_i, \phi_i, m_i).$$

Based on (4.26), the conditional logit for a dead foetus given the number of deaths in the group of remaining foetuses can be written as a linear function of ψ_i and ϕ_i:

$$\text{logit}[P(\text{foetus } j \text{ dead} \mid r_i \text{ of the other foetuses also dead})]$$
$$= \psi_i - \phi_i(m_i - 2r_i - 1),$$

where $j = 1, \dots, m_i$. This implies that if the number of implants is odd, then the parameter ψ_i equals the logit for a dead foetus given that one half of the remaining foetuses are dead as well. Also, (4.26) results in

$$\psi_i = \frac{1}{m_i} \ln \left(\frac{P(R_i = m_i)}{P(R_i = 0)} \right). \quad (4.27)$$

From (4.26) and (4.27), it follows that the parameter $\psi_i = 0$ if and only if the distribution of R_i is symmetric around $m_i/2$. Furthermore, it can be shown that the parameter ϕ_i is one half of the log odds ratio for a pair of foetuses given the number of deaths in the remaining group of foetuses. Thus, clearly, the parameters in the model of Molenberghs and Ryan have a conditional interpretation.

A special case of the conditional model (4.26) is obtained when the association parameter $\phi_i = 0$:

$$f(r_i|\psi_i, m_i) = \binom{m_i}{r_i} \exp(\psi_i r_i - A_i). \tag{4.28}$$

Let

$$\psi_i = \ln\left\{\frac{\omega_i}{1 - \omega_i}\right\}.$$

Then, formula (4.28) can be re-expressed as

$$
\begin{aligned}
f(r_i|\psi_i, m_i) &= \binom{m_i}{r_i}\left\{\frac{\omega_i}{1 - \omega_i}\right\}^{r_i} \exp(-A_i) \\
&= \binom{m_i}{r_i}\omega_i^{r_i}(1 - \omega_i)^{m_i - r_i}.
\end{aligned}
$$

Hence, if in the conditional model with parameters ψ_i, ϕ_i and m_i, the parameter ϕ_i is set equal to zero, then this model reduces to the logistic regression model with parameters m_i and

$$\omega_i = \frac{1}{1 + \exp(-\psi_i)}.$$

Furthermore, one notices from (4.26) that positive and negative values of ϕ_i correspond to overdispersion and underdispersion respectively. Also, there are no restrictions on the parameter space of the conditional model, even in case of underdispersion (Molenberghs and Ryan 1999).

The parameters ψ_i and ϕ_i can be modeled as

$$\begin{pmatrix} \psi_i \\ \phi_i \end{pmatrix} = X_i\boldsymbol{\beta},$$

with X_i and $\boldsymbol{\beta}$ as in (4.4). Estimation of these model parameters can easily be carried out using maximum likelihood techniques. Grouping the summary statistics in

$$\boldsymbol{W}_i = \begin{pmatrix} R_i \\ -R_i(m_i - R_i) \end{pmatrix},$$

the contribution of the ith cluster to the log-likelihood is given by $\ell_i = \boldsymbol{w}_i^t X_i\boldsymbol{\beta} - A_i$, whence the score function becomes

$$\boldsymbol{U}(\boldsymbol{\beta}) = \sum_{i=1}^{C} X_i^t(\boldsymbol{w}_i - E(\boldsymbol{W}_i)).$$

The expectation π_i of R_i/m_i, the marginal death probability in a cluster of m_i implants, is clearly a (non-linear) function of m_i:

$$\pi_i = \frac{\sum_{r_i=0}^{m_i} r_i \binom{m_i}{r_i} \exp\left\{(\beta_0 + \beta_d d_i)r_i - \beta_2 r_i(m_i - r_i)\right\}}{\sum_{r_i=0}^{m_i} m_i \binom{m_i}{r_i} \exp\left\{(\beta_0 + \beta_d d_i)r_i - \beta_2 r_i(m_i - r_i)\right\}}. \tag{4.29}$$

Methods similar to those of Cox and Wermuth (1994) could be invoked to develop approximate expressions for the marginal means and odds ratios. Because the model is conditional in nature, the marginal parameter (4.29) does not simplify in general. As a consequence, the conditional model implies a natural dependence of π_i on the number of implants, in contrast to marginal models. Furthermore, only when the clustering parameters are equal to zero, the conditional model, the Bahadur model and the beta-binomial model reduce to logistic regression. In Section 4.1.2, it has been shown that the general expression of the George-Bowman model reduces to the logistic regression model too if the foetuses of a cluster are independent. The previous discussion implies that the parameters of the conditional model are not directly comparable to their counterparts in the Bahadur, George-Bowman and beta-binomial models.

4.3 Cluster-specific Models

Cluster-specific models are differentiated from population-averaged models by the inclusion of parameters that are specific to the cluster. Unlike for correlated Gaussian outcomes, the parameters of the cluster-specific and of the population-averaged models for correlated binary data describe different types of effects of the covariates on the response probabilities (Neuhaus 1992).

The choice between population-averaged and cluster-specific strategies may heavily depend on the scientific goals. Population-averaged models evaluate the overall risk as a function of covariates; the conditionally specified models and marginal models, described above, belong to this class. With the cluster-specific approach, the response rates are modeled as a function of covariates and parameters, specific to a cluster. In such models, interpretation of fixed-effect parameters is conditional on a constant level of the cluster-specific parameter (e.g., random effect). Population-averaged comparisons, on the other hand, make no use of within cluster comparisons for cluster varying covariates and substantially underestimate within cluster risks. Neuhaus, Kalbfleisch and Hauck (1991) discuss parameter interpretations of these models. They also draw the analogy with omitted covariates; i.e., unless the included and omitted covariates are uncorrelated (conditional on the response), the effect of a randomly assigned treatment will be biased towards zero. Thus, from these papers, population-averaged effects would be expected to be closer to zero than cluster-specific effects.

Cluster-specific parameters can be dealt with in essentially three ways: (1) as fixed effects, (2) as random-effects, and (3) by conditioning upon them. The first approach is seemingly simplest but in many cases flawed since the number of parameters then increases with a rate proportional to the sample size, thereby invalidating most standard inferential results. The second approach is very popular. There are two routes to introduce randomness into the model parameters. Stiratelli, Laird and Ware (1984) assume the parameter vector to be normally distributed. This idea has been carried further in the work on so-called *generalized linear mixed models* (Breslow and Clayton 1993) which is closely related to linear and non-linear mixed models. Alternatively, Skellam (1948) introduced the beta-binomial model, in which the adverse event probability of any foetus of a particular cluster comes from a beta distribution. Hence, this model can also be viewed as a random effects model. The third approach is well known in epidemiology, more precisely in the context of matched case-control studies. In particular, conditional logistic regression is then often considered (Breslow and Day 1987). In general, with so-called conditional likelihood methods, one conditions on the sufficient statistics for the cluster-specific effects (Ten Have, Landis and Weaver 1995, Conaway 1989).

In the remainder of this section we will consider the beta-binomial model and classical mixed-effects models. To facilitate introduction and understanding of generalized linear mixed model, we first introduce the linear and non-linear mixed-model family. A somewhat different way of introducing mixed-effects models is through the concept of multilevel models. These will be studied in some detail in Chapter 15.

It is implicit in the treatment here that the models will be fitted using maximum likelihood or related estimation methods. Alternatively, a fully Bayesian treatment could be envisaged (Carlin and Louis 1996). However, the study of these methods is outside the scope of this text.

4.3.1 The Beta-binomial Model

Rather than modeling marginal functions directly, a popular approach is to assume a random effects model in which each litter has a random parameter (vector). Skellam (1948), Kleinman (1973) and Williams (1975) assume the malformation probability P_i of any foetus in litter i to come from a beta distribution with parameters α_i and β_i:

$$\frac{p^{\alpha_i-1}(1-p)^{\beta_i-1}}{B(\alpha_i,\beta_i)}, \qquad 0 \leq p \leq 1,$$

where $B(.,.)$ denotes the beta function. Conditional on P_i, the number of malformations Z_i in the ith cluster follows a binomial distribution. This leads to the well-known beta-binomial model. The mean of this distribution is

$$\mu_i = n_i\pi_i = n_i\frac{\alpha_i}{\alpha_i + \beta_i} \tag{4.30}$$

and the variance is $\sigma_i^2 = n_i\pi_i(1-\pi_i)\frac{1+n_i\theta_i}{1+\theta_i}$ with $\theta_i = 1/(\alpha_i + \beta_i)$. It can be shown that the intra-litter correlation can be expressed as

$$\rho_i = \frac{1}{\alpha_i + \beta_i + 1}. \tag{4.31}$$

In a litter of size n_i, the probability mass function of Z_i can be expressed directly in terms of the mean and correlation parameters (4.30) and (4.31), i.e., $f(z_i \mid \pi_i, \rho_i, n_i)$ can be expressed as

$$\binom{n_i}{z_i} \frac{B(\pi_i(\rho_i^{-1} - 1) + z_i, (1 - \pi_i)(\rho_i^{-1} - 1) + n_i - z_i)}{B(\pi_i(\rho_i^{-1} - 1), (1 - \pi_i)(\rho_i^{-1} - 1))}, \qquad (4.32)$$

where $B(.,.)$ denotes the beta function. The only association parameter of this model is ρ_i, which is the correlation between two binary responses of litter i. The higher order correlations of the beta-binomial model can be expressed as a function of the mean malformation probability π_i and ρ_i. The association in both the beta-binomial and the Bahadur model is expressed by means of the intraclass correlation. It turns out that both models have the same first and second moments. As a consequence, the parameter ρ_i of the beta-binomial model equals $\rho_{i(2)}$ of the Bahadur model. The parameters π_i and ρ_i of the beta-binomial model have a marginal interpretation and, therefore, they are the parameters in the derived marginal model as well. This results in similarities between the beta-binomial and marginal models, such as the Bahadur model.

It can be shown (Williams 1975) that the contribution of the ith cluster to the log-likelihood, $\ln f(z_i | \pi_i, \rho_i, n_i) \equiv \ell_i$, can be written as

$$\ell_i = \sum_{r=0}^{z_i - 1} \ln \left(\pi_i + \frac{r\rho_i}{1 - \rho_i} \right) + \sum_{r=0}^{n_i - z_i - 1} \ln \left(1 - \pi_i + \frac{r\rho_i}{1 - \rho_i} \right)$$
$$- \sum_{r=0}^{n_i - 1} \ln \left(1 + \frac{r\rho_i}{1 - \rho_i} \right), \qquad (4.33)$$

with $i = 1, ..., N$. It follows from (4.33) that if the association parameter ρ_i equals zero, then the beta-binomial model reduces to the logistic regression model.

Assuming the same generalized linear regression relations (4.3) and (4.4) for π_i and ρ_i, the maximum likelihood estimator $\hat{\beta}$ is the solution to $U(\beta) = 0$ with the score function for β defined as in (4.5).

Kupper and Haseman (1978) compare the Bahadur model to the beta-binomial model. They conclude that the models perform similarly in three clustered data experiments, whereas they both outperform the (naive) binomial model.

4.3.2 Mixed Models

Perhaps the most commonly encountered subject-specific (or cluster-specific model) is the generalized linear mixed model. It is best to first introduce linear mixed models and non-linear mixed models as a basis for the introduction of generalized linear mixed models. To emphasize they fit within a single common framework, we first give a general formulation.

General Formulation

Let y_{ij} denote the jth measurement available for the ith unit, $i = 1, \ldots, N$, $j = 1, \ldots, n_i$, and let $\boldsymbol{y_i}$ denote the vector of all measurements for the ith unit, i.e., $\boldsymbol{y_i}' = (y_{i1}, \ldots, y_{in_i})$. Our general model assumes that $\boldsymbol{y_i}$ (possibly appropriately transformed) satisfies

$$\boldsymbol{y_i}|\boldsymbol{b_i} \quad \sim \quad F_i(\boldsymbol{\theta}, \boldsymbol{b_i}), \tag{4.34}$$

i.e., conditional on $\boldsymbol{b_i}$, $\boldsymbol{y_i}$ follows a pre-specified distribution F_i, possibly depending on covariates, and parameterized through a vector $\boldsymbol{\theta}$ of unknown parameters, common to all subjects. Further, $\boldsymbol{b_i}$ is a q-dimensional vector of subject-specific parameters, called random effects, assumed to follow a so-called mixing distribution G which may depend on a vector $\boldsymbol{\psi}$ of unknown parameters, i.e., $\boldsymbol{b_i} \sim G(\boldsymbol{\psi})$. The $\boldsymbol{b_i}$ reflect the between-unit heterogeneity in the population with respect to the distribution of $\boldsymbol{y_i}$. Different factorizations of F_i will lead to different models. For example, considering the factors made up of the outcomes y_{ij} given its predecessors $(y_{i1}, \ldots, y_{i,j-1})'$ leads to a so-called transitional model. A model without any random effects $\boldsymbol{b_i}$ is called a marginal model for the response vector $\boldsymbol{y_i}$. In the presence of random effects, conditional independence is often assumed, under which the components y_{ij} in $\boldsymbol{y_i}$ are independent, conditional on $\boldsymbol{b_i}$. The distribution function F_i in (4.34) then becomes a product over the n_i independent elements in $\boldsymbol{y_i}$.

In general, unless a fully Bayesian approach is followed, inference is based on the marginal model for $\boldsymbol{y_i}$ which is obtained from integrating out the random effects, over their distribution $G(\boldsymbol{\psi})$. Let $f_i(\boldsymbol{y_i}|\boldsymbol{b_i})$ and $g(\boldsymbol{b_i})$ denote the density functions corresponding to the distributions F_i and G, respectively; we have that the marginal density function of $\boldsymbol{y_i}$ equals

$$f_i(\boldsymbol{y_i}) \quad = \quad \int f_i(\boldsymbol{y_i}|\boldsymbol{b_i})g(\boldsymbol{b_i})d\boldsymbol{b_i}, \tag{4.35}$$

which depends on the unknown parameters $\boldsymbol{\theta}$ and $\boldsymbol{\psi}$. Assuming independence of the units, estimates of $\widehat{\boldsymbol{\theta}}$ and $\widehat{\boldsymbol{\psi}}$ can be obtained from maximizing the likelihood function built from (4.35), and inferences immediately follow from classical maximum likelihood theory.

Obviously, the random-effects distribution G is crucial in the calculation of the marginal model (4.35). One approach is to leave G completely unspecified and to use non-parametric maximum likelihood (NPML) estimation, which maximizes the likelihood over all possible distributions G. The resulting estimate $\widehat{G}$ is then always discrete with finite support. Depending on the context, this may or may not be a realistic reflection of the true heterogeneity between units. One therefore often assumes G to be of a specific parametric form, such as a (multivariate) normal. Depending on F_i and G, the integration in (4.35) may or may not be possible analytically. Proposed solutions are based on Taylor series expansions of $f_i(\boldsymbol{y_i}|\boldsymbol{b_i})$, or on numerical approximations of the integral, such as (adaptive) Gaussian quadrature.

Although in practice one is usually primarily interested in estimating the parameters in the marginal model, it is often useful to calculate estimates for the random effects $\boldsymbol{b_i}$ as well. They reflect between-subject variability, which makes them helpful for detecting special profiles (i.e., outlying individuals) or groups of individuals evolving differently in time. Also, estimates for the random effects are needed whenever interest is in prediction of subject-specific evolutions. Inference for the random effects is often based on their so-called posterior distribution $f_i(\boldsymbol{b_i}|\boldsymbol{y_i})$, given by

$$f_i(\boldsymbol{b_i}|\boldsymbol{y_i}) = \frac{f_i(\boldsymbol{y_i}|\boldsymbol{b_i})\, g(\boldsymbol{b_i})}{\int f_i(\boldsymbol{y_i}|\boldsymbol{b_i})\, g(\boldsymbol{b_i})\, d\boldsymbol{b_i}}, \tag{4.36}$$

in which the unknown parameters $\boldsymbol{\theta}$ and $\boldsymbol{\psi}$ are replaced by their estimates obtained earlier from maximizing the marginal likelihood. The mean or mode corresponding to (4.36) can be used as point estimates for $\boldsymbol{b_i}$, yielding empirical Bayes (EB) estimates.

Linear Mixed Models

When continuous (normally distributed) hierarchical data are considered (repeated measures, clustered data, geographical data, longitudinal data, ...), a general, and very flexible, class of parametric covariance models is obtained from introducing random effects $\boldsymbol{b_i}$ in the multivariate linear regression model. Linear mixed models assume the outcome vector $\boldsymbol{y_i}$ follows a multivariate normal distribution, with mean vector $X_i\boldsymbol{\beta} + Z_i\boldsymbol{b_i}$ and some covariance matrix Σ_i, and assume that the random effects $\boldsymbol{b_i}$ also follow a (multivariate) normal distribution, i.e., it is assumed that the n_i-dimensional vector $\boldsymbol{y_i}$ satisfies

$$\boldsymbol{y_i}|\boldsymbol{b_i} \sim N(X_i\boldsymbol{\beta} + Z_i\boldsymbol{b_i}, \Sigma_i), \tag{4.37}$$
$$\boldsymbol{b_i} \sim N(\boldsymbol{0}, D), \tag{4.38}$$

where X_i and Z_i are $(n_i \times p)$ and $(n_i \times q)$ dimensional matrices of known covariates, $\boldsymbol{\beta}$ is a p-dimensional vector of regression parameters, called the fixed effects, D is a general $(q \times q)$ covariance matrix and Σ_i is a $(n_i \times n_i)$ covariance matrix that depends on i only through its dimension n_i, i.e., the set of unknown parameters in Σ_i will not depend upon i.

The above model can be interpreted as a linear regression model for the vector $\boldsymbol{y_i}$ of repeated measurements for each unit separately, where some of the regression parameters are specific (random effects, $\boldsymbol{b_i}$), while others are not (fixed effects, $\boldsymbol{\beta}$). The distributional assumptions in (4.38) with respect to the random effects can be motivated as follows. First, $E(\boldsymbol{b_i}) = \boldsymbol{0}$ implies that the mean of $\boldsymbol{y_i}$ still equals $X_i\boldsymbol{\beta}$, such that the fixed effects in the random-effects model (4.37) can also be interpreted marginally. Not only do they reflect the effect of changing covariates within specific units, they also measure the marginal effect in the population of changing the same covariates. As will be discussed further, this important property only holds for very specific random-effects models, one of which is the linear mixed model considered here. Second, the normality assumption immediately implies that, marginally, $\boldsymbol{y_i}$ also follows a normal distribution with mean vector $X_i\boldsymbol{\beta}$ and with covariance matrix $V_i = Z_i D Z_i' + \Sigma_i$. Hence, no numerical approximation to the integral in (4.35) is needed. Apart from this mathematical convenience, the normality assumption for the b_i is further supported by noticing that the b_i express how unit-specific trends deviate from the population-averaged trends, which suggests that they can be interpreted as residuals.

Note that the random effects in (4.37) implicitly imply the marginal covariance matrix V_i of y_i to be of the very specific form $V_i = Z_i D Z_i' + \Sigma_i$. Let us consider two examples under the assumption of conditional independence, i.e., assuming $\Sigma_i = \sigma^2 I_{n_i}$. First, consider the case where the random effects are univariate and represent unit-specific intercepts. This corresponds to covariates Z_i which are n_i-dimensional vectors containing only ones. The implied covariance matrix can then easily be shown to have the compound symmetry structure which makes the strong assumption that the variance remains constant over all repeated measures and that the correlation between any two measures within a specific unit is also constant. Second, for longitudinal data, suppose that the b_i represent unit-specific intercepts as well as linear time effects. The corresponding Z_i are then of the form

$$
Z_i \;=\; \begin{pmatrix} 1 & t_{i1} \\ 1 & t_{i2} \\ \vdots & \vdots \\ 1 & t_{in_i} \end{pmatrix},
$$

where t_{ij} is the time point at which the jth measurement was taken for the ith subject. Denoting the (k,l) element in D as d_{kl}, we have that the covariance between two repeated measures within a single unit is given by

$$
\begin{aligned}
\mathrm{Cov}(y_{ik}, y_{il}) &=\; \begin{pmatrix} 1 & t_{ik} \end{pmatrix} D \begin{pmatrix} 1 \\ t_{il} \end{pmatrix} + \sigma^2 \\
&=\; d_{22}\, t_{ik}\, t_{il} + d_{12}(t_{ik} + t_{il}) + d_{11} + \sigma^2.
\end{aligned}
$$

Note how the model now implies the variance function of the response to be quadratic over time, with positive curvature d_{22}.

When time is replaced by dose level, this modeling approach can also be used for clustered data with foetus-specific exposure, such as in the heatshock studies (Section 2.2).

The marginal model implied by expressions (4.37) and (4.38) is

$$
y_i \;\sim\; N(X_i \boldsymbol{\beta}, V_i), \quad V_i = Z_i D Z_i' + \Sigma_i,
$$

that can be viewed as another multivariate linear regression model, with a very particular parameterization of the covariance matrix V_i. Hence, our earlier remarks with respect to the fitting of the marginal model remain valid. The vector $\boldsymbol{\alpha}$ of variance components then consists of the variances and covariances in D as well as all unknown parameters in Σ_i.

With respect to the estimation of unit-specific parameters b_i, the posterior distribution of b_i given the observed data y_i can be shown to be (multivariate) normal with mean vector equal to $D Z_i' V_i^{-1}(\boldsymbol{\alpha})(y_i - X_i \boldsymbol{\beta})$. Replacing $\boldsymbol{\beta}$ and

α by their maximum likelihood estimates, we obtain the EB estimates $\widehat{b_i}$ for the b_i, introduced at the start of this section. A key property of these EB estimates is shrinkage, which is best illustrated by considering the prediction $\widehat{y_i} \equiv X_i\widehat{\beta} + Z_i\widehat{b_i}$ of the ith profile. It can easily be shown that

$$\widehat{y_i} = \Sigma_i V_i^{-1} X_i\widehat{\beta} + \left(I_{n_i} - \Sigma_i V_i^{-1}\right) y_i,$$

which can be interpreted as a weighted average of the population-averaged profile $X_i\widehat{\beta}$ and the observed data y_i, with weights $\Sigma_i V_i^{-1}$ and $I_{n_i} - \Sigma_i V_i^{-1}$, respectively. Note that the "numerator" of $\Sigma_i V_i^{-1}$ represents within-unit variability and the "denominator" is the overall covariance matrix V_i. Hence, much weight will be given to the overall average profile if the within-unit variability is large in comparison to the between-unit variability (modeled by the random effects), whereas much weight will be given to the observed data if the opposite is true. This phenomenon is referred to as shrinkage toward the average profile $X_i\widehat{\beta}$. An immediate consequence of shrinkage is that the EB estimates show less variability than actually present in the random-effects distribution, i.e., for any linear combination λ of the random effects,

$$\text{var}(\lambda'\widehat{b_i}) \leq \text{var}(\lambda'b_i) = \lambda'D\lambda.$$

This is also the main reason why, in practice, EB estimates cannot be used to check the normality assumption of the random effects. For example, histograms of elements of the $\widehat{b_i}$ do not necessarily reflect their correct underlying distribution.

Non-linear Mixed Models

An extension of model (4.37) which allows for non-linear relationships between the responses in y_i and the covariates in X_i and/or Z_i is

$$y_i|b_i \sim N(h(X_i, Z_i, \beta, b_i), \Sigma_i) \tag{4.39}$$

for some known 'link' function h. The definition of X_i, Z_i, β, and b_i remains unchanged, the random effects b_i are again assumed to be normally distributed with mean vector $\mathbf{0}$ and covariance matrix D and inference can proceed as explained for the general model.

There are at least two major differences in comparison to the linear mixed model discussed in the previous section. First, the marginal distribution of y_i can no longer be calculated analytically, such that numerical approximations to the marginal density (4.35) come into play, seriously complicating the computation of the maximum likelihood estimates of the parameters in the marginal model, i.e., β, D and the parameters in all Σ_i. A consequence is that the marginal covariance structure does not immediately follow from the model formulation, such that it is not always clear in practice what assumptions a specific model implies with respect to the underlying variance function and the underlying correlation structure in the data.

A second important difference is with respect to the interpretation of the fixed effects β. Under the linear model (4.37), we have that $E(y_i)$ equals $X_i\beta$, such that the fixed effects have a subject-specific as well as a population-averaged interpretation. Indeed, the elements in $\boldsymbol{\beta}$ reflect the effect of specific covariates, conditionally on the random effects $\boldsymbol{b_i}$, as well as marginalized over these random effects. Under non-linear mixed models, however, this does no longer hold in general. The fixed effects now only reflect the conditional effect of covariates, and the marginal effect is not easily obtained anymore as $E(\boldsymbol{y_i})$ is given by

$$E(\boldsymbol{y_i}) \;\; = \;\; \int \boldsymbol{y_i} \int f_i(\boldsymbol{y_i}|\boldsymbol{b_i})g(\boldsymbol{b_i})d\boldsymbol{b_i}d\boldsymbol{y_i},$$

which, in general, is not of the form $h(X_i, Z_i, \boldsymbol{\beta}, \boldsymbol{0})$.

Only for very particular models, (some of) the fixed effects can still be interpreted as marginal covariate effects. For example, consider the model where, apart from an exponential link function, the mean is linear in the covariates, and the only random effects in the model are intercepts. More specifically, this corresponds to the model with $h(X_i, Z_i, \boldsymbol{\beta}, \boldsymbol{b_i}) = \exp(X_i\boldsymbol{\beta} + Z_i b_i)$, in which Z_i is now a vector containing only ones. The expectation of $\boldsymbol{y_i}$ is now given by

$$
\begin{aligned}
E(\boldsymbol{y_i}) \;\; &= \;\; E\left[\exp(X_i\boldsymbol{\beta} + Z_i b_i)\right] \\
&= \;\; \exp(X_i\boldsymbol{\beta}) \; Z_i E\left[\exp(b_i)\right], \quad\quad\quad (4.40)
\end{aligned}
$$

which shows that, except for the intercept, all parameters in $\boldsymbol{\beta}$ have a marginal interpretation.

The Generalized Linear Mixed Model

The generalized linear mixed model is the most frequently used random-effects model for discrete outcomes. A general formulation is as follows. Conditionally on random effects $\boldsymbol{b_i}$, it assumes that the elements y_{ij} of $\boldsymbol{y_i}$ are independent, with density function of the form

$$f_i(y_{ij}|\boldsymbol{b_i}) \;\; = \;\; \exp\left[(y_{ij}\eta_{ij} - a(\eta_{ij}))/\phi + c(y_{ij}, \phi)\right],$$

with mean $E(y_{ij}|\boldsymbol{b_i}) = a'(\eta_{ij}) = \mu_{ij}(\boldsymbol{b_i})$ and variance $Var(y_{ij}|\boldsymbol{b_i}) = \phi a''(\eta_{ij})$, and where, apart from a link function h, a linear regression model with parameters $\boldsymbol{\beta}$ and $\boldsymbol{b_i}$ is used for the mean, i.e., $h(\boldsymbol{\mu_i}(\boldsymbol{b_i})) = X_i\boldsymbol{\beta} + Z_i\boldsymbol{b_i}$. Note that the linear mixed model is a special case, with identity link function. The random effects $\boldsymbol{b_i}$ are again assumed to be sampled from a (multivariate) normal distribution with mean $\boldsymbol{0}$ and covariance matrix D. Usually, the canonical link function is used, i.e., $h = a'^{-1}$, such that $\boldsymbol{\eta_i} = X_i\boldsymbol{\beta} + Z_i\boldsymbol{b_i}$.

The non-linear nature of the model again implies that the marginal distribution of $\boldsymbol{y_i}$ is, in general, not easily obtained, such that model fitting requires approximation of the marginal density function. An exception to this occurs when the probit link is used. Further, as was also the case for non-linear mixed models, the parameters $\boldsymbol{\beta}$ have no marginal interpretation, except for some very particular models. An example where the marginal interpretation does hold is the Poisson model for count data, for which the logarithm is the canonical link function. In case the model only includes random intercepts, it immediately follows from the calculations in (4.40) that the only element in β which has no marginal interpretation is the intercept.

As another example, consider the binomial model for binary data, with the logit canonical link function, and where the only random effects are intercepts b_i. It can then be shown that the marginal mean $\mu_i = \mathrm{E}(y_{ij})$ satisfies $h(\mu_i) \approx X_i \beta^*$ with $\beta^* = (c^2 \mathrm{Var}(b_i) + 1)^{-1/2}\beta$, in which c equals $16\sqrt{3}/15\pi$. Hence, although the parameters β in the generalized linear mixed model have no marginal interpretation, they do show a strong relation to their marginal counterparts. Note that, as a consequence of this relation, larger covariate effects are obtained under the random-effects model in comparison to the marginal model.

Several approaches have been developed to the fit of generalized linear mixed models. One approach, proposed by Wolfinger and O'Connell (1993), is based on an extension of the method of Nelder and Wedderburn (1972) (see also McCullagh and Nelder 1989) to fit fixed-effects generalized linear models. Let us briefly recall this procedure. Dropping the subject-specific index i, the basic form of a generalized linear model is:

$$\boldsymbol{\eta} = X\boldsymbol{\beta},$$

where $\boldsymbol{\eta} = g(\boldsymbol{\mu})$, $\boldsymbol{\mu} = E(\boldsymbol{Y})$ and g is an appropriate link function. Nelder and Wedderburn (1972) showed that maximum likelihood estimates for β can be obtained by iteratively solving

$$X'WX\boldsymbol{\beta} = X'W\boldsymbol{y}^*, \tag{4.41}$$

where

$$
\begin{aligned}
W &= D\Sigma^{-1}D, \\
\boldsymbol{y}^* &= \hat{\eta} + (\boldsymbol{y} - \hat{\mu})D^{-1}, \\
D &= (\partial\mu/\partial\eta), \\
\Sigma &= \Sigma_{\mu}^{1/2} A \Sigma_{\mu}^{1/2}.
\end{aligned}
$$

Here, Σ_{μ} is a diagonal matrix of variances and A is a correlation matrix. McCullagh and Nelder (1989) note that the "working" dependent variable in these estimating equations is not $\boldsymbol{y}$ but $\boldsymbol{y}^*$, a linearized version of $\boldsymbol{y}$.

As indicated earlier, likelihood inference for generalized linear mixed models requires evaluation of integrals (Breslow and Clayton 1993), where the integral's dimension is equal to the number of random effects. Zeger and Karim (1991) avoid the need for numerical integration by casting the generalized linear random-effects model in a Bayesian framework and by resorting to the Gibbs sampler. Wolfinger and O'Connell (1993) circumvent numerical integration by using pseudo-likelihood (and restricted pseudo-likelihood) procedures. The latter approach is implemented in the SAS macro GLIMMIX and is essentially a random-effects extension of (4.41). The GLIMMIX macro is known to have some drawbacks such as, for example, downward biases in fixed-effects and covariance parameters. In contrast, the MLWIN software, the MIXOR software package (Hedeker and Gibbons 1994) and the SAS procedure NLMIXED use either better approximations or numerical integration and are known to have better properties.

CHAPTER 5

Generalized Estimating Equations

Helena Geys, Geert Molenberghs

transnationale Universiteit Limburg, Diepenbeek–Hasselt, Belgium

Louise M. Ryan

Harvard School of Public Health, Boston, MA

Generalized estimating equations play an important role in the analysis of repeated or clustered outcomes of a non-normally distributed type. In this work, it will be used, together with pseudo-likelihood methodology, as non-likelihood based method for the analysis of clustered binary data. A comparison between both will be made in Chapter 6. Also, the use of generalized estimating equations will be illustrated in the contexts of individual-level covariates and combined continuous and discrete outcomes, in Chapters 13 and 14, respectively. Further applications of the GEE technology can be found in Section 9.2.6 and Chapter 11.

When we are mainly interested in first order marginal mean parameters and pairwise interactions, a full likelihood procedure can be replaced by quasi-likelihood methods (McCullagh and Nelder 1989). In quasi-likelihood, the mean response is expressed as a parametric function of covariates; the variance is assumed to be a function of the mean up to possibly unknown scale parameters. Wedderburn (1974) first noted that likelihood and quasi-likelihood theories coincide for exponential families and that the quasi-likelihood "estimating equations" provide consistent estimates of the regression parameters β in any generalized linear model, even for choices of link and variance functions that do not correspond to exponential families.

For clustered and repeated data, Liang and Zeger (1986) proposed so-called *generalized estimating equations* (GEE or GEE1) which require only the correct specification of the univariate marginal distributions provided one is willing to adopt "working" assumptions about the association structure. They estimate the parameters associated with the expected value of an individual's vector of binary responses and phrase the working assumptions about the association between pairs of outcomes in terms of marginal correlations.

Prentice (1988) extended their results to allow joint estimation of probabil-

ities and pairwise correlations. Lipsitz, Laird and Harrington (1991) modified
the estimating equations of Prentice (1988) to allow modeling of the association through marginal odds ratios rather than marginal correlations. When
adopting GEE1 one does not use information of the association structure to
estimate the main effect parameters. As a result, it can be shown that GEE1
yields consistent main effect estimators, even when the association structure
is misspecified. However, severe misspecification may seriously affect the efficiency of the GEE1 estimators. In addition, GEE1 should be avoided when
some scientific interest is placed on the association parameters.

A second order extension of these estimating equations (GEE2) that include
the marginal pairwise association as well has been studied by Liang, Zeger and
Qaqish (1992). They note that GEE2 is nearly fully efficient though bias may
occur in the estimation of the main effect parameters when the association
structure is misspecified. A variation to this theme, using conditional probability ideas, has been proposed by Carey, Zeger and Diggle (1993). It is referred
to as *alternating logistic regressions.*

In Section 5.1 we present general GEE theory, whereas several applications
and specializations to the case of clustered binary data are presented in Section 5.2.

5.1 General Theory

Usually, when confronted with the analysis of clustered or otherwise correlated
data, conclusions based on mean parameters (e.g., dose effect) are of primary
interest. When inferences for the parameters in the mean model $E(\boldsymbol{y_i})$ are
based on classical maximum likelihood theory, full specification of the joint
distribution for the vector $\boldsymbol{y_i}$ of repeated measurements within each unit i
is necessary. For discrete data, this implies specification of the first-order
moments, as well as all higher-order moments and, depending on whether
marginal or random-effects models are used, assumptions are either explicitly
made or implicit in the random-effects structure. For Gaussian data, full-
model specification reduces to modeling the first- and second-order moments
only. However, even then can inappropriate covariance models seriously invali-
date inferences for the mean structure. Thus, a drawback of a fully parametric
model is that incorrect specification of nuisance characteristics can lead to in-
valid conclusions about key features of the model.

A very flexible approach, frequently used in practice, is so-called generalized estimating equations (GEEs). The GEE methodology is based on two
perceptions. First, the score equations to be solved when computing maxi-
mum likelihood estimates under a marginal normal model $\boldsymbol{y_i} \sim N(X_i\boldsymbol{\beta}, V_i)$
are given by

$$\sum_{i=1}^{N} X_i'(A_i^{1/2} R_i A_i^{1/2})^{-1}(\boldsymbol{y_i} - X_i\boldsymbol{\beta}) = \boldsymbol{0}, \qquad (5.1)$$

in which the marginal covariance matrix V_i has been decomposed in the form

$A_i^{1/2} R_i A_i^{1/2}$, with A_i the matrix with the marginal variances on the main diagonal and zeros elsewhere, and with R_i equal to the marginal correlation matrix. Second, the score equations to be solved when computing maximum likelihood estimates under a marginal generalized linear model (omitting the random effects b_i from the model formulation of Section 4.3.2), assuming independence of the responses within units (i.e., ignoring the repeated measures structure), are given by

$$\sum_{i=1}^{N} \frac{\partial \mu_i}{\partial \beta'} (A_i^{1/2} I_{n_i} A_i^{1/2})^{-1} (y_i - \mu_i) \;=\; 0, \tag{5.2}$$

where A_i is again the diagonal matrix with the marginal variances on the main diagonal.

Note that expression (5.1) is of the form (5.2) but with the correlations between repeated measures taken into account. A straightforward extension of (5.2) that accounts for the correlation structure is

$$S(\beta) \;=\; \sum_{i=1}^{N} \frac{\partial \mu_i}{\partial \beta'} (A_i^{1/2} R_i A_i^{1/2})^{-1} (y_i - \mu_i) \;=\; 0, \tag{5.3}$$

that is obtained from replacing the identity matrix I_{n_i} by a correlation matrix $R_i = R_i(\alpha)$, often referred to as the *working* correlation matrix. Usually, the marginal covariance matrix $V_i = A_i^{1/2} R_i A_i^{1/2}$ contains a vector α of unknown parameters which is replaced for practical purposes by a consistent estimate.

Assuming that the marginal mean μ_i has been correctly specified as $h(\mu_i) = X_i \beta$, it can be shown that, under mild regularity conditions, the estimator $\widehat{\beta}$ obtained from solving (5.3) is asymptotically normally distributed with mean β and with covariance matrix

$$I_0^{-1} I_1 I_0^{-1}, \tag{5.4}$$

where

$$I_0 \;=\; \left(\sum_{i=1}^{N} \frac{\partial \mu_i'}{\partial \beta} V_i^{-1} \frac{\partial \mu_i}{\partial \beta'} \right),$$

$$I_1 \;=\; \left(\sum_{i=1}^{N} \frac{\partial \mu_i'}{\partial \beta} V_i^{-1} \mathrm{Var}(y_i) V_i^{-1} \frac{\partial \mu_i}{\partial \beta'} \right).$$

In practice, $\mathrm{Var}(y_i)$ in (5.4) is replaced by $(y_i - \mu_i)(y_i - \mu_i)'$, which is unbiased on the sole condition that the mean was again correctly specified.

Note that valid inferences can now be obtained for the mean structure, only assuming that the model assumptions with respect to the first-order moments are correct. Note also that, although arising from a likelihood approach, the GEE equations in (5.3) cannot be interpreted as score equations corresponding to some full likelihood for the data vector y_i.

Liang and Zeger (1986) proposed moment-based estimates for the working correlation. To this end, first define deviations:

$$e_{ij} = \frac{y_{ij} - \mu_{ij}}{\sqrt{v(\mu_{ij})}}$$

and decompose the variance slightly more generally as above in the following way:

$$V_i = \phi A_i^{1/2} R_i A_i^{1/2},$$

where ϕ is an overdispersion parameter.

Some of the more popular choices for the working correlations are:

- Independence:

$$\text{Corr}(Y_{ij}, Y_{ik}) = 0 \qquad (j \neq k).$$

  There are no parameters to be estimated.

- Exchangeable:

$$\text{Corr}(Y_{ij}, Y_{ik}) = \alpha \qquad (j \neq k).$$

$$\hat{\alpha} = \frac{1}{N} \sum_{i=1}^{N} \frac{1}{n_i(n_i - 1)} \sum_{j \neq k} e_{ij} e_{ik}.$$

- AR(1):

$$\text{Corr}(Y_{ij}, Y_{i,j+t}) = \alpha^t \qquad (t = 0, 1, \ldots, n_i - j).$$

$$\hat{\alpha} = \frac{1}{N} \sum_{i=1}^{N} \frac{1}{n_i - 1} \sum_{j \leq n_i - 1} e_{ij} e_{i,j+1}.$$

- Unstructured:

$$\text{Corr}(Y_{ij}, Y_{ik}) = \alpha_{jk} \qquad (j \neq k).$$

$$\hat{\alpha}_{jk} = \frac{1}{N} \sum_{i=1}^{N} e_{ij} e_{ik}.$$

A dispersion parameter can be estimated by

$$\hat{\phi} = \frac{1}{N} \sum_{i=1}^{N} \frac{1}{n_i} \sum_{j=1}^{n_i} e_{ij}^2.$$

The standard iterative procedure to fit GEE, based on Liang and Zeger (1986), is then as follows:

1. Compute initial estimates for $\boldsymbol{\beta}$, using a univariate GLM (i.e., assuming independence).

2. Compute the quantities needed in the estimating equation:

   - Compute Pearson residuals e_{ij}.
   - Compute estimates for $\boldsymbol{\alpha}$.
   - Compute $R_i(\boldsymbol{\alpha})$.
   - Compute an estimate for ϕ.
   - Compute $V_i(\boldsymbol{\beta}, \boldsymbol{\alpha}) = \phi A_i^{1/2}(\boldsymbol{\beta}) R_i(\boldsymbol{\alpha}) A_i^{1/2}(\boldsymbol{\beta})$.

3. Update the estimate for $\boldsymbol{\beta}$:

$$\boldsymbol{\beta}^{(t+1)} = \boldsymbol{\beta}^{(t)} - \left[\sum_{i=1}^{N} D_i^T V_i^{-1} D_i\right]^{-1} \left[\sum_{i=1}^{N} D_i^T V_i^{-1}(\boldsymbol{y}_i - \boldsymbol{\mu}_i)\right].$$

Iterate the second and third steps until convergence.

Standard procedures, such as the SAS/STAT procedure GENMOD (1997) and the Oswald functions in Splus (Smith, Robertson and Diggle 1996), that include GEE1 capabilities use an iterative fitting process, where estimation of the parameters $\boldsymbol{\alpha}$ is based on standardized residuals. The model based estimator of $\text{Cov}(\hat{\boldsymbol{\beta}})$ is given by I_0^{-1}, where

$$I_0 = \sum_{i=1}^{N} \frac{\partial \boldsymbol{\pi}_i'}{\partial \boldsymbol{\beta}} V_i^{-1} \frac{\partial \boldsymbol{\mu}_i}{\partial \boldsymbol{\beta}}.$$

The empirically corrected variance estimator (Liang and Zeger 1986) takes the form $I_0^{-1} I_1 I_0^{-1}$, where

$$I_1 = \sum_{i=1}^{N} \frac{\partial \boldsymbol{\mu}_i'}{\partial \boldsymbol{\beta}} V_i^{-1} \text{Cov}(\boldsymbol{Y}_i) V_i^{-1} \frac{\partial \boldsymbol{\mu}_i}{\partial \boldsymbol{\beta}}.$$

Williamson, Lipsitz and Kim (1997) wrote a SAS macro for GEE1 that is based on Prentice's approach. The latter considered an extension of the GEE1 approach of Liang and Zeger (1986) that allows joint estimation of the parameters $\boldsymbol{\beta}$ and $\boldsymbol{\alpha}$ in both the marginal response probabilities and the pairwise correlations. A GEE1 estimator for $\boldsymbol{\beta}$ and $\boldsymbol{\alpha}$ may be defined as a solution to:

$$\sum_{i=1}^{N} \boldsymbol{D}_i^T \boldsymbol{V}_i^{-1}(\boldsymbol{Y}_i - \boldsymbol{\mu}_i) = 0$$

$$\sum_{i=1}^{N} \boldsymbol{E}_i^T \boldsymbol{W}_i^{-1}(\boldsymbol{Z}_i - \boldsymbol{\delta}_i) = 0,$$

where

$$Z_{ijk} = \frac{(Y_{ij} - \mu_{ij})(Y_{ik} - \mu_{ik})}{\sqrt{\mu_{ij}(1 - \mu_{ij})\mu_{ik}(1 - \mu_{ik})}}$$

and $\delta_{ijk} = E(Z_{ijk})$. Under exchangeability we have $\delta_{ijk} = \rho_i$, the correlation between any two outcomes of the same cluster i. This can be reparametrized in terms of α, using Fisher's z-transformation: $\alpha = \ln(1 + \rho) - \ln(1 - \rho)$. The joint asymptotic distribution of $\sqrt{N}(\hat{\boldsymbol{\beta}} - \boldsymbol{\beta})$ and $\sqrt{N}(\hat{\boldsymbol{\alpha}} - \boldsymbol{\alpha})$ is Gaussian with mean zero and with variance-covariance matrix consistently estimated by N times

$$\begin{pmatrix} \boldsymbol{A} & \boldsymbol{0} \\ \boldsymbol{B} & \boldsymbol{C} \end{pmatrix} \begin{pmatrix} \Lambda_{11} & \Lambda_{12} \\ \Lambda_{21} & \Lambda_{22} \end{pmatrix} \begin{pmatrix} \boldsymbol{A} & \boldsymbol{B}^T \\ \boldsymbol{0} & \boldsymbol{C} \end{pmatrix},$$

where

$$A = \left(\sum_{i=1}^{N} \boldsymbol{D}_i^T \boldsymbol{V}_i^{-1} \boldsymbol{D}_i \right)^{-1},$$

$$B = \left(\sum_{i=1}^{N} \boldsymbol{E}_i^T \boldsymbol{W}_i^{-1} \boldsymbol{E}_i \right)^{-1} \left(\sum_{i=1}^{N} \boldsymbol{E}_i^T \boldsymbol{W}_i^{-1} \frac{\partial \boldsymbol{Z}_i}{\partial \beta} \right) \left(\sum_{i=1}^{N} \boldsymbol{D}_i^T \boldsymbol{V}_i^{-1} \boldsymbol{D}_i \right)^{-1},$$

$$C = \left(\sum_{i=1}^{N} \boldsymbol{E}_i^T \boldsymbol{W}_i^{-1} \boldsymbol{E}_i \right)^{-1},$$

$$\Lambda_{11} = \sum_{i=1}^{N} \boldsymbol{D}_i^T \boldsymbol{V}_i^{-1} \text{Cov}(\boldsymbol{Y}_i) \boldsymbol{V}_i^{-1} \boldsymbol{D}_i,$$

$$\Lambda_{12} = \sum_{i=1}^{N} \boldsymbol{D}_i^T \boldsymbol{V}_i^{-1} \text{Cov}(\boldsymbol{Y}_i, \boldsymbol{Z}_i) \boldsymbol{W}_i^{-1} \boldsymbol{E}_i,$$

$$\Lambda_{21} = \Lambda_{12},$$

$$\Lambda_{22} = \sum_{i=1}^{N} \boldsymbol{E}_i^T \boldsymbol{W}_i^{-1} \text{Cov}(\boldsymbol{Z}_i) \boldsymbol{W}_i^{-1} \boldsymbol{E}_i,$$

and $\text{Var}(\boldsymbol{Y}_i)$, $\text{Cov}(\boldsymbol{Y}_i, \boldsymbol{Z}_i)$ and $\text{Var}(\boldsymbol{Z}_i)$ respectively estimated by the quantities $(\boldsymbol{Y}_i - \boldsymbol{\mu}_i)(\boldsymbol{Y}_i - \boldsymbol{\mu}_i)^T$, $(\boldsymbol{Y}_i - \boldsymbol{\mu}_i)(\boldsymbol{Z}_i - \boldsymbol{\delta}_i)^T$ and $(\boldsymbol{Z}_i - \boldsymbol{\delta}_i)(\boldsymbol{Z}_i - \boldsymbol{\delta}_i)^T$. It is convenient to define:

$$\boldsymbol{Z}_i = \begin{pmatrix} Y_{i1} Y_{i2} \\ Y_{i1} Y_{i3} \\ \vdots \\ Y_{in_i} Y_{i(n_i-1)} \end{pmatrix}.$$

Hence, under exchangeability,

$$E(Z_{ijk}) = \mu_{ijk} = \rho \sqrt{\mu_{ij}(1 - \mu_{ij})\mu_{ik}(1 - \mu_{ik})} + \mu_{ij}\mu_{ik},$$

$$\text{Var}(Z_{ijk}) = \mu_{ijk}(1 - \mu_{ijk}),$$

$$\frac{\partial E(Z_{ijk})}{\partial \alpha} = \frac{2 \exp(\alpha)}{(\exp(\alpha) + 1)^2} \sqrt{\mu_{ij}(1 - \mu_{ij})\mu_{ik}(1 - \mu_{ik})}.$$

The matrix C then reduces to:

$$C = \left(\frac{2 \exp(\alpha)}{(\exp(\alpha) + 1)^2} \sqrt{\mu_{ij}(1 - \mu_{ij})\mu_{ik}(1 - \mu_{ik})} \right)^2 \frac{1}{\mu_{ijk}(1 - \mu_{ijk})}.$$

To obtain the variance-covariance matrix of the correlation parameters $\boldsymbol{\rho}$, one can apply the delta method. In the case of exchangeability we multiply the standard error of α with a factor $2 \exp(\alpha)/(\exp(\alpha) + 1)^2$ to obtain the standard error of ρ.

GEE2

The GEE2 approach naturally accommodates individual-level covariates in the estimation of marginal response probabilities. For each cluster, define

$$\boldsymbol{w}_i = (y_{i1}, \ldots, y_{in_i}, y_{i1}y_{i2}, \ldots, y_{in_i-1}y_{in_i})^T,$$

a vector of $n_i + \binom{n_i}{2}$ components. Further, let $\boldsymbol{\Theta}_i = (\boldsymbol{\mu}_i^T, \boldsymbol{\rho}_i^T)^T$ which depends on a $p \times 1$ vector of regression parameters $\boldsymbol{\beta}$ through a generalized linear model. Estimation of $\boldsymbol{\beta}$ is accomplished by solving the following second order estimating equations:

$$\boldsymbol{U}(\boldsymbol{\beta}) = \sum_{i=1}^{N} X_i^T \left(T_i^{-1} \right)^T V_i^{-1} (\boldsymbol{W}_i - E(\boldsymbol{W}_i)) = \boldsymbol{0},$$

with $X_i = \partial \boldsymbol{\eta}_i / \partial \boldsymbol{\beta}$, $T_i = \partial \boldsymbol{\eta}_i / \partial \boldsymbol{\Theta}_i$ and $V_i = \text{Cov}(\boldsymbol{W}_i)$. Calculation of all matrices involved is straightforward with the exception of the covariance matrix, which contains third and fourth order probabilities. To this end, the three-way and higher order correlations are set equal to zero. As before, the parameter estimates $\hat{\boldsymbol{\beta}}$ can then be calculated using, for example, a Fisher scoring algorithm. Provided the first and second order models have been correctly specified, $\hat{\boldsymbol{\beta}}$ is consistent for $\boldsymbol{\beta}$ and has an asymptotic multivariate normal distribution with mean vector $\boldsymbol{\beta}$ and variance-covariance matrix consistently estimated by:

$$V(\hat{\boldsymbol{\beta}}) = \left(\sum_{i=1}^{N} X_i^T \hat{T}_i^{-T} \hat{V}_i^{-1} \hat{T}_i^{-1} X_i \right)^{-1} \sum_{i=1}^{N} \boldsymbol{U}_i(\hat{\boldsymbol{\beta}}) \boldsymbol{U}_i(\hat{\boldsymbol{\beta}})^T$$

$$\left(\sum_{i=1}^{N} X_i^T \hat{T}_i^{-T} \hat{V}_i^{-1} \hat{T}_i^{-1} X_i \right)^{-1}.$$

5.2 Clustered Binary Data

In addition to Prentice's (1988) proposal to use a second set of estimating equations for the correlation nuisance parameters, Lipsitz, Laird and Harrington (1991) suggested the use of odds ratios rather than correlations to capture the within-cluster covariance. Adopting their ideas, we first consider a GEE1 approach that allows joint estimation of regression parameters $(\boldsymbol{\beta}^T, \boldsymbol{\alpha}^T)^T$ in, respectively, the marginal means and pairwise associations, using two sets of estimating equations. Both extended the GEE1 approach of Liang and Zeger (1986), where estimators for $(\boldsymbol{\beta}^T, \boldsymbol{\alpha}^T)^T$ were obtained using iteratively reweighted least squares calculations and moment-based estimation of $\boldsymbol{\alpha}$.

Earlier in this chapter, GEE1 was studied, together with its alternative proposed by Prentice (1988), as well as GEE2. We will now specifically consider the case of clustered binary data. To this effect, let us introduce the notation π_{i11} for the joint observation for two successes in cluster i and π_{i00} for the joint observation of two failures. The probability of one success and one failure is then $\pi_{i10} \equiv \pi_{i01}$. It is insightful to study this case in a bit more detail.

If we let the marginal means π_{i10} and pairwise probabilities π_{i11} depend on a vector of regression parameters $(\boldsymbol{\beta}^T, \boldsymbol{\alpha}^T)^T$ through the following generalized linear model:

$$\boldsymbol{\eta}_i = \left(\begin{array}{c} \ln(\pi_{i10}) - \ln(1 - \pi_{i10}) \\ \ln(\pi_{i11}) + \ln(1 - 2\pi_{i10} + \pi_{i11}) - 2\ln(\pi_{i10} - \pi_{i11}) \end{array} \right) = X_i \left(\begin{array}{c} \boldsymbol{\beta} \\ \boldsymbol{\alpha} \end{array} \right),$$

then the two sets of estimating equations for, respectively, $\boldsymbol{\beta}$ and $\boldsymbol{\alpha}$ can be combined into:

$$\sum_{i=1}^{N} \left(\begin{array}{cc} D_i^T & 0 \\ 0 & C_i^T \end{array} \right) \left(\begin{array}{cc} \mathrm{Var}(Z_i) & 0 \\ 0 & \mathrm{Var}(\binom{Z_i}{2}) \end{array} \right)^{-1} \left(\begin{array}{c} Z_i - n_i \pi_{i10} \\ \binom{Z_i}{2} - \binom{n_i}{2} \pi_{i11} \end{array} \right),$$

where $D_i = n_i \partial \pi_{i10} / \partial \boldsymbol{\beta}$ and $C_i = \binom{n_i}{2} \partial \pi_{i11} / \partial \boldsymbol{\alpha}$. An iterative procedure for calculating $\boldsymbol{\beta}$ and $\boldsymbol{\alpha}$ begins with starting values $\boldsymbol{\beta}_0$ and $\boldsymbol{\alpha}_0$ and produces updated values $\boldsymbol{\beta}_{s+1}, \boldsymbol{\alpha}_{s+1}$ from values $\boldsymbol{\beta}_s, \boldsymbol{\alpha}_s$ by means of

$$\boldsymbol{\beta}_{s+1} = \boldsymbol{\beta}_s + \left(\sum_{i=1}^{N} D_i^T V_i^{-1} D_i \right)^{-1} \sum_{i=1}^{N} D_i^T V_i^{-1} (Z_i - n_i \pi_{i10})$$

$$\boldsymbol{\alpha}_{s+1} = \boldsymbol{\alpha}_s + \left(\sum_{i=1}^{N} C_i^T W_i^{-1} C_i \right)^{-1} \sum_{i=1}^{N} C_i^T W_i^{-1} \left(\binom{Z_i}{2} - \binom{n_i}{2} \pi_{i11} \right),$$

where $V_i = \mathrm{Var}(Z_i)$ and $W_i = \mathrm{Var}(\binom{Z_i}{2}) = \mathrm{Var}(\sum_{j<k} Y_{ij} Y_{ik})$. Here, W_i is a function of third and fourth order probabilities, which are nuisance parameters we would rather not estimate. Assuming three- and higher-order independence, in the spirit of Lipsitz, Laird and Harrington (1991), and taking into account the exchangeability assumption, W_i reduces to:

$$\binom{n_i}{2} \pi_{i11}(1 - \pi_{i11}).$$

Prentice (1988) and Lipsitz, Laird and Harrington (1991) have shown that the joint asymptotic covariance matrix of $(\hat{\boldsymbol{\beta}}^T, \hat{\boldsymbol{\alpha}}^T)^T$ equals:

$$V_{\beta,\alpha} = \lim_{N \to \infty} \left(\begin{array}{cc} B_{11}^{-1} & 0 \\ B_{21} & B_{22}^{-1} \end{array} \right) \left(\begin{array}{cc} \Sigma_{11} & \Sigma_{12} \\ \Sigma_{12}^T & \Sigma_{22} \end{array} \right) \left(\begin{array}{cc} B_{11}^{-1} & 0 \\ B_{21} & B_{22}^{-1} \end{array} \right)^T,$$

where

$$B_{11} = N^{-1} \sum_{i=1}^{N} D_i^T V_i^{-1} D_i,$$

$$B_{22} = N^{-1} \sum_{i=1}^{N} C_i^T W_i^{-1} C_i,$$

$$B_{21} = B_{22}^{-1} (\sum_{i=1}^{N} C_i^T W_i^{-1} \partial(\binom{n_i}{2}\pi_{i11})/\partial\boldsymbol{\beta}) B_{11}^{-1},$$

$$\Sigma_{11} = N^{-1} \sum_{i=1}^{N} D_i^T V_i^{-1} \mathrm{Var}(Z_i) V_i^{-1} D_i,$$

$$\Sigma_{22} = N^{-1} \sum_{i=1}^{N} C_i^T W_i^{-1} \mathrm{Var}(\binom{Z_i}{2}) W_i^{-1} C_i,$$

$$\Sigma_{12} = N^{-1} \sum_{i=1}^{N} D_i^T V_i^{-1} \mathrm{Cov}(Z_i, \binom{Z_i}{2}) W_i^{-1} C_i.$$

The matrix $V_{\beta,\alpha}$ can be consistently estimated by replacing $\boldsymbol{\beta}$ and $\boldsymbol{\alpha}$ by their estimates, and also

$$\mathrm{Var}(Z_i) \text{ by } (Z_i - n_i\pi_{i10})(Z_i - n_i\pi_{i10})^T,$$

$$\mathrm{Var}(\binom{Z_i}{2}) \text{ by } ((\binom{Z_i}{2}) - \binom{n_i}{2}\pi_{i11})((\binom{Z_i}{2}) - \binom{n_i}{2}\pi_{i11})^T,$$

$$\mathrm{Cov}(Z_i, \binom{Z_i}{2}) \text{ by } (Z_i - n_i\pi_{i10})((\binom{Z_i}{2}) - \binom{n_i}{2}\pi_{i11})^T.$$

Note that GEE1 operates as if $\boldsymbol{\beta}$ and $\boldsymbol{\alpha}$ are orthogonal to one another even when they actually are not. The effect is that GEE1 gives consistent estimators of $\boldsymbol{\beta}$ whether or not the association structure is correctly specified. On the other hand, GEE1 can be extremely inefficient for the estimation of $\boldsymbol{\alpha}$.

A second order extension of these estimating equations that includes marginal pairwise associations as well has been studied by Liang, Zeger and Qaqish (1992), Molenberghs and Ritter (1996) and Heagerty and Zeger (1996). Liang, Zeger and Qaqish (1992) point out the connection of the quasi-likelihood theories with second order generalized estimating equations, GEE2. In fact, GEE2 can be simply regarded as a multivariate extension of quasi-likelihood. As in quasi-likelihood, GEE2 requires specification of first and second order moments, which are usually of great scientific interest. Indeed, even when there is considerable association between outcomes, three-way and higher order interactions tend to be negligible and are certainly more difficult to interpret. Therefore, a working higher order independence assumption is often plausible. We will develop a second-order estimating equations procedure (GEE2), following the ideas of Liang, Zeger and Qaqish (1992) and adopting a working higher order independence assumption. It is very appealing that such a procedure closely corresponds to the way in which the pseudo-likelihood function was represented. Recall that the pseudo-likelihood function also limits its attention to pairwise interactions, since it is constructed as a product of pairwise probabilities. In the GEE2 framework the following set of estimating equations can be considered:

$$U(\boldsymbol{\beta}) = \sum_{i=1}^{N} X_i^T (T_i^{-1})^T V_i^{-1} (\boldsymbol{Z}_i - \boldsymbol{\pi}_i) = 0$$

with

$$\boldsymbol{Z}_i = \left(\begin{pmatrix} Z_i \\ Z_i \\ 2 \end{pmatrix} \right) \quad \text{and} \quad \pi_i = \left(\begin{pmatrix} n_i \pi_{i10} \\ n_i \\ 2 \end{pmatrix} \pi_{i11} \right).$$

Furthermore, $T_i = \partial \boldsymbol{\eta}_i / \partial \boldsymbol{\pi}_i$ and V_i is the covariance matrix of $\boldsymbol{Z}_i$. The computation of T_i presents no difficulties and is analogous to the calculations performed in Section 6.3.1. We obtain, for T_i:

$$\left(\begin{array}{cc} \frac{1}{n_i} \left(\frac{1}{\pi_{i10}} + \frac{1}{1-\pi_{i10}} \right) & 0 \\ \frac{1}{n_i} \left(\frac{-2}{1-2\pi_{i10}+\pi_{i11}} - \frac{2}{\pi_{i10}-\pi_{i11}} \right) & \frac{2}{n_i(n_i-1)} \left(\frac{1}{\pi_{i11}} + \frac{1}{1-2\pi_{i10}+\pi_{11}} + \frac{2}{\pi_{i10}-\pi_{i11}} \right) \end{array} \right).$$

However, the matrix V_i contains third and fourth order probabilities, which can be found using either the iterative proportional fitting (IPF) algorithm, outlined in Molenberghs and Lesaffre (1999), or alternatively by the procedure given in Molenberghs and Lesaffre (1994), which we use here. This is an important difference with both PL and GEE1, as will be indicated in Section 6.3. Indeed, these only need first and second order probabilities, which are straightforward to implement. Probabilities of order n can be computed, provided all lower-dimensional probabilities together with the odds-ratio of dimension n are known. At this point we introduce the higher order independence working assumption. Let us denote the so-obtained three and four way probabilities $P(y_{ij} = 1, y_{ik} = 1, y_{il} = 1)$ and $P(y_{ij} = 1, y_{ik} = 1, y_{il} = 1, y_{im} = 1)$ by $\pi_{i1}^{(3)}$ resp. $\pi_{i1}^{(4)}$, then we can calculate the different components of V_i:

$$\begin{aligned}
\text{Var}(Z_{i1}) &= E(Z_i^2) - E(Z_i)^2 \\
&= 2E\left[\sum_{j=1}^{n_i} \sum_{k>j} Y_{ij} Y_{ik} \right] + E\left[\sum_{j=1}^{n_i} Y_{ij}^2 \right] - E\left[\sum_{j=1}^{n_i} Y_{ij} \right]^2 \\
&= 2\binom{n_i}{2} \pi_{i11} + n_i \pi_{i10}(1 - n_i \pi_{i10}). \quad (5.5)
\end{aligned}$$

Note that (5.5) reduces to $n_i \pi_{i10}(1 - \pi_{i10})$, under independence. Similarly, we calculate:

$$
\mathrm{Cov}(Z_{i1}, Z_{i2}) = 3\sum_{j=1}^{n_i}\sum_{k>j}\sum_{l>k} E\left[Y_{ij}Y_{ik}Y_{il}\right] + 2E\left[\sum_{j=1}^{n_i}\sum_{l>k} Y_{ik}^2 Y_{il}\right]
$$

$$
- n_i \binom{n_i}{2} \pi_{i10}\pi_{i11}
$$

$$
= 3\binom{n_i}{3}\pi_1^{(3)} + 2\binom{n_i}{2}\pi_{i11} - n_i\binom{n_i}{2}\pi_{i10}\pi_{i11},
$$

$$
\mathrm{Var}(Z_{i2}) = E\left[\sum_{j=1}^{n_i}\sum_{k>j} Y_{ij}Y_{ik}\sum_{r=1}^{n_i}\sum_{s>r} Y_{ir}Y_{is}\right] - \binom{n_i}{2}^2 \pi_{i11}^2
$$

$$
= 6\binom{n_i}{4}\pi_1^{(4)} + 6\binom{n_i}{3}\pi_1^{(3)} + 2\binom{n_i}{2}\pi_{i11} - \binom{n_i}{2}^2 \pi_{i11}^2.
$$

A Fisher scoring algorithm can now be applied to calculate the parameter estimates. The empirically corrected version of the asymptotic covariance matrix proposed by Liang and Zeger (1986) is similar to the one described in Section 6.3.1 and is estimated by:

$$
\left(\sum_{i=1}^{N} X_i^T \hat{T}_i^{-T} \hat{V}_i^{-1} \hat{T}_i^{-1} X_i\right)^{-1} \left(\sum_{i=1}^{N} \boldsymbol{U}_i(\hat{\boldsymbol{\beta}}) \boldsymbol{U}_i(\hat{\boldsymbol{\beta}})^T\right)
$$

$$
\times \left(\sum_{i=1}^{N} X_i^T \hat{T}_i^{-T} \hat{V}_i^{-1} \hat{T}_i^{-1} X_i\right)^{-1}. \tag{5.6}
$$

Thus, provided the model is correctly specified, $\hat{\boldsymbol{\beta}}$ is consistent for $\boldsymbol{\beta}$ and is asymptotically normally distributed with the covariance matrix estimated by (5.6). If the model for the association structure is misspecified, bias may follow in first order parameters (Liang, Zeger and Qaqish 1992). This contrasts with the classical first order estimating equations, GEE1, which yield consistent estimates even if the association structure is misspecified.

CHAPTER 6

Pseudo-likelihood Estimation

Helena Geys, Geert Molenberghs

transnationale Universiteit Limburg, Diepenbeek–Hasselt, Belgium

Louise M. Ryan

Harvard School of Public Health, Boston, MA

It is well known that full maximum likelihood estimation can become prohibitive for many models. For example, in the framework of a marginally specified odds ratio model (Lipsitz, Laird and Harrington 1991, Dale 1986, Molenberghs and Lesaffre 1994, Glonek and McCullagh 1995, Lang and Agresti 1994) for multivariate, clustered binary data, full maximum likelihood estimation is prohibitive, especially with large within-unit representation. Conditional models such as the Molenberghs and Ryan (1999) models, introduced in Section 4.2, are based on an exponential family model for multivariate binary data and exhibit a high flexibility to capture different patterns of non-linear dependencies of the marginal probabilities on the cluster size. Like most exponential family models, the Molenberghs and Ryan (1999) model enjoys well known properties, such as linearity of the log-likelihood in the minimal sufficient statistics, unimodality, etc. This implies a high numerical stability of iterative procedures to determine maximum likelihood estimators. In multivariate settings (with 3 or more outcomes), however, where the normalizing constant takes a complicated form, all of these advantages can be lost as this leads to excessive computational requirements. This is especially true for clusters of variable length, because the normalizing constant depends on the cluster size. Hence, alternative estimation methods, which do not require the explicit calculation of the normalizing constant, are in demand.

In this chapter, we introduce the pseudo-likelihood estimation method. Strictly speaking this is a non-likelihood method. The principal idea is to replace a numerically challenging joint density by a simpler function that is a suitable product of ratios of likelihoods of subsets of the variables. For example, when a joint density contains a computationally intractable normalizing constant, one might calculate a suitable product of conditional densities which does not involve such a complicated function. A bivariate distribution

$f(y_1, y_2)$, for example, can be replaced by the product of both conditionals $f(y_1|y_2)f(y_2|y_1)$. While the method achieves important computational economies by changing the method of estimation, it does not affect model interpretation. Model parameters can be chosen in the same way as with full likelihood and retain their meaning. This method converges quickly with only minor efficiency losses, especially for a range of realistic parameter settings.

6.1 Pseudo-likelihood: Definition and Asymptotic Properties

To formally introduce pseudo-likelihood, we will use the convenient general definition given by Arnold and Strauss (1991). Without loss of generality we can assume that the vector $\boldsymbol{Y}_i$ of binary outcomes for subject i ($i = 1, \ldots, N$) has constant dimension L. The extension to variable lengths of $\boldsymbol{Y}_i$ is straightforward.

6.1.1 Definition

Define S as the set of all $2^L - 1$ vectors of length L, consisting solely of zeros and ones, with each vector having at least one non-zero entry. Denote by $\boldsymbol{y}_i^{(s)}$ the subvector of $\boldsymbol{y}_i$ corresponding to the components of s that are non-zero. The associated joint density is $f_s(\boldsymbol{y}_i^{(s)}; \boldsymbol{\Theta}_i)$. In order to define a pseudo-likelihood function, one chooses a set $\delta = \{\delta_s | s \in S\}$ of real numbers, with at least one non-zero component. The log of the pseudo-likelihood is then defined as

$$p\ell = \sum_{i=1}^{N} \sum_{s \in S} \delta_s \ln f_s(\boldsymbol{y}_i^{(s)}; \boldsymbol{\Theta}_i). \tag{6.1}$$

Adequate regularity conditions have to be assumed to ensure that (6.1) can be maximized by solution of the pseudo-likelihood (score) equations, the latter obtained by differentiation of the logarithm of the pseudo-likelihood and the setting of the derivative to zero.

The classical log-likelihood function is found by setting $\delta_s = 1$ if s is the vector consisting solely of ones, and 0 otherwise. Subsequently we will present some examples of pseudo-likelihood functions that satisfy (6.1).

6.1.2 Consistency and Asymptotic Normality

Before stating the main asymptotic properties of the PL estimators, we first list the required regularity conditions on the density functions $f_s(\boldsymbol{y}^{(s)}; \boldsymbol{\Theta})$.

A0 The densities $f_s(\boldsymbol{y}^{(s)}; \boldsymbol{\Theta})$ are distinct for different values of the parameter $\boldsymbol{\Theta}$.

A1 The densities $f_s(\boldsymbol{y}^{(s)}; \boldsymbol{\Theta})$ have common support, which does not depend on $\boldsymbol{\Theta}$.

A2 The parameter space Ω contains an open region ω of which the true parameter value $\boldsymbol{\Theta}_0$ is an interior point.

A3 ω is such that for all s, and almost all $\boldsymbol{y}^{(s)}$ in the support of $\boldsymbol{Y}^{(s)}$, the densities admit all third derivatives

$$\frac{\partial^3 f_s(\boldsymbol{y}^{(s)}; \boldsymbol{\Theta})}{\partial \theta_j \partial \theta_k \partial \theta_\ell}.$$

A4 The first and second logarithmic derivatives of f_s satisfy

$$E_{\boldsymbol{\Theta}} \left(\frac{\partial \ln f_s(\boldsymbol{y}^{(s)}; \boldsymbol{\Theta})}{\partial \theta_k} \right) = 0, \qquad k = 1, \ldots, q,$$

and

$$0 < E_{\boldsymbol{\Theta}} \left(\frac{-\partial^2 \ln f_s(\boldsymbol{y}^{(s)}; \boldsymbol{\Theta})}{\partial \theta_k \partial \theta_\ell} \right) < \infty, \qquad k, \ell = 1, \ldots, q.$$

A5 The matrix J, defined in (6.2), is positive definite.

A6 There exist functions M_{klr} such that

$$\sum_{s \in S} \delta_s E_{\boldsymbol{\Theta}} \left| \frac{\partial^3 \ln f_s(\boldsymbol{y}^{(s)}; \boldsymbol{\Theta})}{\partial \theta_k \partial \theta_\ell \partial \theta_r} \right| < M_{k\ell r}(\boldsymbol{y})$$

for all $\boldsymbol{y}$ in the support of f and for all $\boldsymbol{\theta} \in \omega$ and $m_{k\ell r} = E_{\boldsymbol{\Theta}_0}(M_{k\ell r}(Y)) < \infty$.

Theorem 6.1.1, proven by Arnold and Strauss (1991), guarantees the existence of at least one solution to the pseudo-likelihood equations, which is consistent and asymptotically normal. Without loss of generality, we can assume $\boldsymbol{\Theta}$ is constant. Replacing it by $\boldsymbol{\Theta}_i$, and modeling it as a function of covariates is straightforward.

Theorem 6.1.1 (Consistency and Asymptotic Normality) *Assume that* $(\boldsymbol{Y}_1, \ldots, \boldsymbol{Y}_N)$ *are i.i.d. with common density that depends on* $\boldsymbol{\Theta}_0$*. Then under regularity conditions (A1)–(A6):*

1. *the pseudo-likelihood estimator* $\tilde{\boldsymbol{\Theta}}_N$*, defined as the maximizer of (6.1), converges in probability to* $\boldsymbol{\Theta}_0$*.*

2. $\sqrt{N}(\tilde{\boldsymbol{\Theta}}_N - \boldsymbol{\Theta}_0)$ *converges in distribution to* $N_p(\boldsymbol{0}, J(\boldsymbol{\Theta}_0)^{-1} K(\boldsymbol{\Theta}_0) J(\boldsymbol{\Theta}_0)^{-1})$ *with* $J(\boldsymbol{\Theta})$ *defined by*

$$J_{k\ell}(\boldsymbol{\Theta}) = -\sum_{s \in S} \delta_s E_{\boldsymbol{\Theta}} \left(\frac{\partial^2 \ln f_s(\boldsymbol{y}^{(s)}; \boldsymbol{\Theta})}{\partial \theta_k \partial \theta_\ell} \right) \tag{6.2}$$

and $K(\boldsymbol{\Theta})$ *by*

$$K_{k\ell}(\boldsymbol{\Theta}) = \sum_{s,t \in S} \delta_s \delta_t E_{\boldsymbol{\Theta}} \left(\frac{\partial \ln f_s(\boldsymbol{y}^{(s)}; \boldsymbol{\Theta})}{\partial \theta_k} \frac{\partial \ln f_t(\boldsymbol{y}^{(t)}; \boldsymbol{\Theta})}{\partial \theta_\ell} \right). \tag{6.3}$$

Similar in spirit to generalized estimating equations (Liang and Zeger 1986), the asymptotic normality result provides an easy way to estimate consistently the asymptotic covariance matrix. Indeed, the matrix J is found from evaluating the second derivative of the log PL function at the PL estimate. The expectation in K can be replaced by the cross-products of the observed scores. We will refer to J^{-1} as the model based variance estimator (which should not be used since it overestimates the precision), to K as the empirical correction, and to $J^{-1}KJ^{-1}$ as the empirically corrected variance estimator. In the context of generalized estimating equations, this is also known as the sandwich estimator.

As discussed by Arnold and Strauss (1991), the Cramèr-Rao inequality implies that $J^{-1}KJ^{-1}$ is greater than the inverse of I (the Fisher information matrix for the maximum likelihood case), in the sense that $J^{-1}KJ^{-1} - I^{-1}$ is positive semi-definite. Strict inequality holds if the PL estimator fails to be a function of a minimal sufficient statistic. Therefore, a PL estimator is always less efficient than a ML estimator.

6.1.3 Applied to Exponential Family Models with a Single Clustered Binary Outcome

A convenient pseudo-likelihood function for exponential family models such as (4.16) with a single clustered outcome is found by replacing the joint density $f_Y(y_i; \Theta_i)$ by the product of univariate "full" conditional densities $f(y_{ij}|\{y_{ij'}\}, j' \neq j; \Theta_i)$ for $j = 1, \ldots, L$, obtained by conditioning each observed outcome on all others. This idea can be put into the framework (6.1) by choosing $\delta_{1_L} = L$ and $\delta_{s_j} = -1$ for $j = 1, \ldots, L$ where 1_L is a vector of ones and s_j consists of ones everywhere, except for the jth entry. For all other vectors s, δ_s equals zero. We refer to this particular choice as the *full conditional pseudo-likelihood function*. This pseudo-likelihood has the effect of replacing a joint mass function with a complicated normalizing constant by L univariate functions.

If we can assume that outcomes within a cluster are exchangeable, there are only two types of contributions: (1) the conditional probability of an additional success, given there are $z_i - 1$ successes and $n_i - z_i$ failures (this contribution occurs with multiplicity z_i):

$$p_{is} = \frac{\exp\{\theta_i - \delta_i(n_i - 2z_i + 1)\}}{1 + \exp\{\theta_i - \delta_i(n_i - 2z_i + 1)\}},$$

and (2) the conditional probability of an additional failure, given there are z_i successes and $n_i - z_i - 1$ failures (with multiplicity $n_i - z_i$):

$$p_{if} = \frac{\exp\{-\theta_i + \delta_i(n_i - 2z_i - 1)\}}{1 + \exp\{-\theta_i + \delta_i(n_i - 2z_i - 1)\}}.$$

The log PL contribution for cluster i can then be expressed as $p\ell_i = z_i \ln p_{is} + (n_i - z_i) \ln p_{if}$. The contribution of cluster i to the pseudo-likelihood score vector is of the form

$$\begin{pmatrix} z_i(1 - p_{is}) - (n_i - z_i)(1 - p_{if}) \\ -z_i(n_i - 2z_i + 1)(1 - p_{is}) + (n_i - z_i)(n_i - 2z_i - 1)(1 - p_{if}) \end{pmatrix}.$$

Note that, if $\delta_i \equiv 0$, then $p_{is} \equiv 1 - p_{if}$ and the first component of the score vector is a sum of terms $z_i - n_i p_{is}$, i.e., standard logistic regression follows. In the general case, we have to account for the association, but this non-standard system of equations can be solved using logistic regression software as follows. Represent the contribution for cluster i by two separate records, with repetition counts z_i for the "success case" and $n_i - z_i$ for the "failure case", respectively. All interaction covariates need to be multiplied by $-(n_i - 2z_i + 1)$ in the success case and $-(n_i - 2z_i - 1)$ in the failure case.

6 1.4 Applied to Exponential Family Models With Clustered Multivariate Binary Data

For clustered multivariate binary data, several formulations can be adopted. One convenient PL function is found by replacing the joint density (4.21) by the product of Mn_i univariate conditional densities describing outcome j for the kth individual in a cluster, given all other outcomes in that cluster:

$$PL(1) = \prod_{i=1}^{N} \prod_{j=1}^{M} \prod_{k=1}^{n_i} f(y_{ijk}|y_{ij'k'}, j' \neq j \text{ or } k' \neq k; \Theta_i). \qquad (6.4)$$

This fits into framework (6.1) by choosing $\delta_{1_{Mn_i}} = Mn_i$ and $\delta_{s_{kj}} = -1$ for $k = 1, \ldots, n_i$ and $j = 1, \ldots, M$ where 1_{Mn_i} is a vector of ones and s_{kj} is a $Mn_i \times 1$ vector, obtained by applying the vec operator to an $n_i \times M$ matrix, consisting of ones everywhere, except for entry (k, j), which is 0. If the members of each cluster are assumed to be exchangeable on every outcome separately, there are only $M2^M$ different contributions. Subsequently one can model components of Θ as a function of covariates, and take derivatives of the log PL function with respect to the regression parameters β to derive the score functions.

Equation (6.4) is one convenient definition of the PL function but certainly not the only one. For example, one might want to preserve the multivariate nature of the data on each cluster member by considering the product of n_i conditional densities of the M outcomes for subject k, given the outcomes for the other subjects:

$$PL(2) = \prod_{i=1}^{N} \prod_{k=1}^{n_i} f(y_{ijk}, j = 1, \ldots, M|y_{ijk'}, k \neq k', j = 1, \ldots, M). \qquad (6.5)$$

This satisfies (6.1) by taking $\delta_{1_{Mn_i}} = n_i$ and $\delta_{s_k} = -1$ for $k = 1, \ldots, n_i$. Here, 1_{Mn_i} denotes the Mn_i dimensional vector of ones, while s_k is the $(Mn_i \times 1)$ vector, obtained by applying the vec operator to an $(n_i \times M)$ matrix, consisting of ones everywhere, except for the kth row which consists of zeros.

Computational convenience may be the primary reason for choosing one PL definition over another. Let us discuss the relative merits of definitions (6.4) and (6.5). The former procedure is straightforward and natural when interest is focused on the estimation of main effect parameters. Furthermore, it is slightly easier to evaluate. If, however, interest lies in the estimation of

Table 6.1 *NTP Data. Maximum likelihood estimates (model based standard errors; empirically corrected standard errors) of univariate outcomes.*

Study	Par.	External	Visceral	Skeletal	Collapsed
DEHP	β_0	-2.81 (0.58;0.52)	-2.39 (0.50;0.52)	-2.79 (0.58;0.77)	-2.04 (0.35;0.42)
	β_d	3.07 (0.65;0.62)	2.45 (0.55;0.60)	2.91 (0.63;0.82)	2.98 (0.51;0.66)
	β_a	0.18 (0.04;0.04)	0.18 (0.04;0.04)	0.17 (0.04;0.05)	0.16 (0.03;0.03)
EG	β_0	-3.01 (0.79;1.01)	-5.09 (1.55;1.51)	-0.84 (0.17;0.18)	-0.81 (0.16;0.16)
	β_d	2.25 (0.68;0.85)	3.76 (1.34;1.20)	0.98 (0.20;0.20)	0.97 (0.20;0.20)
	β_a	0.25 (0.05;0.06)	0.23 (0.09;0.09)	0.20 (0.02;0.02)	0.20 (0.02;0.02)
DYME	β_0	-5.78 (1.13;1.23)	-3.32 (0.98;0.89)	-1.62 (0.35;0.48)	-2.90 (0.43;0.51)
	β_d	6.25 (1.25;1.41)	2.88 (0.93;0.83)	2.45 (0.51;0.82)	5.08 (0.74;0.96)
	β_a	0.09 (0.06;0.06)	0.29 (0.05;0.05)	0.25 (0.03;0.03)	0.19 (0.03;0.03)

multivariate associations then approach (6.5) would be more natural. Geys, Molenberghs and Ryan (1999) have shown that both procedures are roughly equally efficient.

While we have now exemplified the definition on pseudo-likelihood functions for conditional models, one can also develop pseudo-likelihood functions for marginal models that satisfy (6.1). These will be considered in Section 6.3.

Further, it should be noted that, in general, it is not guaranteed that a $p\ell$ function corresponds to an existing and uniquely defined probability mass function. However, since PL(1) and PL(2) are derived from (4.21), existence is guaranteed. In addition, both definitions (6.4) and (6.5) satisfy the conditions of the theorem presented in Gelman and Speed (1993), and hence uniqueness is guaranteed as well.

6.1.5 Illustration: NTP Data

To illustrate our findings, we apply the proposed method to three developmental toxicity studies in mice (DEHP, EG, DYME) conducted by the Research Triangle Institute under contract to the National Toxicology Program (NTP). These studies were described in Section 2.1. We will adopt the pseudo-likelihood method both for a univariate and multivariate Molenberghs and Ryan (1999) model.

Single Clustered Outcome

We fitted Model (4.18) to 4 outcomes in each of the 3 datasets: external, visceral, and skeletal malformation, as well as a collapsed outcome, defined to be 1 if any malformation occurred and -1 otherwise. Parameters were estimated

Table 6.2 *NTP Data. Pseudo-likelihood estimates (standard errors) of univariate outcomes.*

Study	Par.	External	Visceral	Skeletal	Collapsed
DEHP	β_0	-2.85 (0.53)	-2.30 (0.50)	-2.41 (0.73)	-1.80 (0.35)
	β_d	3.24 (0.60)	2.55 (0.53)	2.52 (0.81)	2.95 (0.56)
	β_a	0.18 (0.04)	0.20 (0.04)	0.21 (0.05)	0.20 (0.03)
EG	β_0	-2.61 (0.88)	-5.10 (1.55)	-1.18 (0.14)	-1.11 (0.14)
	β_d	2.14 (0.71)	3.79 (1.18)	1.43 (0.19)	1.41 (0.19)
	β_a	0.30 (0.06)	0.23 (0.10)	0.21 (0.01)	0.21 (0.01)
DYME	β_0	-5.04 (0.94)	-3.34 (0.99)	-2.20 (0.27)	-3.08 (0.47)
	β_d	5.52 (1.01)	2.91 (0.91)	3.22 (0.49)	5.20 (0.97)
	β_a	0.13 (0.05)	0.29 (0.06)	0.25 (0.02)	0.19 (0.02)

by both maximum likelihood (Table 6.1) and pseudo-likelihood (Table 6.2). The empirically corrected standard errors are commonly referred to as "robust" standard errors (Liang and Zeger 1986; see also Chapter 5). The fitting procedure has been implemented in GAUSS. The natural parameters were modeled as follows: $\theta_i = \beta_0 + \beta_d d_i$ where d_i is the dose level applied to the ith cluster, and $\delta_i = \beta_a$, i.e., a constant association model.

An attractive feature of the proposed approach is that the parameters can also be obtained using standard and readily available software, such as the SAS procedures LOGISTIC or GENMOD. As an illustration, the parameters for the external outcome in the DEHP study were also determined with the LOGISTIC procedure. An implementation and selected output is presented in Figures 6.1 and 6.2. Each cluster is represented by a two-line record. The first line corresponds with the "success" case so that the variable ASSOC represents $-(n_i - 2z_i + 1)$; the second line corresponds with the "failure" case so that ASSOC represents $-(n_i - 2z_i - 1)$.

While the estimates are identical to those obtained in Table 6.2, the standard errors are incorrect since they are based on the assumption of independence. To obtain a correct estimate of the variability, a short macro could be written.

The methods can be compared based on the parameter estimates, their standard errors (model based likelihood, empirically corrected likelihood, and pseudo-likelihood), or a combination of both (e.g., the Z statistic, defined as the ratio of estimate and standard error). Obviously, the development of methods to assess the fit of the proposed methods is necessary. However, classical tools cannot be used within the pseudo-likelihood framework without modification. Of course, one can always assess the fit by fitting an extended model and testing whether the additional parameters are significant. The ex-

```
data pseudo;
input success failure dose assoc;
total=success+failure;
cards;
      0.0000      0.0000      0.0000    -10.0000
      0.0000      9.0000      0.0000     -8.0000
      0.0000      0.0000      0.1667    -11.0000
      0.0000     10.0000      0.1667     -9.0000
      1.0000      0.0000      0.3333     -6.0000
      0.0000      6.0000      0.3333     -4.0000


      . . .


      2.0000      0.0000      0.6667    -10.0000
      0.0000     11.0000      0.6667     -8.0000
;
run;


proc logistic data=pseudo;
model success/total = dose assoc;
run;
```

Figure 6.1 *DEHP Study. Implementation using the SAS procedure PROC LOGIS-TIC.*

Analysis of Maximum Likelihood Estimates

Variable	DF	Parameter Estimate	Standard Error	Wald Chi-Square	Pr > Chi-Square	Standardized Estimate
INTERCPT	1	-2.8520	0.5621	25.7456	0.0001	.
DOSE	1	3.2369	0.6501	24.7921	0.0001	0.474261
ASSOC	1	0.1833	0.0429	18.2737	0.0001	0.393847

Figure 6.2 *DEHP Study. Selected output of the SAS procedure PROC LOGISTIC.*

tension of flexible tools such as likelihood ratio and score tests to the PL framework has been proposed by Geys, Molenberghs and Ryan (1999) and will be described in Chapter 7.

Maximum likelihood and pseudo-likelihood dose parameter estimates agree fairly closely, except for the EG outcomes skeletal and collapsed. No method

Table 6.3 *NTP Data. Pseudo-likelihood estimates (standard errors) for trivariate outcomes (different main dose effects).*

Par.	DEHP	EG	DYME
β_{01}	-2.13 (0.64)	-1.64 (1.04)	-5.67 (1.16)
β_{02}	-2.38 (0.63)	-5.04 (1.75)	-2.34 (1.26)
β_{03}	-2.76 (0.72)	-0.39 (0.51)	-2.97 (0.90)
δ_1	0.14 (0.07)	0.18 (0.13)	0.15 (0.04)
δ_2	0.18 (0.04)	0.12 (0.17)	0.30 (0.06)
δ_3	0.29 (0.06)	0.20 (0.01)	0.25 (0.02)
ω_{12}	0.06 (0.25)	-0.05 (0.57)	-0.45 (0.20)
ω_{13}	0.60 (0.20)	0.11 (0.31)	0.25 (0.31)
ω_{23}	0.36 (0.29)	0.86 (0.34)	0.35 (0.31)
γ_{12}	0.11 (0.06)	0.14 (0.13)	0.07 (0.04)
γ_{13}	-0.06 (0.05)	0.08 (0.04)	-0.11 (0.05)
γ_{23}	-0.14 (0.06)	-0.09 (0.04)	0.01 (0.05)
β_{d1}	2.70 (0.66)	1.12 (0.86)	6.48 (1.26)
β_{d2}	2.63 (0.66)	3.63 (1.04)	1.66 (1.36)
β_{d3}	2.70 (0.76)	1.42 (0.19)	4.29 (0.99)

systematically leads to larger parameter estimates (each one yields the largest value in about half of the cases).

Rather than comparing estimated standard errors directly, one could also consider the derived Z statistics (not shown) and their associated significance levels. Pairwise comparisons of the test statistics reveal again that no procedure systematically yields larger values. Indeed, in all three comparisons, the magnitude of one statistic is larger than the other in approximately 50% of the cases.

These results are promising because a loss of efficiency of pseudo-likelihood versus maximum likelihood could be anticipated. However, even though in Section 6.2 it will be shown that the asymptotic relative efficiency (ARE) is in general strictly less than 1 (except for saturated models), the data analysis suggests that the efficiency loss is moderate.

Clustered Multivariate Outcomes

When considering all three outcomes (external, visceral, and skeletal, respectively indexed by 1, 2, and 3) jointly, ML becomes prohibitively difficult to fit. Some analyses are very sensitive to initial values and take more than 10 hours to converge. Therefore, we abandoned ML and concentrated solely on the PL method, which took less than 3 minutes to converge.

Table 6.4 *NTP Data. Pseudo-likelihood estimates (standard errors) for trivariate outcomes (common main dose effects).*

Par.	DEHP	EG	DYME
β_{01}	-2.10 (0.51)	-1.97 (0.56)	-3.89 (0.83)
β_{02}	-2.42 (0.50)	-2.96 (0.87)	-4.77 (0.87)
β_{03}	-2.74 (0.49)	-0.27 (0.55)	-3.21 (0.81)
δ_1	0.14 (0.07)	0.18 (0.13)	0.22 (0.03)
δ_2	0.18 (0.04)	0.17 (0.17)	0.25 (0.06)
δ_3	0.29 (0.05)	0.20 (0.01)	0.25 (0.02)
ω_{12}	0.06 (0.24)	-0.05 (0.57)	-0.46 (0.19)
ω_{13}	0.60 (0.20)	0.11 (0.30)	0.29 (0.30)
ω_{23}	0.36 (0.28)	0.97 (0.37)	0.28 (0.31)
γ_{12}	0.11 (0.06)	0.13 (0.13)	0.05 (0.04)
γ_{13}	-0.06 (0.05)	0.06 (0.04)	-0.09 (0.04)
γ_{23}	-0.14 (0.06)	-0.07 (0.03)	-0.03 (0.05)
β_d	2.67 (0.48)	1.50 (0.20)	4.31 (0.85)

For all three NTP studies, we considered (1) a model with a different dose effect per outcome and (2) a common dose effect model, both of which are tested for the null hypothesis of no dose effect. In both cases all association parameters are held constant. Results of these analyses are tabulated in Tables 6.3 and 6.4 and indicate, based on Wald tests, that all dose effect parameters are significant (except for External outcomes in EG and for Visceral malformations in DYME). In addition, Tables 6.3 and 6.4 show that by fitting a relatively simple model with different dose effects for each outcome and constant association parameters, the three different main dose effect parameters in the DEHP study all seem to be relevant and of similar magnitude. This suggests that the use of a common main dose parameter is desirable, hereby increasing the efficiency (Lefkopoulou and Ryan 1993). The estimated clustering parameters δ_j ($j = 1, 2, 3$) are all significant, except for External and Visceral malformation outcomes in the EG study. In contrast, the other association parameters often do not reach the 5% significance level.

6.2 Relative Efficiency of Pseudo-likelihood versus Maximum Likelihood

6.2.1 Asymptotic Relative Efficiency for the Saturated Model

The price for computational ease usually consists of some efficiency loss. In this section we will however show that the ARE equals one for all saturated models, i.e., models of the form (4.15) without covariates and where all subvectors of W are included. The ARE for non-saturated models will be discussed in Section 6.2.2.

Consider the PL contribution for a single cluster, consisting of the product of all univariate conditional densities. Like in Section 4.2.1 the cluster index i is kept fixed and dropped from notation:

$$PL = \prod_{j=1}^{M} f_j(y_j | \boldsymbol{y}_{(j)}),$$

where $\boldsymbol{y}_{(j)}$ indicates omission of the jth component. Extending the notational conventions, the logit of the conditional probability that y_j equals 1 given all others can be written as:

$$\text{logit}\,(\mu_j(y_1, \ldots, y_{j-1}, y_{j+1}, \ldots, y_M)) = \theta_j + \sum_{k \neq j} \omega_{jk} y_k$$
$$+ \sum_{k < k'; k, k' \neq j} \omega_{jkk'} y_k y_{k'} + \cdots + \omega_{12\ldots M} y_1 \cdots y_{j-1} y_{j+1} \cdots y_M. \quad (6.6)$$

In short, we denote the logit in (6.6) by logit μ_j. In general, the pseudo-likelihood score contributions of the rth $(r = 1, \ldots, M)$ association parameter for a single subject can then be derived as:

$$\sum_{\ell=1}^{r} (y_{k_\ell} - \mu_{k_\ell}) y_{k_1} \cdots y_{k_{\ell-1}} y_{k_{\ell+1}} \cdots y_{k_r}, \quad (6.7)$$

$(1 \leq k_1 < k_2 < \cdots < k_r \leq M)$. For the main effect and the pairwise interactions, these contributions reduce to

$$y_j - \mu_j, \qquad\qquad 1 \leq j \leq M,$$
$$(y_j - \mu_j) y_k + (y_k - \mu_k) y_j, \qquad 1 \leq j < k \leq M.$$

We will now show that the maximum likelihood estimator satisfies (6.7) in the sense that it solves for this equation summed over all subjects.

Organize the data into an M dimensional contingency table with cell counts $z_{j_1 \ldots j_M} (j_p = 0, 1; p = 1, \ldots, M)$. Obviously, it may be more convenient to introduce an alternative notation for these cell counts. Rather than giving a sequence of M zeros and ones, we can present the subscripts for which $j_p = 1$. Thus, $z.$ is the number of individuals with failures on all variables, z_j refers to those having a success on outcome j and a failure on all others, $z_{j_1 j_2}$ refers to those having successes on both outcomes j_1 and j_2 and a failure on all others, etc. With straightforward notation, the maximum likelihood estimates for the corresponding cell probabilities are given by:

$$\hat{\pi}_{j_1 \ldots j_p} = \frac{z_{j_1 \ldots j_p}}{N}.$$

Now, simple relations exist between these cell probabilities and the natural parameters:

$$\hat{\pi}_j = e^{\hat{\theta}_j} / A(\widehat{\boldsymbol{\Theta}})$$

and hence

$$e^{\hat{\theta}_j} = z_j / z.,$$

which is the classical relationship between the main effect parameters and the conditional odds associated with outcome j, given failures on all others. Similarly,

$$\hat{\pi}_{j_1 j_2} = e^{\hat{\theta}_{j_1} + \hat{\theta}_{j_2} + \hat{\omega}_{j_1 j_2}} / A(\hat{\Theta})$$

and thus

$$e^{\hat{\omega}_{j_1 j_2}} = (z_{j_1 j_2} z.)/(z_{j_1} z_{j_2}).$$

Using the notation introduced above, the PL score contribution for the main effect θ_j, combined over all subjects, can be written as:

$$\sum_{(t_1, \ldots, t_{j-1}, t_{j+1}, \ldots, t_M)} z_{t_1 \ldots t_{j-1} 1 t_{j+1} \ldots t_M} \{1 - \mu_j(t_1, \ldots, t_{j-1}, t_{j+1}, \ldots, t_M)\}$$

$$+ \sum_{(t_1, \ldots, t_{j-1}, t_{j+1}, \ldots, t_M)} z_{t_1 \ldots t_{j-1} 0 t_{j+1} \ldots t_M} \{-\mu_j(t_1, \ldots, t_{j-1}, t_{j+1}, \ldots, t_M)\} = 0,$$

where the summation is over all $M - 1$ vectors (no jth component) of zeros and ones. Rewriting this equation as

$$\sum_{(t_1, \ldots, t_{j-1}, t_{j+1}, \ldots, t_M)} z_{t_1 \ldots t_{j-1} 1 t_{j+1} \ldots t_M} - \sum_{(t_1, \ldots, t_{j-1}, t_{j+1}, \ldots, t_M)} z_{t_1 \ldots t_{j-1} + t_{j+1} \ldots t_M} \mu_j(t_1, \ldots, t_{j-1}, t_{j+1}, \ldots, t_M) = 0,$$

it is easily seen that the MLE satisfies this equation, since on the one hand

$$\mu_j(t_1, \ldots, t_{j-1}, t_{j+1}, \ldots, t_M)$$

is the probability of observing a success on outcome j, given the value of the other outcomes, and on the other hand its MLE is given by

$$\hat{\mu}_j(t_1, \ldots, t_{j-1}, t_{j+1}, \ldots, t_M) = \frac{z_{t_1 \ldots t_{j-1} 1 t_{j+1} \ldots t_M}}{z_{t_1 \ldots t_{j-1} + t_{j+1} \ldots t_M}}.$$

Similar calculations can be carried out for the equations pertaining to the association parameters. This shows that the maximum likelihood estimator and the pseudo-likelihood estimator coincide in this case. A trivial consequence of this result is that ARE$\equiv$1.

6.2.2 Asymptotic Relative Efficiency for Clustered Outcomes

Although explicit formulae for the ARE were derived for unclustered outcomes in previous sections, similar expressions in the clustered case are difficult to obtain. We will focus on a single clustered binary outcome. Results for clustered multivariate binary data are similar (Geys, Molenberghs and Ryan 1999). To study the ARE, we will follow the recommendations of Rotnitzky and Wypij (1994). In order to compute asymptotic bias or efficiency, an artificial sample can be constructed, where each possible realization is weighted according to its true probability. In our case, we need to consider all realizations of the form (n_i, z_i, d_i), and hence have to specify: (1) $f(d_i)$, the relative frequencies of the dose groups, as prescribed by the design; (2) $f(n_i|d_i)$, the probability

Table 6.5 *Local linear smoothed cluster frequencies.*

n_i	$f(n_i)$	n_i	$f(n_i)$
1	0.0046	11	0.1179
2	0.0057	12	0.1529
3	0.0099	13	0.1605
4	0.0139	14	0.1424
5	0.0147	15	0.0975
6	0.0148	16	0.0542
7	0.0225	17	0.0207
8	0.0321	18	0.0086
9	0.0475	19	0.0030
10	0.0766		

with which each cluster size can occur, possibly depending on the dose level; and (3) $f(z_i|n_i, d_i)$, the actual model probabilities.

Throughout, we assume that there are 4 dose groups, with one control ($d_i = 0$) and three exposed groups ($d_i = 0.25, 0.5, 1.0$). The number n_i of viable foetuses per cluster is chosen at random, using a local linear smoothed version of the relative frequency distribution given in Table 1 of Kupper *et al.* (1986) (which is considered representative of that encountered in actual experimental situations). Least squares cross-validation has been used to choose the bandwidth. The smoothed frequencies are presented in Table 6.5. Guided by the analysis of the examples, we identified three values for each of the three parameters: $\beta_0 = -5, -3, 0$, $\beta_d = 0, 3, 5$, and $\beta_a = 0, 0.15, 0.30$, with notation as defined in Section 6.2. The full grid of 27 parameter combinations has been explored. Results are displayed in Table 6.6.

No AREs are exactly equal to one, although some appear to be due to rounding. The AREs are very high for the lowest background rate ($\beta_0 = -5$) and they are almost all above 90% for the medium background rate ($\beta_0 = -3$). We can notice the non-monotone relationship of the ARE with β_d and β_a. While still high in some areas of the (β_d, β_a) space for $\beta_0 = 0$, a dramatic decrease is observed when β_a increases and/or β_d decreases. PL performs very poorly when there is no dose effect together with a reasonably high association. Unless background malformation probabilities or dose effects are extreme, large associations diminish the contribution to the information of a full conditional. As a limiting case it can even be reduced to zero when the association parameter approaches infinity. This phenomenon is further illustrated in Figure 6.3.

The parameter estimates found from the data analysis are all in regions of the parameter space with a high ARE.

Table 6.6 *Simulation Results. Asymptotic relative efficiencies of pseudo-likelihood versus maximum likelihood.*

β_0	β_d	β_a		
		0.00	0.15	0.30
-5	0	1.000	1.000	1.000
	3	0.982	0.999	1.000
	5	0.940	0.978	0.966
-3	0	1.000	1.000	1.000
	3	0.938	0.938	0.897
	5	0.921	0.959	0.907
0	0	1.000	0.725	0.055
	3	0.958	0.895	0.792
	5	0.943	0.928	0.890

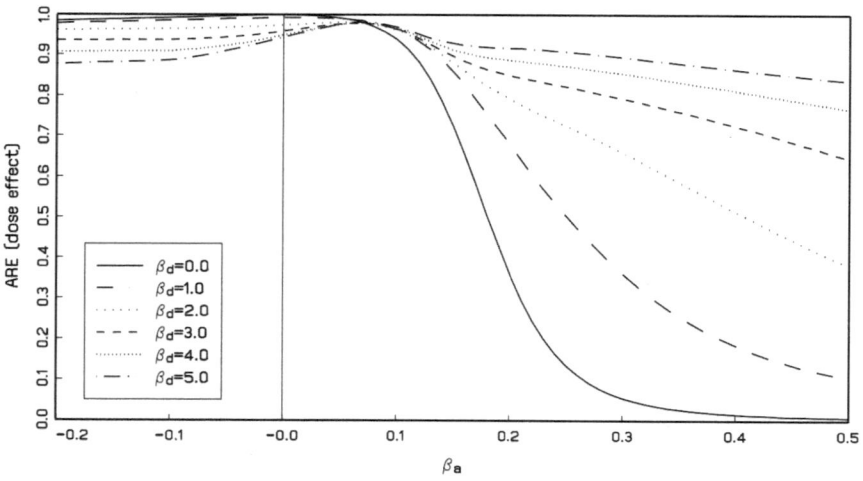

Figure 6.3 *Simulation Results. Asymptotic relative efficiency of pseudo-likelihood versus maximum likelihood for the dose effect parameter in the clustered data model.*

In order to investigate whether these conclusions also hold for random samples, a small simulation study has been performed, to be described next.

6.2.3 Small Sample Relative Efficiency of Pseudo-likelihood versus Maximum Likelihood

The same 27 parameter combinations of the previous sections are investigated, for samples of size 30. For each setting, 500 simulations were run. The estimated covariance matrices were kept and averaged at the end of the run. The relative efficiencies for the dose-effect parameters are displayed in Table 6.7. For the maximum likelihood procedure, both the purely model based as well as the empirically corrected version are considered.

The results of the asymptotic study and the small sample study are remarkably well in agreement, except for the small sample relative efficiency (SSRE), which tends to be slightly higher in certain regions of the grid, such as $\beta_0 = -3$ or 0 and $\beta_a = 0.3$. Also, the SSRE is larger for the model based than for the empirically corrected likelihood version, which is in line with knowledge about the sandwich estimator. The only major discrepancies, deserving further explanation, are seen for $\beta_0 = -5$, no dose effect ($\beta_d = 0$), and $\beta_a \neq 0$. First, observe that these parameter settings correspond to a very low background rate (the background probability of observing no malformation in a single foetus being 0.9933). It can be calculated that the marginal probability of sampling a cluster without malformations is 0.9229, 0.9851, and 0.9966 for the corresponding association parameters 0.0, 0.15, and 0.30. Correspondingly, the number of datasets without malformations (and thus with parameters at infinity) in a batch of 500 runs is on average 0.03, 83, and 332, respectively. In our simulation study, we actually encountered 0, 83, 331 of such datasets. All 83 (331) of these datasets were ignored, along with 98 (76) other problematic sets of data, mainly because the latter contain merely a single malformation, which renders the association parameter inestimable. Still, the remaining 319 and 93 datasets are not free of difficulties. Let us consider variances and relative efficiencies for the dose effect in the 0.30 association case. The asymptotic variances are all about 17.8, while the simulation result for the small sample variances is smaller (8.84 for model based likelihood, 0.94 for empirically corrected likelihood, and 0.99 for PL). This is most likely due to the fact that omitting the problematic datasets truncates the sampling space and effectively reduces the variability. In particular, these problematic datasets contain no events, yielding an estimate for the intercept of $-\infty$, the dose effect being inestimable. Typically, samples with extreme parameter values are excluded, leading to still smaller sample variances. This effect is more pronounced in the empirically corrected estimators than in the purely model based one.

For the other, often more realistic parameter settings, the asymptotic and small sample variances are in fairly good agreement. This leads to SSREs and AREs that are very close. Further, the observed variances in these settings, whether asymptotic or small sample, are much smaller than in the problematic settings described earlier, e.g., when $\beta_0 = -3.0$, $\beta_d = 3.0$, and $\beta_a = 0.15$, the asymptotic variances of the dose effect are all close to 0.13, while the small sample versions are about 0.14.

Table 6.7 *Simulation Results. Small sample relative efficiencies (500 replications) of pseudo-likelihood versus maximum likelihood.*

		β_a								
		0.00			0.15			0.30		
β_0	β_d	model	emp.	runs	model	emp.	runs	model	emp.	runs
-5	0	1.073	0.973	500	2.592	0.990	319	8.919	0.945	93
	3	0.976	0.955	500	1.086	0.995	485	1.613	1.017	254
	5	0.959	0.918	500	1.003	0.982	499	1.123	0.986	411
-3	0	1.027	0.994	500	1.116	1.018	500	1.966	1.009	425
	3	0.929	0.915	500	0.970	0.938	500	1.005	0.921	500
	5	0.931	0.905	500	0.979	0.938	500	1.058	0.951	498
0	0	1.000	0.995	500	0.746	0.732	500	0.055	0.055	500
	3	0.948	0.942	500	0.925	0.903	500	0.912	0.831	500
	5	0.934	0.909	500	0.951	0.932	500	1.064	0.950	500

6.3 Pseudo-likelihood and Generalized Estimating Equations

In the framework of a marginally specified odds ratio model (Lipsitz, Laird and Harrington 1991, Dale 1986, Molenberghs and Lesaffre 1994, Glonek and Mc-Cullagh 1995, Lang and Agresti 1994) for multivariate, clustered binary data, full maximum likelihood estimation can also become prohibitive, especially with large within-unit replication. In this section, we compare generalized estimating equations with pseudo-likelihood, to gain insight in both.

Note that, while GEE is typically aimed at marginal models, PL can be used for both marginal (le Cessie and van Houwelingen 1994, Geys, Molenberghs and Lipsitz 1998) and conditional models. Here, we discuss the relative merits of PL and GEE, which will be illustrated using data from NTP studies. As before, we will only pay attention to exchangeable association structures and cluster-level covariates, since this simplifies comparison and covers the setting encountered in the data. While our findings can be applied to some longitudinal settings, the assumption of exchangeability is frequently not tenable, so that more complex association structures are needed. An extension of these results to longitudinal data could be of interest, but would need further investigation.

6.3.1 Pseudo-likelihood Estimating Equations

In this section we first present a general PL form, accommodating clustered responses. Next, we concentrate on the special case of exchangeability leading to an elegant formulation of the PL. Again, we assume there are N clusters with $k = 1, \ldots, n_i$ indexing the individuals in the ith cluster. If we denote the binary outcome for subject k in cluster i by Y_{ik} then the exchangeability assumption allows us to introduce the summary statistic $Z_i = \sum_{k=1}^{n_i} Y_{ik}$: the total number of successes within the ith cluster.

Classical Representation

First Definition   le Cessie and van Houwelingen (1994) replace the true contribution of a vector of correlated binary data to the full likelihood, written as $f(y_{i1}, \ldots, y_{in_i})$, by the product of all pairwise contributions $f(y_{ij}, y_{ik})$ $(1 \leq j < k \leq n_i)$, to obtain a *pseudo-likelihood function*. Grouping the outcomes for subject i into a vector $\boldsymbol{Y}_i$, the contribution of the ith cluster to the log pseudo-likelihood is

$$p\ell_i = \sum_{j<k} \ln f(y_{ij}, y_{ik}), \tag{6.8}$$

if it contains more than one observation. Otherwise $p\ell_i = f(y_{i1})$. In the sequel we restrict our attention to clusters of size larger than 1. Clusters of size 1 contribute to the marginal parameters only.

Using a bivariate Plackett distribution (Plackett 1965) the joint probabilities $f(y_{ij}, y_{ik})$, denoted by π_{ijk}, can be specified in terms of marginal probabilities and pairwise odds ratios. For individuals j and k (or for measurement occasions j and k in a longitudinal study), the pairwise odds ratio ψ_{ijk} is defined as (Fitzmaurice, Molenberghs and Lipsitz 1995):

$$\psi_{ijk} = \frac{P(Y_{ij} = 1, Y_{ik} = 1)P(Y_{ij} = 0, Y_{ik} = 0)}{P(Y_{ij} = 1, Y_{ik} = 0)P(Y_{ij} = 0, Y_{ik} = 1)}.$$

Dale (1986) refers to this quantity as the *global cross ratio*.

The univariate marginal means π_{ij}, as well as the pairwise odds ratios ψ_{ijk}, can be modeled in terms of regression parameters, using (for example) logit and log links, respectively, whence the bivariate marginal means π_{ijk} satisfy:

$$\pi_{ijk} = \begin{cases} \frac{1+(\pi_{ij}+\pi_{ik})(\psi_{ijk}-1)-S(\pi_{ik},\pi_{ij},\psi_{ijk})}{2(\psi_{ijk}-1)} & \text{if } \psi_{ijk} \neq 1, \\ \pi_{ij}\pi_{ik} & \text{if } \psi_{ijk} = 1, \end{cases}$$

with

$$S(\pi_{ij}, \pi_{ik}, \psi_{ijk}) = \sqrt{[1 + (\pi_{ij} + \pi_{ik})(\psi_{ijk} - 1)]^2 + 4\psi_{ijk}(1 - \psi_{ijk})\pi_{ij}\pi_{ik}}.$$

Under Exchangeability   For binary data and taking the exchangeability assumption into account, the log pseudo-likelihood contribution $p\ell_i$ can be formulated as:

$$p\ell_i = \binom{z_i}{2} \ln \pi_{i11}^* + \binom{n_i - z_i}{2} \ln \pi_{i00}^* + z_i(n_i - z_i) \ln \pi_{i10}^*. \tag{6.9}$$

In this formulation, π_{i11}^* and π_{i00}^* denote the bivariate probabilities of observing two *successes* or two *failures*, respectively, while π_{i10}^* is the probability for the first component being 1 and the second being 0. Under exchangeability, this is identical to the probability π_{i01}^* for the first being 0 and the second being 1. If we consider the following reparameterization:

$$\begin{aligned}
\pi_{i11} &= \pi_{i11}^*, \\
\pi_{i10} &= \pi_{i11}^* + \pi_{i10}^* = \pi_{01}, \\
\pi_{i00} &= \pi_{i11}^* + \pi_{i10}^* + \pi_{i01}^* + \pi_{i00}^* = 1,
\end{aligned}$$

then this one-to-one reparameterization maps the three, common within-cluster, two-way marginal probabilities $(\pi_{i11}^*, \pi_{i10}^*, \pi_{i00}^*)$ to two one-way marginal probabilities (which under exchangeability are both equal to π_{i10}) and one two-way probability $\pi_{i11} = \pi_{i11}^*$. Hence, equation (6.9) can be reformulated as:

$$\begin{aligned}
p\ell_i &= \binom{z_i}{2} \ln \pi_{i11} + \binom{n_i - z_i}{2} \ln(1 - 2\pi_{i10} + \pi_{i11}) \\
&\quad + z_i(n_i - z_i) \ln(\pi_{i10} - \pi_{i11}),
\end{aligned} \tag{6.10}$$

and the pairwise odds ratio ψ_{ijk} reduces to:

$$\psi_i = \frac{\pi_{i11}(1 - 2\pi_{i10} + \pi_{i11})}{(\pi_{i10} - \pi_{i11})^2}.$$

To enable model specification, we assume a composite link function $\boldsymbol{\eta}_i = (\eta_{i1}, \eta_{i2})^T$ with a mean and an association component:

$$\begin{aligned}
\eta_{i1} &= \ln(\pi_{i10}) - \ln(1 - \pi_{i10}), \\
\eta_{i2} &= \ln(\psi_i) = \ln(\pi_{i11}) + \ln(1 - 2\pi_{i10} + \pi_{i11}) - 2\ln(\pi_{i10} - \pi_{i11}).
\end{aligned}$$

From these links, the univariate and pairwise probabilities are easily derived (Plackett 1965):

$$\pi_{i10} = \frac{\exp(\eta_{i1})}{1 + \exp(\eta_{i1})}$$

and

$$\pi_{i11} = \begin{cases} \frac{1 + 2\pi_{i10}(\psi_i - 1) - S_i}{2(\psi_i - 1)}, & \text{if } \psi_i \neq 1 \\ \pi_{i10}^2 & \text{if } \psi_i = 1, \end{cases}$$

with

$$S_i = \sqrt{[1 + 2\pi_{i10}(\psi_i - 1)]^2 + 4\psi_i(1 - \psi_i)\pi_{i10}^2}.$$

Next, we can assume a linear model $\boldsymbol{\eta}_i = X_i\boldsymbol{\beta}$, with X_i a known design matrix and $\boldsymbol{\beta}$ a vector of unknown regression parameters. The maximum pseudo-likelihood estimator $\hat{\boldsymbol{\beta}}$ of $\boldsymbol{\beta}$ is then defined as the solution to the pseudo-score equations $\boldsymbol{U}(\boldsymbol{\beta}) = \mathbf{0}$. Using the chain rule, $\boldsymbol{U}(\boldsymbol{\beta})$ can be written as:

$$\boldsymbol{U}(\boldsymbol{\beta}) = \sum_{i=1}^{N} X_i^T (T_i^{-1})^T \frac{\partial p\ell_i}{\partial \boldsymbol{\pi}_i} \tag{6.11}$$

with $\boldsymbol{\pi}_i = (\pi_{i10}, \pi_{i11})^T$ and $T_i = \partial\boldsymbol{\eta}_i/\partial\boldsymbol{\pi}_i$. Two frequently used fitting algorithms are the Newton-Raphson and the Fisher scoring algorithms. Newton-Raphson starts with a vector of initial estimates $\boldsymbol{\beta}^{(0)}$ and updates the current value of the parameter vector $\boldsymbol{\beta}^{(s)}$ by

$$\boldsymbol{\beta}^{(s+1)} = \boldsymbol{\beta}^{(s)} + W(\boldsymbol{\beta}^{(s)})^{-1}\boldsymbol{U}(\boldsymbol{\beta}^{(s)}).$$

Here, $W(\boldsymbol{\beta})$ is the matrix of the second derivatives of the log pseudo-likelihood with respect to the regression parameters $\boldsymbol{\beta}$:

$$W(\boldsymbol{\beta}) = \sum_{i=1}^{N} X_i^T \left(\boldsymbol{F}_i + (T_i^{-1})^T \frac{\partial^2 p\ell_i}{\partial \boldsymbol{\pi}_i \partial \boldsymbol{\pi}_i^T}(T_i^{-1}) \right) X_i,$$

and $\boldsymbol{F}_i$ is defined by (McCullagh 1987, p. 5; Molenberghs and Lesaffre 1999):

$$(F_i)_{jk} = \sum_s \sum_{r,t,u} \frac{\partial^2 \eta_{ir}}{\partial \pi_{it}\partial \pi_{iu}} \frac{\partial \pi_{is}}{\partial \eta_{ir}} \frac{\partial \pi_{it}}{\partial \eta_{ij}} \frac{\partial \pi_{iu}}{\partial \eta_{ik}} \frac{\partial p\ell_i}{\partial \pi_{is}}.$$

The Fisher scoring algorithm is obtained by replacing the matrix $W(\boldsymbol{\beta})$ by its expected value:

$$W(\boldsymbol{\beta}) = \sum_{i=1}^{N} X_i^T (T_i^{-1})^T A_i (T_i^{-1}) X_i,$$

with A_i the expected value of the matrix of second derivatives of the log pseudo-likelihood $p\ell_i$ with respect to $\boldsymbol{\pi}_i$.

Similar in spirit to generalized estimating equations, the asymptotic covariance matrix of the regression parameters $\hat{\boldsymbol{\beta}}$ is consistently estimated by (Arnold and Strauss 1991, Geys, Molenberghs and Ryan 1997):

$$W(\hat{\boldsymbol{\beta}})^{-1} \left(\sum_{i=1}^{N} \boldsymbol{U}_i(\hat{\boldsymbol{\beta}})\boldsymbol{U}_i(\hat{\boldsymbol{\beta}})^T \right) W(\hat{\boldsymbol{\beta}})^{-1}.$$

In the context of generalized estimating equations, this estimator is also known as the empirically corrected or sandwich estimator.

Second Definition   A non-equivalent specification of the pseudo-likelihood contribution (6.8) is:

$$p\ell_i^* = p\ell_i/(n_i - 1).$$

The factor $1/(n_i - 1)$ corrects for the effect that each response Y_{ij} occurs $n_i - 1$ times in the ith contribution to the PL and it ensures that the PL reduces to full likelihood under independence. Indeed, under independence, (5.10) simplifies to:

$$p\ell_i = (n_i - 1)\left[z_i \ln(\pi_{i10}) + (n_i - z_i)\ln(1 - \pi_{i10})\right].$$

We can replace $p\ell_i$ by $p\ell_i^*$ everywhere in this discussion. However, if $(n_i - 1)$ is considered random it is not obvious that the expected value of $U_i(\boldsymbol{\beta})/(n_i - 1)$ equals zero. To ensure that the solution to the new pseudo-score equation is consistent, we have to assume that n_i is independent of z_i given the dose level d_i for the ith cluster. When all clusters are equal in size, the PL estimator $\boldsymbol{\beta}$ and its variance-covariance matrix remain the same, no matter whether we use $p\ell_i$ or $p\ell_i^*$ in the definition of the log pseudo-likelihood.

Generalized Linear Model Representation

To obtain the pseudo-likelihood function described in Section 6.3.1 we replaced the true contribution $f(y_{i1}, \ldots, y_{in_i})$ of the ith cluster to the full likelihood, by the product of all pairwise contributions $f(y_{ij}, y_{ik})$ with $1 \leq j < k \leq n_i$. This implies that a particular response y_{ij} occurs $n_i - 1$ times in $p\ell_i$. Therefore, it is useful to construct for each response y_{ij}, $n_i - 1$ replicated $y_{ij}^{(k)}$ with $k \neq j$. The dummy response $y_{ij}^{(k)}$ is to be interpreted as the particular replicate of y_{ij} that is paired with the replicate $y_{ik}^{(j)}$ of y_{ik} in the pseudo-likelihood function. Using this specific device we are able to rewrite the gradient of the log pseudo-likelihood $p\ell$ in an appealing generalized linear model type representation. With notation introduced in the previous section the gradient can now be written as

$$U(\boldsymbol{\beta}) = \sum_{i=1}^{N} X_i^T (T_i^{-1})^T V_i^{-1} (\boldsymbol{Z}_i - \boldsymbol{\pi}_i),$$

or, using the second representation $p\ell_i^*$, as

$$U(\boldsymbol{\beta}) = \sum_{i=1}^{N} X_i^T (T_i^{-1})^T V_i^{-1} (\boldsymbol{Z}_i - \boldsymbol{\pi}_i)/(n_i - 1),$$

where

$$\boldsymbol{Z}_i = \begin{pmatrix} \sum_{j=1}^{n_i} \sum_{k \neq j} Y_{ij}^{(k)} \\ \frac{1}{2} \sum_{j=1}^{n_i} \sum_{k \neq j} Y_{ij}^{(k)} Y_{ik}^{(j)} \end{pmatrix}, \quad \pi_i = \begin{pmatrix} n_i(n_i - 1)\pi_{i10} \\ \binom{n_i}{2}\pi_{i11} \end{pmatrix}$$

and V_i is the covariance matrix of $\boldsymbol{Z}_i$. Geys, Molenberghs and Lipsitz (1998) have shown that the elements of V_i take appealing expressions and are easy to implement. One only needs to evaluate first and second order probabilities. Under independence, the variances reduce to well-known quantities. To obtain a suitable PL estimator for $\boldsymbol{\beta}$ we can use the Fisher-scoring algorithm where the matrix A_i in the previous section is now replaced by the inverse of V_i. The asymptotic covariance matrix of $\hat{\boldsymbol{\beta}}$ is estimated in a similar fashion as before.

6.3.2 Comparison with Generalized Estimating Equations

In the previous sections we described one alternative estimating procedure for full maximum likelihood estimation in the framework of a marginally specified odds ratio model, which is easier and much less time consuming. Another popular alternative approach is generalized estimating equations, described in detail in Chapter 5. Several questions arise such as to how the different methods compare in terms of efficiency and in terms of computing time and what the mathematical differences and similarities are. At first glance, there is a fundamental difference. A pseudo-likelihood function is constructed by modifying a joint density. Parameters are estimated by setting the first derivatives of this function equal to zero. On the contrary, generalized estimating equations follow from specification of the first few moments and by adopting assumptions about the higher order moments. One could also consider them as resulting from modifying the score equations from the likelihood function. In that respect, McCullagh and Nelder (1989) note that these estimating equations need not necessarily integrate to a so-called *quasi-likelihood*.

The close connection of PL to likelihood is an attractive feature. Indeed, it enabled Geys, Molenberghs and Ryan (1999) to construct pseudo-likelihood ratio test statistics that have easy-to-compute expressions and intuitively appealing limiting distributions. In contrast, likelihood ratio test statistics for GEE (Rotnitzky and Jewell 1990) are slightly more complicated.

In Section 6.3.1 we have rewritten the PL score equations as contrasts of observed and fitted frequencies, establishing some agreement between PL and GEE2. Both procedures lead to similar estimating equations. The most important difference is in the evaluation of the matrix $V_i = \text{Cov}(Z_i)$. This only involves first and second order probabilities for the pseudo-likelihood procedure. In that respect, PL resembles GEE1. In contrast, GEE2 also requires evaluation of third and fourth order probabilities. This makes the GEE2 score equations harder to evaluate and also more time consuming.

Both pseudo-likelihood and generalized estimating equations yield consistent and asymptotically normally distributed estimators, provided an empirically corrected variance estimator is used and provided the model is correctly specified. This variance estimator is similar for both procedures, the main difference being the evaluation of V_i.

If we define the log of the pseudo-likelihood contribution for clusters with size larger than one as $p\ell_i^* = p\ell_i/(n_i - 1)$, the first component of the PL vector contribution $S_i = Z_i - \pi_i$ equals that of GEE2. On the contrary, the association component differs by a factor of $1/(n_i - 1)$. Yet, if we would define the log pseudo-likelihood as $p\ell = \sum_{i=1}^{N} p\ell_i$, then the second components would be equal, while the first components would differ by a factor of $n_i - 1$. Therefore, in studies where the main interest lies in the marginal mean parameters one would prefer $p\ell^*$ over $p\ell$. However, if main interest lies in the estimation of the association parameters we advocate the use of $p\ell$ instead. GEE1 in that case should be avoided, since its goal is limited to estimation of the mean model parameters, while GEE2 is computationally more complex.

The price to pay for computational ease is usually efficiency. Therefore, we will study the asymptotic relative efficiencies (AREs) of the different estimation procedures. For clusters of fixed size, $p\ell$ and $p\ell^*$ are equally efficient. For variable sized clusters, the loss of efficiency for main effects of $p\ell$ will turn out to be very small compared to $p\ell^*$. On the contrary, $p\ell$ will turn out to be superior for estimation of association parameters. We follow the suggestion of Rotnitzky and Wypij (1994), described in Section 6.2.2. In our case, we need to consider all realizations of the form $(n_i, d_i, y_{i1}, \ldots, y_{in_i})$, and have to specify: (1) $f(d_i)$, the relative frequencies of the dose groups, as prescribed by the design; (2) $f(n_i|d_i)$, the probability with which each cluster size can occur, possibly depending on the dose level (we will assume $f(n_i|d_i) = f(n_i)$); and

(3) $f(y_{i1}, \ldots, y_{in_i} | n_i, d_i)$, the actual model probabilities. These can be derived from the cumulative Dale model probabilities. For instance, let $\pi^{(k)}$ denote the cumulative Dale probability of observing at least k successes and $\pi^{(k)*}$ the probability of observing exactly k successes, then

$$\pi^{(k)*} = \binom{n_i}{k} \sum_{j=0}^{(n_i-k)} (-1)^j \binom{n_i - k}{n_i - k - j} \pi^{(k+j)}.$$

As before, we assume that there are 4 dose groups, with one control ($d_i = 0$) and three exposed groups ($d_i = 0.25, 0.5, 1.0$). The number n_i of viable foetuses per cluster can be chosen at random using a local linear smoothed version of the relative frequency distribution given in Table 6.5. Due to excessive time requirements for the maximum likelihood procedure, the calculations are restricted to clusters of size 4. The ML estimating equations are:

$$U(\beta) = \sum_{i=1}^{N} \frac{\partial \pi}{\partial \beta} V_i^{-1} (Z_i - \pi) = 0,$$

where

$$Z_i = \begin{pmatrix} Z_i \\ \binom{Z_i}{2} \\ \binom{Z_i}{3} \\ \binom{Z_i}{4} \end{pmatrix} \quad \text{and} \quad \pi_i = \begin{pmatrix} n_i \pi_{i10} \\ \binom{n_i}{2} \pi_{i11} \\ \binom{n_i}{3} \pi_i^{(3)} \\ \binom{n_i}{4} \pi_i^{(4)} \end{pmatrix}.$$

This involves the evaluation of third and fourth order probabilities, which is computationally laborious, though feasible. Data are generated from a univariate model where the parameters of interest are modeled as follows: $\text{logit}(\pi_{i10}) = \beta_0 + \beta_d d_i$ with d_i, the dose level applied to the ith cluster, and $\ln \psi_i = \beta_a$, i.e., a constant marginal odds ratio model. The background rate parameters (β_0) equal either 0 or -5 and dose effect parameters (β_d) are chosen from $0, 3, 5$. The second order association parameters (β_a) are chosen from $0, 0.3, 1$. The third and fourth order associations are assumed to be zero. The AREs will decrease for increasing higher order associations.

Table 6.8 shows that, when main interest lies in the estimation of the dose effect, the AREs are highest for GEE2, followed by GEE1 and PL. Since the cluster sizes are assumed to be constant and equal to 4, it does not matter whether we use $p\ell$ or $p\ell^*$ to define the log of the pseudo-likelihood. This result shows that GEE1 has some advantage when interest lies in the estimation of main effect parameters. ML and GEE2 are computationally more complex. GEE1 is the easiest one to fit and the loss of efficiency for the main effect parameters is very small compared to GEE2 and ML. Similar results were found by Liang, Zeger and Qaqish (1992). The PL estimation procedure proposed by le Cessie and van Houwelingen (1994) is also computationally easy but is slightly less efficient than GEE1. The differences in ARE between GEE1 and PL are minor.

When main interest lies in the estimation of the association parameters,

Table 6.8 *Simulation Studies. Asymptotic relative efficiencies for dose effect parameter of GEE1, GEE2 and PL versus ML.*

β_0	β_d	β_a	PL	GEE1	GEE2
-5	5	0.0	1.000	1.000	1.000
		0.3	0.999	0.999	0.999
		1.0	0.995	0.999	0.999
-5	3	0.0	1.000	1.000	1.000
		0.3	0.999	0.999	0.999
		1.0	0.998	0.999	0.999
-5	0	0.0	1.000	1.000	1.000
		0.3	0.999	0.999	0.999
		1.0	0.999	0.999	0.999
0	0	0.0	1.000	1.000	1.000
		0.3	0.999	0.999	0.999
		1.0	0.999	0.999	0.999

Table 6.9 *Simulation Results. Asymptotic relative efficiencies for association parameter of GEE1, GEE2 and PL versus ML.*

β_0	β_d	β_a	PL	GEE1	GEE2
-5	5	0.0	1.000	0.865	1.000
		0.3	0.998	0.888	0.999
		1.0	0.995	0.862	0.999
-5	3	0.0	1.000	0.992	1.000
		0.3	0.999	0.992	0.999
		1.0	0.993	0.992	0.999
-5	0	0.0	1.000	1.000	1.000
		0.3	1.000	1.000	1.000
		1.0	1.000	1.000	1.000
0	0	0.0	1.000	1.000	1.000
		0.3	1.000	1.000	1.000
		1.0	1.000	1.000	1.000

Table 6.9 shows that GEE1 can lose considerable efficiency. Moreover, in general, one should not use GEE1 for estimating association parameters, unless confidence in the working assumption is great. Therefore, we would advocate the use of PL. ML and GEE2 are again the most efficient procedures, but

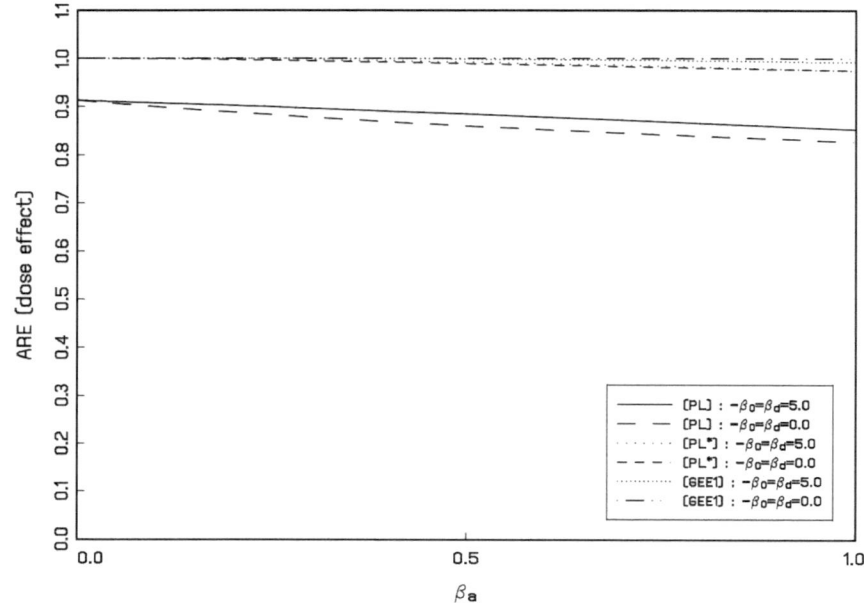

Figure 6.4 *Simulation Results. Asymptotic relative efficiency of GEE2 versus PL and GEE1 for the dose effect parameter in a marginally specified odds ratio model.*

computationally intensive. In case of no dose effect, the three procedures are equally efficient with respect to the association parameter.

As Liang, Zeger and Qaqish (1992) suggested, GEE1, GEE2, and PL may be less efficient when the cluster sizes are unequal. Figures 6.4 and 6.5 show the efficiencies of $p\ell$ and $p\ell^*$ and GEE1 versus GEE2 for *varying* cluster sizes. In that case $p\ell$ and $p\ell^*$ behave differently. Since maximum likelihood is prohibitive, we calculated the AREs of several methods versus the GEE2 method. Since even data generation from the assumed true distribution is rather time consuming, we restricted the calculations to clusters of size less than or equal to 6. Association parameters of order three and higher are assumed to be zero.

Figure 6.4 shows that $p\ell^*$ is much more efficient than $p\ell$ for estimating dose effects. Furthermore, it has the desirable property that the ARE equals 1 under independence. For estimating the second order association parameter, however, Figure 6.5 suggests the use of $p\ell$ rather than $p\ell^*$. Therefore, if main interest lies in the marginal mean parameters we would suggest to use $p\ell^*$ rather than $p\ell$. However, if main interest lies in the estimation of association parameters, the use of $p\ell$ is advised. If interest is combined, and one type of analysis should be chosen, $p\ell$ might be preferable. When using $p\ell^*$ the ARE increases for increasing association. Furthermore, in all cases, AREs are highest for the lowest background rate parameters.

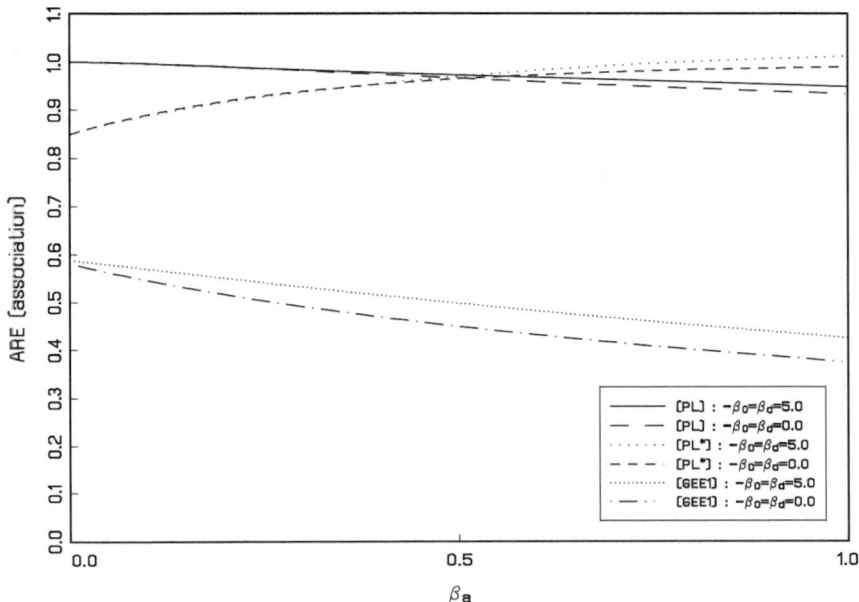

Figure 6.5 *Simulation Results. Asymptotic relative efficiency of GEE2 versus PL and GEE1 for the association parameter in a marginally specified odds ratio model.*

6.3.3 Examples: NTP Data

We apply the PL and first and second order GEE estimating procedures to data from the DEHP and DYME studies, described in Section 2.1. Malformations are classified as being external, visceral, and skeletal. However, we fit the marginal odds ratio model described in the previous sections to a collapsed outcome, defined as 1 if at least one malformation was found and 0 otherwise. The parameters of interest are modeled using $\text{logit}(\pi_{i10}) = \beta_0 + \beta_d d_i$, with d_i the dose level applied to the ith cluster, and $\ln \psi_i = \beta_a$, i.e., a constant marginal odds ratio model is assumed.

Table 6.10 shows that the parameter estimates, obtained by either the pseudo-likelihood or the generalized estimating equations approach, are comparable. Because main interest is focused on the dose effect, we used $p\ell^*$ rather than $p\ell$. Dose effects and association parameters are always significant, except for the GEE1 association estimates. For this procedure, β_a is not significant for the DEHP study and marginally significant for the DYME study. The GEE1 standard errors for β_a are much larger than for their PL and GEE2 counterparts. The GEE2 standard errors are the smallest among the different estimating approaches, which is in agreement with findings in previous sections. Furthermore, it is observed that the standard errors of the Newton-Raphson PL algorithm are generally slightly smaller than those obtained using Fisher scoring, which is in line with other empirical findings. On the other hand, the

Table 6.10 *NTP Data. Parameter estimates (standard errors) for a marginal odds ratio model fitted with PL, GEE1 and GEE2.*

Study	β_0	β_d	β_a
	Newton-Raphson PL Estimates		
DEHP	-3.98 (0.30)	5.57 (0.61)	1.10 (0.27)
DYME	-5.73 (0.46)	8.71 (0.94)	1.42 (0.31)
	Fisher scoring PL Estimates		
DEHP	-3.98 (0.30)	5.57 (0.61)	1.11 (0.27)
DYME	-5.73 (0.47)	8.71 (0.95)	1.42 (0.35)
	GEE2 Estimates		
DEHP	-3.69 (0.25)	5.06 (0.51)	0.97 (0.23)
DYME	-5.86 (0.42)	8.96 (0.87)	1.36 (0.34)
	GEE1 Estimates		
DEHP	-4.02 (0.31)	5.79 (0.62)	0.41 (0.34)
DYME	-5.89 (0.42)	8.99 (0.87)	1.46 (0.75)

Table 6.11 *NTP Data. Time (in seconds) needed for the PL, GEE1 and GEE2 procedures. For PL, Fisher scoring was used.*

Study	GEE2	PL (Fisher scoring)		GEE1
		Classical	GLM	
DEHP	1280	116	72	25
DYME	801	110	76	26

Newton-Raphson procedure is computationally slightly more complex in this case. The time gain of Fisher scoring however is negligible.

Table 6.11 presents the time (in seconds) needed for each procedure. As was expected, GEE2 is relatively time consuming. Then comes the PL estimating approach in its classical form, followed by the generalized linear model type representation, which is computationally less complex. As anticipated, GEE1 is the least complicated fitting procedure.

CHAPTER 7

Pseudo-likelihood Inference

Helena Geys, Geert Molenberghs

transnationale Universiteit Limburg, Diepenbeek–Hasselt, Belgium

In Chapter 6, the pseudo-likelihood estimation procedure was proposed as a viable and attractive alternative to maximum likelihood estimation in the case of clustered (multivariate) binary data. In practice, one will often want to perform a flexible model selection. Therefore, one needs extensions of the Wald, score, or likelihood ratio test statistics to the pseudo-likelihood framework. Rotnitzky and Jewell (1990) examined the asymptotic distributions of generalized Wald and score tests, as well as likelihood ratio tests, for regression coefficients obtained by generalized estimating equations for a class of marginal generalized linear models for correlated data. Using similar ideas, we derive different test statistics, as well as their asymptotic distributions for the pseudo-likelihood framework. Liang and Self (1996) have considered a test statistic, for one specific type of pseudo-likelihood function, which is similar in form to one of the tests we will present below.

7.1 Test Statistics

Suppose we are interested in testing the null hypothesis $H_0 : \gamma = \gamma_0$, where γ is an r-dimensional subvector of the vector of regression parameters β and write β as $(\gamma^T, \delta^T)^T$. Then, several test statistics can be used.

7.1.1 Wald Statistic

Because of the asymptotic normality of the PL estimator $\tilde{\beta}_N$,

$$W^* = N(\tilde{\gamma}_N - \gamma_0)^T \Sigma_{\gamma\gamma}^{-1}(\tilde{\gamma}_N - \gamma_0)$$

has an asymptotic χ_r^2 distribution under the null hypothesis, where $\Sigma_{\gamma\gamma}$ denotes the $r \times r$ submatrix of $\Sigma = J^{-1}KJ^{-1}$, where J is the matrix of minus the second derivatives of the log pseudo-likelihood and K is the matrix of the cross-products of the first derivative vectors. In practice, the matrix Σ can be replaced by a consistent estimator, obtained by substituting the PL estimator $\tilde{\beta}_N$. Although the Wald test is in general simple to apply, it is well known to be sensitive to changes in parameterization. The Wald test statistic is therefore particularly unattractive for conditionally specified models, since marginal effects are likely to depend in a complex way on the model parameters (Diggle, Liang and Zeger 1994).

7.1.2 Pseudo-score Statistics

As an alternative to the Wald statistic, one can propose the *pseudo-score statistic*. A score test has the advantage that it can be obtained by fitting only the null model. Furthermore, it is invariant to reparameterization. Let us define

$$S^*(e.c.) = \frac{1}{N} U_\gamma(\gamma_0, \tilde{\delta}(\gamma_0))^T J^{\gamma\gamma} \Sigma_{\gamma\gamma}^{-1} J^{\gamma\gamma} U_\gamma(\gamma_0, \tilde{\delta}(\gamma_0)),$$

where $\tilde{\delta}(\gamma_0)$ denotes the maximum pseudo-likelihood estimator in the subspace where $\gamma = \gamma_0$, $J^{\gamma\gamma}$ is the $r \times r$ submatrix of the inverse of J, and $J^{\gamma\gamma} \Sigma_{\gamma\gamma}^{-1} J^{\gamma\gamma}$ is evaluated under H_0. Geys, Molenberghs and Ryan (1999) showed that this pseudo-score statistic is asymptotically χ_r^2 distributed under H_0. As discussed by Rotnitzky and Jewell (1990) in the context of generalized estimating equations, such a score statistic may suffer from computational stability problems. A model based test that may be computationally simpler is:

$$S^*(m.b.) = \frac{1}{N} U_\gamma(\gamma_0, \tilde{\delta}(\gamma_0))^T J^{\gamma\gamma} U_\gamma(\gamma_0, \tilde{\delta}(\gamma_0)).$$

However, its asymptotic distribution under H_0 is complicated and given by $\sum_{j=1}^r \lambda_j \chi_{1(j)}^2$ where the $\chi_{1(j)}^2$ are independently distributed as χ_1^2 variables and $\lambda_1 \geq \cdots \geq \lambda_r$ are the eigenvalues of $(J^{\gamma\gamma})^{-1} \Sigma_{\gamma\gamma}$, evaluated under H_0. The score statistic $S^*(m.b.)$ can be adjusted such that it has an approximate χ_r^2 distribution, which is much easier to evaluate. Several types of adjustments have been proposed in the literature (Rao and Scott 1987, Roberts, Rao and Kumar 1987). Similar to Rotnitzky and Jewell (1990), Geys, Molenberghs and Ryan (1999) proposed an adjusted pseudo-score statistic

$$S_a^*(m.b.) = S^*(m.b.)/\overline{\lambda},$$

where $\overline{\lambda}$ is the arithmetic mean of the eigenvalues λ_j. Note that no distinction can be made between $S^*(e.c.)$ and $S_a^*(m.b.)$ for $r = 1$. Moreover, in the likelihood-based case, all eigenvalues reduce to one and thus all three statistics coincide with the model based likelihood score statistic.

7.1.3 Pseudo-likelihood Ratio Statistics

Another alternative is provided by the pseudo-likelihood ratio test statistic, which requires comparison of full and reduced model:

$$G^{*2} = 2 \left[p\ell(\tilde{\beta}_N) - p\ell(\gamma_0, \tilde{\delta}(\gamma_0)) \right].$$

Geys, Molenberghs and Ryan (1999) showed that the asymptotic distribution of G^{*2} can also be written as a weighted sum $\sum_{j=1}^{r} \lambda_j \chi_{1(j)}^2$, where the $\chi_{1(j)}^2$ are independently distributed as χ_1^2 variables and $\lambda_1 \geq \cdots \geq \lambda_r$ are the eigenvalues of $(J^{\gamma\gamma})^{-1} \Sigma_{\gamma\gamma}$. Alternatively, the adjusted pseudo-likelihood ratio test statistic, defined by

$$G_a^{*2} = G^{*2}/\overline{\lambda},$$

is approximately χ_r^2 distributed. Their proof shows that G^{*2} can be rewritten as an approximation to a Wald statistic. The covariance structure of the Wald statistic can be calculated under the null hypothesis, but also under the alternative hypothesis. Both versions of the Wald tests are asymptotically equivalent under H_0 (Rao 1973, p. 418). Therefore, it can be argued that the adjustments in G_a^{*2} can also be evaluated under the null as well as under the alternative hypothesis. These adjusted statistics will then be denoted by $G_a^{*2}(H_0)$ and $G_a^{*2}(H_1)$ respectively. In analogy with the Wald test statistic, we expect $G_a^{*2}(H_1)$ to have high power. A similar reasoning suggests that the score test $S_a^*(m.b.)$ might closely correspond to $G_a^{*2}(H_0)$, since both depend strongly on the fitted null model. Analogous results were obtained by Rotnitzky and Jewell (1990). Section 7.2 briefly compares the small sample behavior of the different test statistics.

The asymptotic distribution of the pseudo-likelihood based test statistics are weighted sums of independent χ_1^2 variables where the weights are unknown eigenvalues. In Aerts and Claeskens (1999) it is shown theoretically that the parametric bootstrap leads to a consistent estimator for the null distribution of the pseudo-likelihood ratio test statistic. The bootstrap approach does not need any additional estimation of unknown eigenvalues and automatically corrects for the incomplete specification of the joint distribution in the pseudo-likelihood. Similar results hold for the robust Wald and robust score test. Their simulation study indicates that the χ^2 tests often suffer from inflated type I error probabilities which are nicely corrected by the bootstrap. This is especially the case for the Wald statistic whereas the asymptotic χ^2 distribution of the robust score statistic test is performing quite well. The parametric bootstrap is expected to break down if the likelihood of the data is grossly misspecified. Chapter 11 presents a more robust semiparametric bootstrap, based on resampling the score and differentiated score values.

7.2 Simulation Results

7.2.1 Asymptotic Simulations

Diggle, Liang and Zeger (1994) note that Wald tests can have poor properties for conditional models. Therefore we advocate the use of score and ratio test statistics.

To explore the performance of these test statistics more thoroughly, we will show some simulation results with asymptotic considerations similar to the ideas of Rotnitzky and Wypij (1994). Section 6.3.2 describes this approach in a

Comparison of Likelihood and Pseudo-likelihood test Statistics

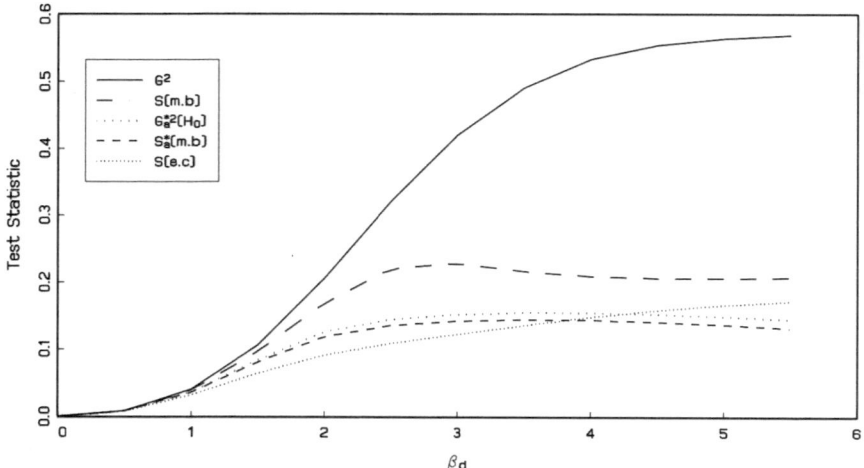

Figure 7.1 *Simulation Results. Comparison of likelihood and pseudo-likelihood test statistics for a common dose trend in the bivariate Molenberghs and Ryan (1999) model.*

univariate context. The extension to a multivariate context is straightforward. As in Chapter 6, we assume that there are 4 dose groups, with one control group ($d_i = 0$) and three active groups ($d_i = 0.25, 0.5, 1.0$) and that the number of viable foetuses (n_i) per cluster is chosen at random from a local linear smoothed version of the relative frequency distribution given in Table 1 of Kupper *et al.*(1986) (which is considered representative of that encountered in actual experimental situations.)

The present study is restricted to clusters of *bivariate* binary data with maximum cluster size of 10, due to prohibitive time requirements of ML. The main effects are modeled as

$$\theta_{ij} = \beta_{0j} + \beta_d d_i (j = 1, 2),$$

i.e., a common main dose effect is assumed, and all association parameters are assumed to be constant. Data are generated from a bivariate model with background rate parameters (β_{01}, β_{02}) = ($-3, -3$) and a zero association vector ($\delta_1, \delta_2, \omega_{12}, \gamma_{12}$). Positive associations yield similar results.

We would like to assess the effect of β_d. Since the Wald test is known to depend on the particular parameterization, it might be a less relevant measure to use. We will therefore concentrate mainly on score and ratio statistics.

Figure 7.1 shows the adjusted pseudo-score and pseudo-likelihood ratio statistics $S_a^*(m.b.)$ and $G_a^{*2}(H_0)$, as well as the model based, $S(m.b.)$, and empirically corrected, $S(e.c.)$, likelihood score tests and the likelihood ratio statistic G^2. We restrict to $G_a^{*2}(H_0)$, since it is similar to $G_a^{*2}(H_1)$ in this case. Note that $S_a^*(m.b.)$ is identical to $S^*(e.c.)$, since we are testing for the effect of a

Table 7.1 *Simulated type I error probabilities for* $\beta_0 = -2.5$ *and dose levels* $0, .25, .50, 1$. *(NC is the number of clusters per dose level.)*

β_a	NC	Likelihood			Pseudo-likelihood		
		G^2	$S(m.b)$	$S(e.c)$	$G_a^*(H_0)$	$G_a^*(H_1)$	$S^*(e.c) = S_a^*(m.b)$
0.10	5	5.09	4.21	3.21	4.29	14.29	2.80
	30	6.00	6.20	5.00	5.40	6.60	5.40
0.25	15	3.63	3.68	1.23	4.70	18.37	2.25
	30	6.01	4.60	5.00	6.63	10.84	5.20

single parameter. In the absence of both a true dose effect and an association between outcomes or between cluster members, likelihood and pseudo-likelihood are equivalent. However, a substantial discrepancy arises between G^2 and $G_a^{*2}(H_0)$ for positive dose effects. Indeed, by ignoring an important effect, we introduce an apparent association, which is given too much weight in the pseudo-likelihood. This leads to a pseudo-likelihood value that is too large under the null. Therefore, the pseudo-deviance is much smaller than the likelihood deviance. As a consequence of the misspecification, the matrix $\Sigma_{\gamma\gamma}(J^{\gamma\gamma})^{-1}$, and hence also the corresponding adjustment, is overestimated, rendering an even greater discrepancy between the test statistics G^2 and $G_a^{*2}(H_0)$. A similar argument explains the discrepancy with the pseudo-score statistic, since this statistic is fully obtained from the null model. As follows from theory, $S_a^*(m.b.)$ and $G_a^{*2}(H_0)$ are comparable. For small to moderate dose effects, both these statistics are situated between $S(m.b.)$ and $S(e.c.)$. However, for larger dose effects, the pseudo-statistics $S_a^*(m.b.)$ and $G_a^{*2}(H_0)$, as well as their adjustments, show a non-monotone behavior, in contrast to the likelihood ratio that increases monotonically with dose.

7.2.2 Small Sample Behavior of the Test Statistics

In this section, we present the results of a small sample simulation study for a single clustered outcome, based upon 500 replications, to illustrate the finite sample behavior of the pseudo-likelihood test statistics with respect to type I error probability and power (Geys, Molenberghs and Ryan 1999). The number n_i of viable foetuses per cluster is again assumed to follow a local linear smoothed version of the relative frequency distribution in Table 1 of Kupper et al. (1986). Data are generated and fitted using a model where the main effect is modeled as $\theta = \beta_0 + \beta_d d$ and the association parameter is held constant ($\delta = \beta_a$). The hypothesis of interest is $\beta_d = 0$. The results are shown in Tables 7.1 and 7.2.

The pseudo-score test statistics as well as $G_a^{*2}(H_0)$ have satisfactory type I error probabilities, in good agreement with their likelihood counterparts. Since

Table 7.2 *Simulated powers for* $\beta_0 = -2.5$, $\beta_a = 0.1$ *and dose levels* $0, .25, .50, 1$. *(NC is the number of clusters per dose level.)*

		Likelihood			Pseudo-likelihood		
β_d	NC	G^2	$S(m.b)$	$S(e.c)$	$G_a^*(H_0)$	$G_a^*(H_1)$	$S^*(e.c) = S_a^*(m.b)$
1.0	5	25.05	24.80	16.80	20.04	29.66	20.20
2.0		96.42	95.79	79.79	90.18	91.38	90.40
2.5		100.00	100.00	92.46	98.20	97.80	98.00
1.0	30	88.40	88.60	83.40	83.80	87.60	84.40
2.0		100.00	100.00	100.00	100.00	100.00	100.00
2.5		100.00	100.00	100.00	100.00	100.00	100.00

we are in the single parameter case, $S^*(e.c.)$ and $S_a^*(m.b.)$ yield identical results. The rejection probabilities for the score statistics tend to be somewhat smaller than for the pseudo-likelihood ratio statistics, which is often observed in the likelihood setting as well. The pseudo-likelihood ratio statistic $G_a^{*2}(H_1)$ shows inflated type I error probabilities, especially for small samples. Consequently its power may be misleadingly high. A bootstrap alternative of the pseudo-likelihood ratio test, constructed by Aerts and Claeskens (1999), seems to nicely correct this towards the nominal size (see also Chapter 11). Similar problems can be remedied by the bootstrap for the Wald statistic (which is also based on the alternative model). The power of $G_a^{*2}(H_0)$ closely corresponds to that of the pseudo-score statistics. For realistic parameter settings such as $(\beta_0, \beta_d, \beta_a) = (-2.5, 2.5, 0.1)$ (based on analyses of National Toxicology Program data; Price *et al.* 1987) and/or large samples, $G_a^{*2}(H_1)$ behaves similarly to the other pseudo-likelihood test statistics. Moreover, powers are then very high for all pseudo-likelihood statistics and comparable to their likelihood counterparts. An analogous result was obtained from asymptotic simulations (not shown).

In summary, the simulations suggest that the pseudo-score statistics as well as $G_a^2(H_0)$ may have lower power than their likelihood counterparts. Calculating the adjusted pseudo-likelihood ratio test under the alternative, $G_a^2(H_1)$ may increase the power, but tends to inflate type I error probabilities in small samples. However, for realistic samples and parameter settings, the pseudo-likelihood ratio tests produce high powers. Therefore, we would suggest the use of the adjusted pseudo-likelihood ratio tests, but recommend caution for small sample sizes.

7.3 Illustration: EG data

In this section, we return to the data, described in Section 2.1, studied in earlier Chapters, and collected to study the toxicity of Ethylene Glycol in mice.
The main goal is to construct an appropriate dose-response model. This will
be achieved by fitting model (4.21) and modeling the natural parameters Θ
in this model as fractional polynomial functions of dose (Royston and Altman 1994), since fractional polynomials provide more flexibly shaped curves
than conventional polynomials. More details on this approach can be read in
Chapter 8. Estimation is by pseudo-likelihood rather than maximum likelihood, due to the latter's excessive computational requirements. To select an
appropriate dose-response model, we need to use the test statistics, developed
in Section 7.1.

Attempts to use conventional low order polynomials of the form $\beta_0 +
\sum_{j=1}^{m} \beta_j d^j$ to express the model parameters Θ as a function of dose (d) are
not successful for the EG data. Royston and Altman (1994) argue that conventional low order polynomials offer only a limited family of shapes and that
high order polynomials may fit poorly at the extreme values of the covariates.
Moreover, polynomials do not have finite asymptotes and cannot fit the data
where limiting behavior is expected. This is a severe limitation when low dose
extrapolation is envisaged. As an alternative, Royston and Altman (1994)
propose an extended family of curves, which they call fractional polynomials.
For a detailed discussion see Section 8.1.

Geys, Molenberghs and Ryan (1999) adopted this approach in their analysis
of the EG data. The following strategy is adopted. First, they select a suitable
set of dose transformations for each of the three developmental outcomes
(skeletal, visceral, and external) separately, using the method described by
Royston and Altman (1994). The resulting set of transformations is then used
to construct more elaborate (multivariate) models that can be scrutinized
further by means of the formal tests proposed in Section 7.1.

Their most complex model we consider (Model 1) allows different $\sqrt{d}$ trends
on the external, visceral, and skeletal main effect parameters, an additional
d trend on the skeletal main effect parameter ($\theta_1 = \beta_{01} + \beta_{\sqrt{d}1}\sqrt{d}$; $\theta_2 =
\beta_{02} + \beta_{\sqrt{d}2}\sqrt{d}$; $\theta_3 = \beta_{03} + \beta_{\sqrt{d}3}\sqrt{d} + \beta_{d3}d$), and different $\sqrt{d}$ trends to the
clustering parameters (δ). All other association parameters (ω and γ) are held
constant. This model can now be scrutinized further by means of the formal
test statistics introduced in Section 7.1.

From Table 7.3 it is clear that the clustering parameters do not depend on
$\sqrt{d}$ (confirming our preliminary, univariate findings). Hence, Model 2 is now
selected. The d trend on the skeletal main effect parameter cannot be removed
(comparing Models 2 and 3), nor can the different $\sqrt{d}$ trends on the external,
visceral, and skeletal main effects be replaced by a common trend (comparing
Models 2 and 4). Therefore we select Model 2 for the time being. Table 7.4
shows parameter estimates for this model.

A key tool to gain insight in this model is the qualitative study of the dose-
response relationship. In the area of developmental toxicity, there is generally

Table 7.3 *EG Data. Model selection. (All effects are constant except the ones mentioned.)*

Model	Description	# Pars.
1	$\neq \sqrt{d}$ trends on θ_1, θ_2, θ_3; d trend on θ_3;	
	$\neq \sqrt{d}$ trends on δ_1, δ_2, δ_3	19
2	$\neq \sqrt{d}$ trends on θ_1, θ_2, θ_3; d trend on θ_3	16
3	$\neq \sqrt{d}$ trends on θ_1, θ_2, θ_3	15
4	$= \sqrt{d}$ trend on θ_1, θ_2, θ_3; d trend on θ_3	14
5	$\neq \sqrt{d}$ trends on θ_1, θ_2, θ_3; d trend on θ_3; No ω, γ pars.	10

Comparison	df	$\overline{\lambda}(H_0)$	$\overline{\lambda}(H_1)$	$S^*(e.c.)$	$S_a^*(m.b.)$	$G_a^{*2}(H_0)$	$G_a^{*2}(H_1)$
1–2	3	1.27	0.89	3.77	2.84	2.84	4.06
2–3	1	0.45	0.78	15.19	15.19	18.55	10.68
2–4	2	0.79	0.70	5.76	8.03	8.05	9.09
2–5	6	1.48	1.44	7.71	9.18	9.68	10.01

little understanding about the complex processes that relate maternal exposure to adverse fetal impacts. For developmental toxicity studies where offspring are clustered within litters, there are several ways to define an adverse effect. A foetus-based approach considers the malformation probability of an individual offspring while a litter-based approach is based on the probability that at least one adverse effect has occurred within a litter. Here, we restrict attention to the litter-based approach. To this end, moment-based methods such as GEE cannot be used, while the Molenberghs and Ryan (1999) model allows flexible modeling for both the main effects and the association structure. Given the number of viable foetuses n_i, the probability of observing at least one abnormal foetus in a cluster is $1 - \exp(-A_{n_i}(\boldsymbol{\Theta}_i))$. Integrating over all possible values of n_i, we obtain the following *risk function*:

$$r(d) = \sum_{n_i=0}^{\infty} P(n_i)[1 - \exp(-A_{n_i}(\boldsymbol{\Theta}_i))], \qquad (7.1)$$

where $P(n_i)$ is the probability of observing n_i viable foetuses in a pregnant dam. (We use the empirical distribution of $P(n_i)$.) One of the major challenges of a teratology study lies in characterizing the relationship between dose and event probability (7.1) by means of a dose-response curve. Here, Model 2 is used to construct dose-response curves representing the probability of observing an adverse event as a function of dose (d). The risk function $r(d)$ is calculated using PL parameter estimates.

Figures 7.2 (a) and (b) show the observed frequencies of malformed litters

Table 7.4 *EG Data. Pseudo-likelihood estimates (standard errors) for the final model.*

Effect	Outcome	Parameter	Est. (s.e.)	
			Model 2	Model 5
θ Main	Ext.	β_{01}	-2.27 (1.16)	-3.58 (1.10)
		$\beta_{\sqrt{d}1}$	1.71 (0.99)	3.07 (0.97)
	Visc.	β_{02}	-6.98 (2.36)	-7.17 (2.26)
		$\beta_{\sqrt{d}2}$	5.54 (1.71)	5.83 (1.96)
	Skel.	β_{03}	-2.81 (0.95)	-3.61 (0.84)
		$\beta_{\sqrt{d}3}$	7.73 (2.32)	7.59 (2.22)
		β_{d3}	-4.01 (1.50)	-3.89 (1.43)
δ Clustering	Ext.	δ_1	0.18 (0.13)	0.29 (0.06)
	Visc.	δ_2	0.12 (0.17)	0.22 (0.09)
	Skel.	δ_3	0.18 (0.01)	0.19 (0.01)
ω Assoc.	Ext.-Visc.	ω_{12}	-0.06 (0.57)	
	Ext.-Skel.	ω_{13}	0.11 (0.29)	
	Skel.-Visc.	ω_{23}	0.81 (0.34)	
γ Assoc.	Ext.-Visc.	γ_{12}	0.14 (0.13)	
	Ext.-Skel.	γ_{13}	0.08 (0.04)	
	Skel.-Visc.	γ_{23}	-0.08 (0.04)	

at the selected dose levels for external and visceral malformations and the (univariate) dose-response curves for models with constant association and $\sqrt{d}$ trends on the main effects. The observed malformation rates are supplemented with pointwise 95% confidence intervals. The dose-response curve for skeletal malformation (Figure 7.2 (c)) is based on the quadratic $(\sqrt{d}, d)$-model for the main effect parameter and constant clustering. Figure 7.2 (d) shows the trivariate dose-response curve based on all three outcomes simultaneously (Model 2). Both the univariate as well as the trivariate fits are excellent. All curves gradually increase when dams are exposed to larger quantities of the toxic substance, before finally reaching an asymptotic. Note that there is a fundamental difference in the dose-response curve for external and visceral outcomes on the one hand, and skeletal malformation on the other, the latter of which shows a much more pronounced dose-response relationship. This is in line with the different functional form for these responses. Further, the joint dose-response curve is clearly driven by skeletal malformation.

These observations suggest to explore additional model simplification. Candidates for removal are the dose trends on the external and visceral outcomes,

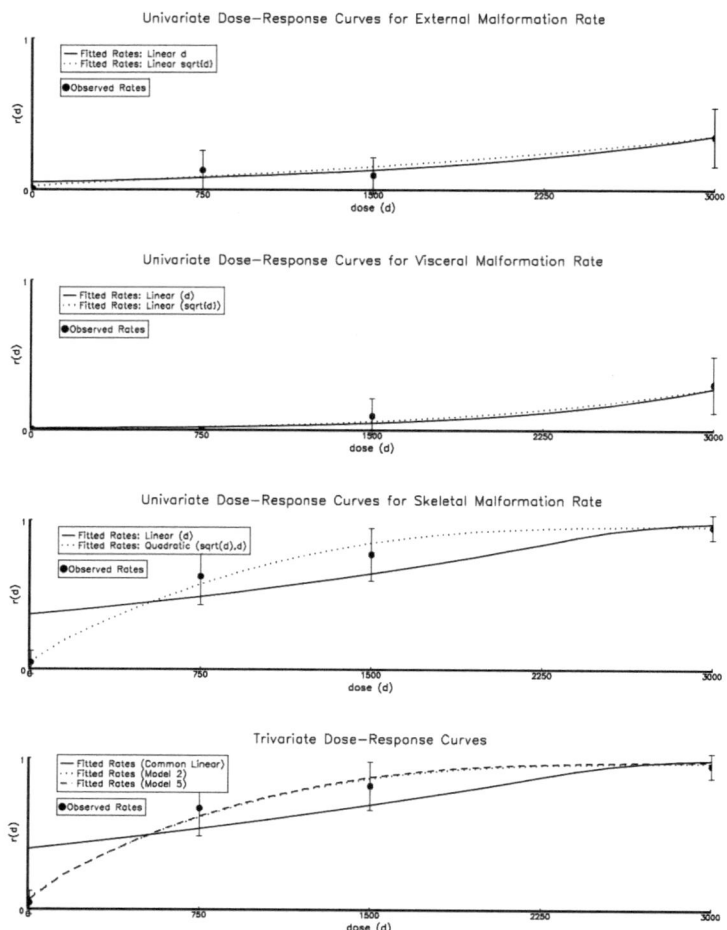

Figure 7.2 *EG Data. Dose-response curves. (a) Univariate dose-response curve for external malformations based on a model with $\sqrt{d}$ trend on main effect parameter θ and constant clustering parameter δ. (b) Univariate dose-response curve for visceral malformations based on a model with $\sqrt{d}$ trend on main effect parameter θ and constant clustering parameter δ. (c) Univariate dose-response curve for skeletal malformations based on the quadratic $(\sqrt{d}, d)$ trend on main effect parameter θ and constant clustering parameter δ. (d) Trivariate dose-response curves based on Models 2 and 5.*

as well as one or more association parameters. Table 7.3 shows that the ω and γ association parameters are redundant (compare Model 2–Model 5). However, the clustering parameters could not be removed from the model without a substantial decrease in fit. Furthermore, the dose trends on the external and visceral main effects are also important. Since the goal of selecting a good-fitting model is to perform risk assessment, merely concentrating on formal model selection criteria is insufficient. Arguably, the excellent fit of the dose-response curves that have been achieved should not be compromised. However, Figure 7.2 shows that the simplified Model 5 produces essentially the same dose-response curve as Model 3. Therefore, Model 5 will be treated as our final model. The parameter estimates are tabulated in Table 7.4. It is important to remember that the model parameters have a conditional interpretation. For example it can be derived from (4.23) that, in Model 5, the main effect parameter θ_{ij} can be interpreted as the conditional logit, associated with an additional malformation of type j in the ith cluster, given the cluster contains already $z_{ij} = (n_i + 1)/2$ foetuses with malformations of that type. Similarly, δ_{ij} can be interpreted as the conditional log odds ratio for a pair of foetuses, exhibiting malformation j, given all other outcomes. Thus, if interest is in marginal quantities, such as the dose-response curve, they have to be obtained as non-linear functions of the parameters. Computationally, this is a very feasible task. In contrast, conditional questions can be answered in terms of linear functions of the parameters.

CHAPTER 8

Flexible Polynomial Models

Marc Aerts, Christel Faes

transnationale Universiteit Limburg, Diepenbeek–Hasselt, Belgium

Gerda Claeskens

Texas A & M University, College Station, TX

Clustered binary data have been analyzed mainly in a parametric way. The selection of the proper functional forms describing the dependence of all main and association parameters in a specific probability model is not always an easy task. There is clearly a need for flexible parametric models and, in case the design allows, for semi- and nonparametric approaches. This chapter illustrates how two popular classes of polynomial models, fractional and local polynomials, offer great flexibility in modeling clustered data.

8.1 Fractional Polynomial Models

Although classical polynomial predictors are still very customary, they are often inadequate. A very elegant alternative approach to classical polynomials, which falls within the realm of (generalized) linear methods, is given by fractional polynomials. They provide a much wider range of functional forms. Let us briefly describe this procedure, advocated by Royston and Altman (1994).

For a given degree m and a univariate argument $x > 0$, fractional polynomials are defined as

$$\beta_0 + \sum_{j=1}^{m} \beta_j x^{[p_j]},$$

where the β_j are regression parameters, $x^{[p]} = x^p$ if $p \neq 0$ and $x^{[0]} = \ln(x)$. The powers $p_1 < \cdots < p_m$ are either positive or negative integers, or fractions. Royston and Altman (1994) argue that polynomials with degree higher than 2 are rarely required in practice and further restrict the powers of x to a small pre-defined set of noninteger values:

$$\Pi = \{-2, -1, -1/2, 0, 1/2, 1, 2, \ldots, \max(3, m)\}.$$

The full definition includes possible "repeated powers" which involve multiplication with $\ln(x)$. For example, a fractional polynomial of degree $m = 3$ with powers $(-1,-1,2)$ is of the form $\beta_0 + \beta_1 x^{-1} + \beta_2 x^{-1} \ln(x) + \beta_3 x^2$ (Royston and Altman 1994, Sauerbrei and Royston 1999). Setting $m = 2$, for example, will generate:

(1) 4 "quadratics" in powers of x, represented by

$$- \beta_0 + \beta_1 1/x + \beta_2 1/x^2,$$
$$- \beta_0 + \beta_1 1/\sqrt{x} + \beta_2 1/x,$$
$$- \beta_0 + \beta_1 \sqrt{x} + \beta_2 x, \text{ and}$$
$$- \beta_0 + \beta_1 x + \beta_2 x^2;$$

(2) a quadratic in $\ln(x)$: $\beta_0 + \beta_1 \ln(x) + \beta_2 \ln^2(x)$; and

(3) several other curves with shapes different from those of low degree polynomials.

For given m, we consider as the best set of transformations, the one producing the highest log (pseudo)-likelihood. For example, the best first degree fractional polynomial is the one with the highest log (pseudo)-likelihood among the eight models with one regressor $(x^{-2}, x^{-1}, \ldots, x^3)$. As with conventional polynomials, the degree m is selected either informally on *a priori* grounds or by increasing m until no worthwhile improvement in the fit of the best fitting fractional polynomial occurs. In the above discussion, it is assumed that x is strictly positive. If x can take zero values, a preliminary transformation of x is needed to ensure positivity (e.g., $x + 1$).

Geys, Molenberghs and Ryan (1999) used fractional polynomials for dose-response modeling as basis for quantitative risk assessment. A similar application is illustrated in the next section.

8.1.1 The EG Data

Consider the EG data from the NTP developmental experiments in mice. In order to describe the dose-response relationship, representing the probability of observing at least one adverse event as a function of dose (d), we use the exponential family likelihood model of Molenberghs and Ryan (1999). See also Section 4.3.

Traditionally, conventional linear predictors are used to describe main effects and associations. Figures 8.1 (a)–(c) show the observed frequencies of malformed litters at the selected dose levels for external, visceral, and skeletal malformations (dots), and the three (univariate) dose-response curves for models with constant association and a linear d trend on the main effect (solid line). Figure 8.1 (d) shows the trivariate dose-response curve based on all three outcomes jointly and with a common linear dose trend on the main effect parameters (solid line). These models are clearly too restricted to adequately describe the underlying dose-response relationship. One can try to further improve the data by adopting the fractional polynomial approach.

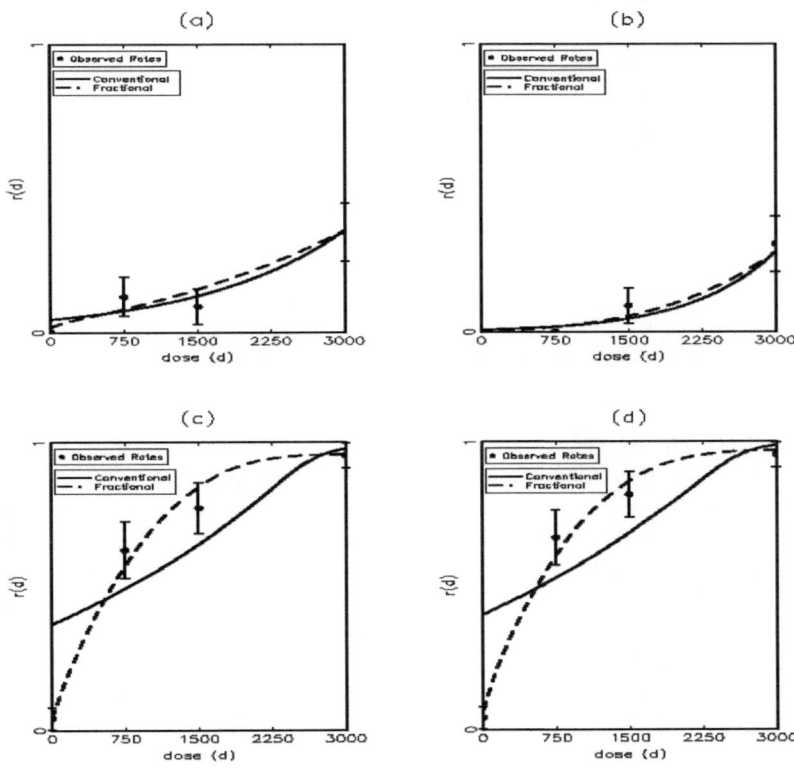

Figure 8.1 *EG Study. Univariate and trivariate dose-response curves for external, visceral and skeletal malformation rates.*

We first select a suitable set of dose transformations for each of the three developmental outcomes separately. We consider as the best set of transformations the one producing the smallest value of Akaike's (1974) information criterion (AIC). The AIC is a measure of model fit, and penalizes for increasing size of the model. AIC is defined as minus twice the log (pseudo)-likelihood value plus twice the number of model parameters. Table 8.1 shows the best first and second degree models for the main effect of skeletal malformation outcomes. Among all models under consideration, the quadratic represented by $(\sqrt{d}, d)$ yields the smallest AIC. A similar approach can be applied to the clustering parameter, but suggests that no dose effect needs to be incorpo-

Table 8.1 *EG Study. Selection procedure for main effect of skeletal malformation outcomes.*

$m = 1$		$m = 2$	
transformation	AIC	transformation	AIC
$1/d^2$	681.18	$(1/d, d^2)$	674.48
$1/d$	680.96	$(1/d, 1/d^2)$	672.88
$1/\sqrt{d}$	682.94	$(1/d, d)$	673.90
$ln(d)$	684.92	$(\sqrt{d}, d)$	671.92
$\sqrt{d}$	677.98	(d, d^2)	675.60
d	688.58	$(ln(d), ln(d)^2)$	675.12
d^2	696.56	$(ln(d), d^2)$	675.18
d^3	699.02	$(d, 1/d)$	673.90

rated. For both the external and visceral malformation outcomes, the main effects are best modeled linearly in $\sqrt{d}$, with constant association.

Based on the univariate selection, we construct more elaborate, trivariate models that can be further scrutinized by means of formal test statistics. This strategy was suggested by Geys, Molenberghs and Ryan (1999). The most complex model we consider allows different $\sqrt{d}$ trends on the external, visceral, and skeletal main effect parameters, an additional d trend on the skeletal main effect parameter, and different $\sqrt{d}$ trends on the clustering parameters. All other associations are assumed constant. Using formal test statistics, we can simplify this model. We can show that a model with different $\sqrt{d}$ trends on the external, skeletal, and visceral main effect parameters, an additional trend on the skeletal main effect parameter, and constant associations is a good fractional model. Figures 8.1 (a)–(d) shows these fractional models (dashed line). Univariate as well as the trivariate dose-response curves fit the data well.

8.1.2 The HIS Data

The HIS data were introduced in Section 2.3. They were collected at three different stages (municipalities, households, and individuals). A multilevel model takes the underlying hierarchical structure into account. We want to study the effects of personal, social, and material characteristics on the body mass index (BMI). The covariates of interest are region, sex, age, education, and households income. The variables region, education, and households income are categorical, with respectively 3, 5, and 5 categories. The variable age is continuous. Figure 8.2 (a) plots ln(BMI) against age, together with the estimated response for the model with a linear age effect. There is clearly a nonlinear trend. We use fractional polynomials to model the effect of age, since these polynomials give very flexible shaped curves.

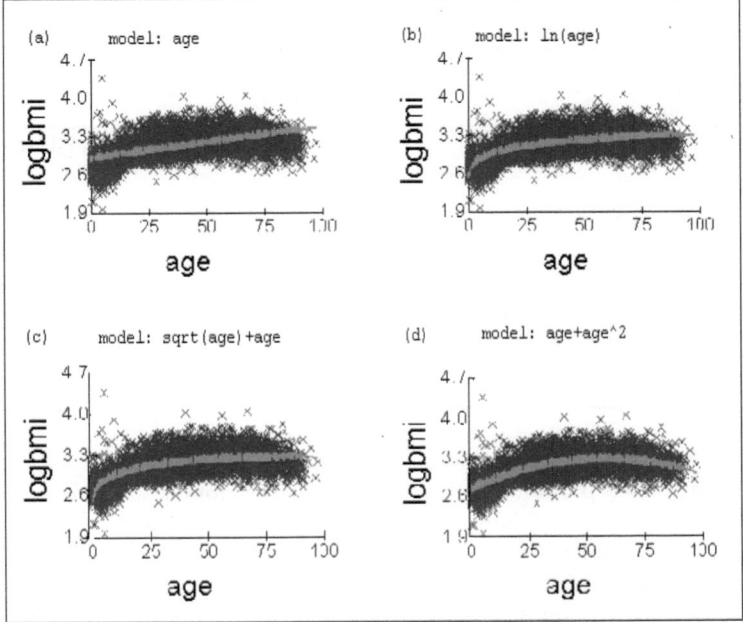

Figure 8.2 *Belgian Health Interview Survey. ln(BMI) versus age.*

Table 8.2 shows that a single effect of dose ($m = 1$) is unacceptable as compared to two effects simultaneously ($m = 2$). This table presents only a few of the models, but none of the other fractional polynomials provided an improvement in fit. The quadratic represented by (age, age^2) yields the smallest AIC. This can also be seen in Figures 8.2 (b)–(d).

In this example, none of the fractional polynomials give a better fit than the conventional quadratic polynomial. It is clear that when the fractional polynomial approach is not necessary, it reduces to a standard polynomial approach.

8.2 Local Polynomial Models

Although flexible, fractional polynomial models are fully parametric and their possible shapes are pre-defined. Fully data driven nonparametric approaches, where there is no need to postulate any assumptions on the curves' shapes, are another very useful modeling tool. In an explorative way, a parametric model can be graphically compared with its nonparametric alternative. In a further stage, a formal test statistic to examine the appropriateness of a cer-

Table 8.2 *Belgian Health Interview Survey. Selection procedure.*

m = 1		m = 2	
transformation	AIC	transformation	AIC
$1/\text{age}^2$	-2760.612	age, age^2	-7737.231
$1/\text{age}$	-3949.483	$\sqrt{\text{age}}, \text{age}$	-7090.243
$1/\sqrt{\text{age}}$	-5551.051	$\sqrt{\text{age}}, \text{age}^2$	-7439.551
$\ln(\text{age})$	-6841.176	$1/\text{age}, \ln(\text{age})$	-7161.566
$\ln(\text{age})^2$	-6823.584	$1/\text{age}^2, \ln(\text{age})$	-7212.649
$\sqrt{\text{age}}$	-6492.688	$\text{age}, \text{age} * \ln(\text{age})$	-7637.586
age	-5368.459	$1/\sqrt{\text{age}}, 1/\text{age}$	-7407.479
age^2	-3741.608	$1/\text{age}, 1/\text{age}^2$	-6272.256

tain hypothesized parametric function can be developed. This chapter focuses on estimation whereas Chapter 9 deals with such lack-of-fit tests.

In the one-parameter case, several authors have examined strategies to implement nonparametric estimation procedures in likelihood based regression models. Recently, many other applications have been introduced and studied. For relevant references on this subject, see, e.g., Fan and Gijbels (1996) and Simonoff (1996). We restrict attention to the local polynomial fitting which has become the standard in kernel smoothing. The corresponding smoothers are known to have several advantages in comparison with other linear smoothers, such as the behavior at the boundary.

8.2.1 Local Likelihood Estimation

Instead of assuming a particular functional form that specifies how the predictor x affects the distribution of the dependent variable Y, one can allow the data to describe this relationship nonparametrically, only requiring some weak smoothness assumptions.

For binomial data (no clustering), kernel estimates of the dose-response curve by locally averaging the sample proportions have been studied by Copas (1983), Staniswalis and Cooper (1988), and Müller and Schmitt (1988). These approaches essentially ignore the categorical nature of the response and do not incorporate clustering effects. Here, the multi-parameter models as introduced in Chapters 4 and 5 are used in a generalization of the concept of local likelihood estimation (introduced by Tibshirani and Hastie 1987). Without loss of generality, we restrict attention to two parameter probability models, where there are two curves of interest, $\theta_1(x)$ and $\theta_2(x)$, for example, representing a dose-response relation and a within cluster correlation, both as a function of a covariate x.

The idea behind local polynomial estimation, which is a pointwise estimation method, is the following. If a covariate value x_i is close to the value x

where we wish to estimate the curve θ_r, then by a Taylor series approximation, $\theta_r(x_i)$ should be well approximated by

$$\theta_r(x) + \theta_r'(x)(x_i - x) + \ldots + \frac{1}{p_r!}\theta_r^{(p_r)}(x)(x_i - x)^{p_r},$$

provided the curve θ_r is smooth enough to possess at least p_r continuous derivatives. Denoting $\theta_{rj}(x) = \theta_r^{(j)}(x)/j!$, $j = 0, \ldots, p_r$, $r = 1, 2$, local polynomial fitting provides estimators for $\boldsymbol{\theta}_r^T(x) = (\theta_{r0}(x), \ldots, \theta_{rp_r}(x))$, that is, the vector containing the values at x of the curve θ_r and its higher order derivatives up to order p_r ($r = 1, 2$). The local polynomial maximum likelihood estimator

$$(\hat{\boldsymbol{\beta}}_1^T, \hat{\boldsymbol{\beta}}_2^T) = (\hat{\beta}_{10}, \ldots, \hat{\beta}_{1p_1}, \hat{\beta}_{20}, \ldots, \hat{\beta}_{2p_2})$$

maximizes the kernel weighted log-likelihood function

$$\mathcal{L}_n(\boldsymbol{\beta}_1, \boldsymbol{\beta}_2) = \frac{1}{nh}\sum_{i=1}^{n}\ln f(Y_i; \sum_{j=0}^{p_1}\beta_{1j}(x_i - x)^j, \sum_{j=0}^{p_2}\beta_{2j}(x_i - x)^j),$$

$$\times K[(x_i - x)/h] \tag{8.1}$$

with respect to $(\boldsymbol{\beta}_1^T, \boldsymbol{\beta}_2^T) = (\beta_{10}, \ldots, \beta_{1p_1}, \beta_{20}, \ldots, \beta_{2p_2})$. The data are centered about x and each individual log-likelihood contribution is multiplied by a weight, governed by the kernel K and the bandwidth h. In this way, those observations x_i close to x have a larger impact on the maximization process. Of course, instead of the smoothing weights $w_{ni}(x) = (1/nh)K[(x_i - x)/h]$, other nonnegative weights satisfying $\lim_{n\to\infty}\sum_{i=1}^{n}w_{ni}(x) = 1$ could be chosen.

Taking a one-parameter family and $p_1 = 0$ (local constant) the aforementioned estimator has been studied by Staniswalis (1989), and Fan, Heckman and Wand (1995) discussed the local polynomial estimator for $f(y; \theta(x))$, a one-parameter exponential family member. Carroll, Ruppert and Welsh (1998) considered the general estimating equations setting.

Although the role of link functions here is less crucial than in parametric models (because the fitting is localized), link functions still can be very useful in this nonparametric setting. Above all, a properly defined link function guarantees that the final estimators $\hat{\theta}_1(x)$ and $\hat{\theta}_2(x)$ will have the correct range of admissible values. Moreover, the choice of the link function affects the computational aspects of the estimation procedure. For example, local polynomial estimates in the beta-binomial model are obtained by defining $\theta_1(x) = \text{logit}[\pi(x)]$, $\theta_2(x) = \ln[(1 + \rho(x))/(1 - \rho(x))]$, and by using the beta-binomial log-likelihood (in terms of $\theta_1(x)$ and $\theta_2(x)$) in the definition of the kernel weighted log-likelihood function (8.1). Since for the beta-binomial model explicit expressions for the maximizer of the function (8.1) are not available, a Newton-Raphson algorithm is used to obtain the estimates in the data examples below.

Assuming that the bandwidth $h \to 0$ in such a way that $nh \to \infty$ as $n \to \infty$, and also the existence of the necessary derivatives of $\boldsymbol{\theta}_r(x)$ and typical likelihood regularity conditions, Aerts and Claeskens (1998b) proved consistency of the local polynomial estimator $(\hat{\boldsymbol{\beta}}_1, \hat{\boldsymbol{\beta}}_2)$ as an estimator for $(\boldsymbol{\theta}_1(x), \boldsymbol{\theta}_2(x))$, derived its asymptotic bias and variance expressions, and proved joint asymptotic normality. Using the delta method, these results can be stated in terms of different scales (e.g., probability instead of logit scale). The main conclusion concerning the asymptotic bias expression is that the leading term is essentially of an order determined by the polynomial of the smallest degree. Hence, it is recommended to use polynomials of the same degree $p_1 = p_2 = p$, preferably odd. Indeed, for odd degree polynomial fits, boundary bias and interior bias are of the same order of magnitude. This automatic incorporation of boundary treatment is one of several nice properties to which local polynomial fitting owes its popularity (Fan and Gijbels 1996).

For $p_1 = p_2 = p$ odd, the asymptotic bias of $\hat{\theta}_\ell(x)$ is given by

$$h^{p+1}\left(\int z^{p+1} K_p(z)dz\right)\theta_\ell^{(p+1)}(x)/(p+1)!$$

and its asymptotic variance by

$$\operatorname{tr}\left(\boldsymbol{I}^{-1}(\theta_1(x), \theta_2(x))\right) \int K_p^2(z)dz/(nhf_X(x)).$$

Here, K_p is a modification of the kernel K (the so-called equivalent kernel), $f_X(x)$ refers to the design density, and $\boldsymbol{I}(\theta_1(x), \theta_2(x))$ is the positive definite Fisher information matrix. The bias expression states explicitly the smoothness condition, namely the existence of the derivative $\theta_\ell^{(p+1)}(x)$. Its asymptotic order $O(h^{p+1})$ further shows that bias decreases according to higher choices for p, the degree of the local polynomial. On the other hand, the order of the variance $O(1/(nh))$ does not depend on p, though a closer inspection of the constant $\int K_p^2(z)dz$ shows that the variance increases with p. A local linear approach has been shown to serve most purposes.

It is known that the choice of the kernel K is not very important for the performance of the resulting estimators, but the choice of the bandwidth is crucial. Indeed, the above expressions show that bias decreases while variance increases as h gets smaller. Minimizing the asymptotic mean squared error or equivalently balancing the squared bias and variance leads to the optimal local bandwidth

$$h_{opt}(x) = C(K, f_X)C(\theta_1(x), \theta_2(x))n^{-1/(2p+3)}$$

where the first constant $C(K, f_X)$ only depends on kernel K and design density $f_X(x)$ but the second contains several unknown quantities. Another more natural approach in likelihood estimation is to consider the bandwidth which maximizes $E[\ln f(Y; \hat{\theta}_1(x), \hat{\theta}_2(x))]$. This yields an optimal bandwidth of the same order $O(n^{-1/(2p+3)})$, but with a different constant $C(\theta_1(x), \theta_2(x))$ also incorporating the (asymptotic) covariance between both estimators $\hat{\theta}_1(x)$ and $\hat{\theta}_2(x)$. These are optimal local bandwidths. Taking a global loss measure such as the asymptotic mean integrated squared error, we get an optimal global,

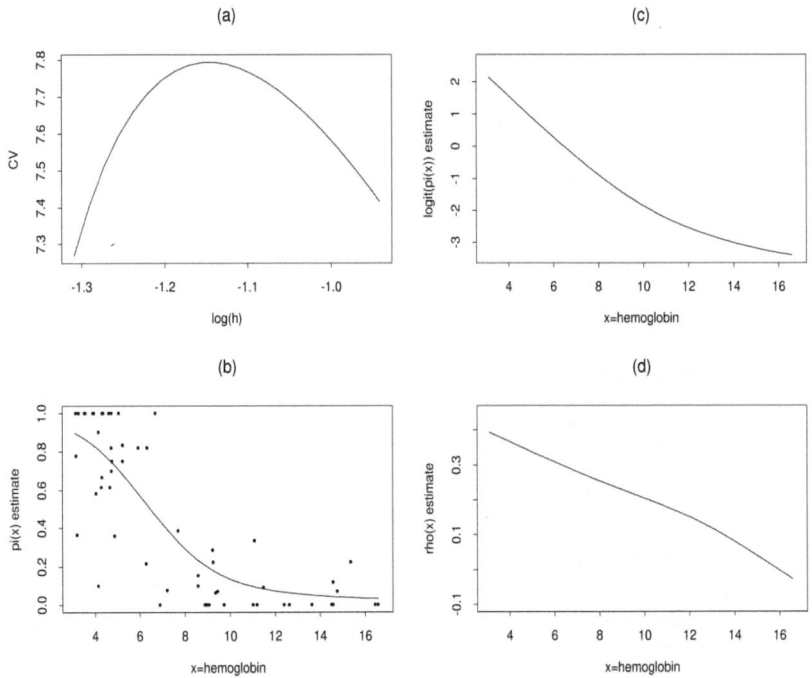

Figure 8.3 *Low-iron Rat Teratology Data. Local linear/linear estimates.*

fixed, bandwidth. Plug-in procedures could be implemented to get a data-driven bandwidth, but this gets very complicated.

An alternative approach is to define a cross-validation criterion

$$\hat{h}_{CV} = \text{argmax}_{h>0} \sum_{i=1}^{n} \ln f(Y_i; \hat{\theta}_1^{[i]}(x_i), \hat{\theta}_2^{[i]}(x_i)) \qquad (8.2)$$

where $\hat{\theta}_1^{[i]}(x_i)$ and $\hat{\theta}_2^{[i]}(x_i)$ are the estimators based on the sample without the ith observation (x_i, Y_i). This cross-validation method is straightforward and fully data-driven. Local linear estimators with data-driven bandwidth choice is applied to the low-iron rat teratology data, introduced in Section 2.5.

8.2.2 The Low-iron Rat Teratology Data

For a grid of hemoglobin levels and with $p_1 = p_2 = 1$ (i.e., local linear), we estimate the proportion of dead animals $\pi(x)$ and the correlation $\rho(x)$. As link functions we choose the logit and Fisher's z-transformations. After

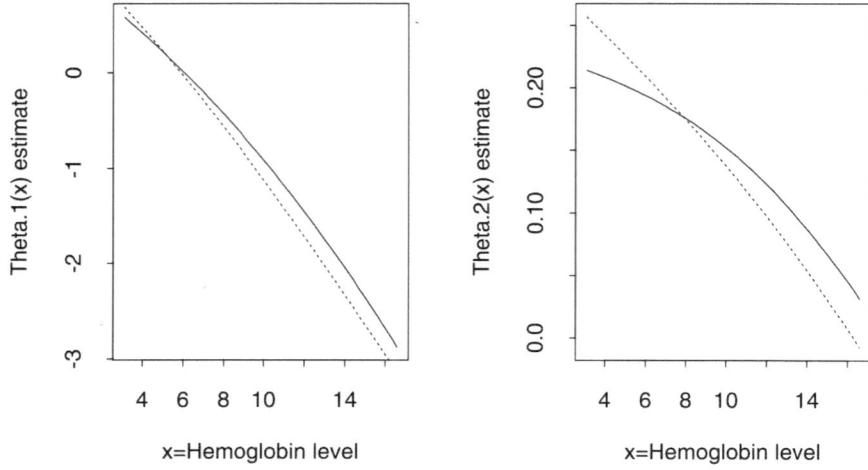

Figure 8.4 *Low-iron Rat Teratology Data. Local linear pseudo-likelihood estimates.*

rescaling the hemoglobin levels to the unit interval, the cross-validation (CV) procedure (8.2) resulted in a bandwidth $\hat{h}_{CV} = 0.319$ (see Figure 8.3 (a) for a plot of the CV-measure as a function of $\log(h)$). Figure 8.3 (b) shows the data together with the fitted $\pi(x)$-curve; the same curve on logit scale is shown in Figure 8.3 (c). The latter picture suggests a possible quadratic type of curvature. Figure 8.3 (d) shows the fitted $\rho(x)$-curve. Apparently, intra-litter correlation decreases linearly as hemoglobin level increases. Inspired by these suggestions, we fitted the data in a parametric way. The quadratic effect of hemoglobin on the proportion dead and the linear effect on the correlation appeared to be statistically significant (the likelihood ratio test producing $p = 0.029$). This illustrates a typical application of a nonparametric estimator as a tool to suggest specific functional relationships.

8.2.3 Local Pseudo-likelihood

Using the notation of Chapter 6 and restricting again to the two-parameter case, the local polynomial pseudo-likelihood estimator is defined, analogously to (8.1), as the maximizer, denoted $(\hat{\boldsymbol{\beta}}_1^T, \hat{\boldsymbol{\beta}}_2^T)$, of the locally weighted log pseudo-likelihood equations; see also equation (6.1),

$$\mathcal{PL}_n(\boldsymbol{\beta}_1, \boldsymbol{\beta}_1) = \frac{1}{nh} \sum_{s \in S} \delta_s \sum_{i=1}^{n} \log f_s(\boldsymbol{Y}_i^{(s)}; \sum_{j=0}^{p_1} \beta_{1j}(x_i - x)^j,$$

$$\sum_{j=0}^{p_2} \beta_{2j}(x_i - x)^j) K[(x_i - x)/h]. \qquad (8.3)$$

Claeskens and Aerts (2000a) prove consistency and asymptotic normality for these estimators. Taking $p_1 = p_2 = p$, the asymptotic bias is identical to that mentioned in Section 8.2.1. In the asymptotic variance expression the Fisher information matrix $I(\theta_1(x), \theta_2(x))$ has to be replaced by the product $J^{-1}KJ^{-1}(\theta_1(x), \theta_2(x))$, where $J(\theta_1(x), \theta_2(x))$ is the matrix with components

$$J_{rr'}(\theta_1(x), \theta_2(x)) =$$

$$\sum_{s \in S} \delta_s E[-\frac{\partial^2}{\partial \theta_r \partial \theta_{r'}} \ln f_s(Y^{(s)}; \theta_1(x), \theta_2(x))], \qquad (8.4)$$

$(r, r' = 1, 2)$, and $K(\theta_1(x), \theta_2(x))$ is the matrix with components

$$K_{rr'}(\theta_1(x), \theta_2(x)) = \sum_{s,t \in S} \delta_s \delta_t E[\frac{\partial}{\partial \delta_s} \ln f_s(Y^{(s)}; \theta_1, \theta_2) \frac{\partial}{\partial \delta_t} \ln f_t(Y^{(t)}; \theta_1, \theta_2)],$$

$(r, r' = 1, 2)$. See also equations (6.2) and (6.3) in the parametric pseudo-likelihood estimation framework. Figure 8.4 shows the local linear estimates of the two-parameter Molenberghs and Ryan (1999) exponential family likelihood model (solid lines) and of the full conditional pseudo-likelihood model (dashed lines). Both estimation approaches result into curves which are in close agreement.

Irrespective of the estimation method chosen, full likelihood or pseudo-likelihood, the asymptotic bias and variance expressions depend in a complicated way on several unknown quantities. Claeskens and Aerts (2000a) discuss several bias and variance estimators. Furthermore, they study a one-step bootstrap method allowing the construction of simultaneous confidence regions. This approach is illustrated in the next section and is a local version of the bootstrap method discussed in more detail in Chapter 11.

8.2.4 The Wisconsin Diabetes Study

Based on the beta-binomial likelihood, Figure 8.5 shows local estimates for the probability of macular edema and the intra-person correlation, together with simultaneous and pointwise 90% confidence intervals (using 1000 bootstrap simulation runs). For this type of application, the correlation structure is often assumed to be constant. In an explorative way, Figure 8.5 already indicates that this assumption might be violated. In Section 9.2.6, we will formally test this assumption using an omnibus lack-of-fit test.

8.3 Other Flexible Polynomial Methods and Extensions

Next to fractional and local polynomial models, there are other interesting approaches using the polynomials as building blocks. One such smoothing technique is the orthogonal series method using orthogonal polynomials as basis functions. In the next chapter this method will be used to construct lack-of-fit tests. Also, smoothing splines or penalized regression splines use

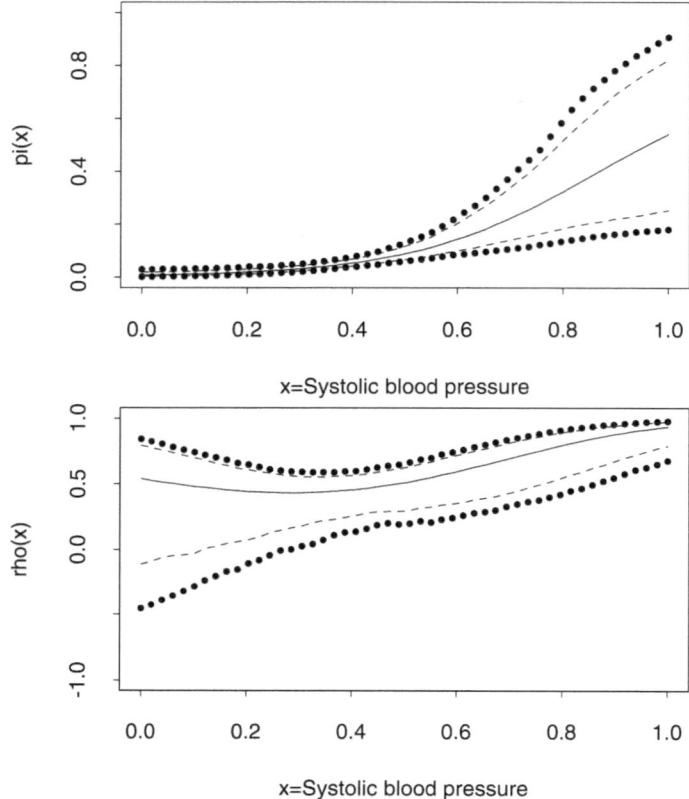

Figure 8.5 *Wisconsin Diabetes Data. Simultaneous and pointwise 90% confidence intervals for the probability of macular edema (left panel) and for the intra-person correlation (right panel), based on the one-step linear bootstrap.*

polynomials and truncated polynomials as basis functions. In case of multiple predictors, these methods can be easily extended but they are all confronted with the curse of dimensionality. One way out is to consider additive models. Some relevant references for clustered data are Wild and Yee (1996), Claeskens and Aerts (2000b), Aerts, Claeskens and Wand (2002), and Gao *et al.* (2001). All of the above mentioned methods are essentially linear. Of course, non-linear models form another interesting family of models. This approach will not be discussed here.

Another important issue in model building is misspecification. Also, here, flexible models can play an important role. A misspecified probability model can be partly corrected by flexible regression models for the parameters involved. More about this topic will be explained in Chapter 11.

CHAPTER 9

Assessing the Fit of a Model

Gerda Claeskens

Texas A & M University, College Station, TX

Marc Aerts

transnationale Universiteit Limburg, Diepenbeek–Hasselt, Belgium

To evaluate how effective a model is in describing the outcome variable, we need to assess the quality of its fit. The lack-of-fit of a regression model is investigated by testing the hypothesis that a function has a prescribed parametric form. The function of interest is typically the mean of the response, but it might also be its variance, or the correlation between different outcomes. In other cases, it might be a complete density function of which we want to investigate the goodness-of-fit. The parametric testing methods of Chapter 7 are designed to detect very specific types of departures from the hypothesized model. For example, likelihood ratio, Wald or score tests are employed to contrast a constant and a linear dose-response curve. While very powerful for this particular class of alternative models, these tests quickly lose power when the truth is more complicated. The omnibus nonparametric methods of this chapter are appealing in that they are consistent against virtually any departure from the hypothesized parametric model.

Section 9.1 describes an adaptation of the Hosmer-Lemeshow (1989) approach, for application to clustered binary data. In the remaining sections of this chapter, we discuss order selection tests based on orthogonal series estimators. In particular, Section 9.2 defines the different testing strategies for simple regression, while Section 9.3 gives extensions to multiple regression and to somewhat more specific alternative models. Section 9.4 applies these ideas to test whether density functions belong to a specific class.

9.1 A Hosmer-Lemeshow Approach for Likelihood Based Models

Several methods for assessing the goodness-of-fit of binary logistic regression models are based on the notion of partitioning the covariate space into groups or regions. Tsiatis (1980) proposed a lack-of-fit statistic for a given partition of the covariate space, but he did not provide a method for partitioning the covariate space into suitable regions. Hosmer and Lemeshow (1989) proposed the partition of subjects into groups or regions on the basis of percentiles of the predicted probabilities from the fitted logistic regression model. To construct a lack-of-fit measure for clustered binary data, an adaptation of the methods proposed by the above authors is necessary. We illustrate this in the context of the heatshock study. First, groups are constructed according to deciles of the predicted malformation probabilities in each temperature-duration combination. Given this partition, the lack-of-fit statistic is formulated by defining $G - 1$ group indicators (in our example, $G = 10$):

$$
I_{ik}^g = \begin{cases} 1 & \text{if } \hat{\pi}_{ik} \text{ is in group } g \quad (g = 1, \ldots, G - 1), \\ 0 & \text{otherwise,} \end{cases}
$$

where $\hat{\pi}_{ik}$ is the estimated malformation probability of the kth individual within the ith cluster, calculated from the model that takes into account the clustering between the individuals. For example, in the context of the heatshock studies, the following model is considered:

$$
\ln \left(\frac{\pi_{ik}}{1 - \pi_{ik}} \right) = \beta_0 + \beta_{t^*} t_{ik}^* + \beta_{dt} dt_{ik} + \sum_{g=1}^{G-1} I_{ik}^g \gamma_g.
$$

The association is modeled similarly as in the model for which the lack-of-fit is assessed. One possible choice would be an exchangeable correlation structure.

If the mean structure in the original model is correctly specified, then all $\gamma_g = 0$. Moore and Spruill (1975) note that, even though II_{ik}^g is based on random quantities $\hat{\pi}_{ik}$, the partition can be treated asymptotically as if it were based on the true π_{ik}. To test the lack-of-fit of the model, one can use either a likelihood ratio, Wald or score statistic to test $H_0 : \gamma_1 = \cdots = \gamma_{G-1} = 0$. For large samples, each of these statistics has approximately a χ^2 distribution with $G - 1$ degrees of freedom, if the model under the null hypothesis is correctly specified. We suggest the use of the likelihood ratio statistic, since it is simple to calculate and is fairly powerful. For large samples, all estimated expected frequencies should typically be greater than 1 and at least 80 percent should be greater than 5. Otherwise, one can collapse some frequencies, reducing the number of groups G (Lipsitz, Fitzmaurice and Molenberghs 1996). Hosmer and Lemeshow (1989) noted that $G = 6$ should be a minimum, since a test statistic calculated from fewer than 6 groups will usually have low power and thus indicates that the model fits well. Note that in the lack-of-fit assessment described above, correlation is essentially treated as a nuisance parameter and interest is focused on the relationship between the covariates and the probability of response. Hosmer, Hosmer, Lemeshow and le Cessie (1997) have shown that there may be disadvantages in the use of lack-of-fit tests based on the ones proposed by Hosmer and Lemeshow. Decisions on model fit may depend more on choice of cut points than on lack-of-fit and their test statistic may have relatively low power with small sample sizes. This goodness-of-fit method will be illustrated on the heatshock data in detail in Section 13.4. Finally, we mention that the method as proposed above has been studied for GEE for repeated binary outcomes by Horton *et al.* (1999).

9.2 Order Selection Tests

The basic idea of the construction of omnibus lack-of-fit tests based on orthogonal series expansions comes from Eubank and Hart (1992) who proposed the order selection test for checking the fit of a linear model in fixed-design ordinary regression. The tests in the current section, as proposed by Aerts, Claeskens and Hart (1999), are applicable to a much wider range of statistical problems. Robust versions of the tests have an asymptotic significance level which is correct even when the likelihood is misspecified, or when quasi-likelihood, pseudo-likelihood, or generalized estimating equations methods are being used.

Suppose we have independent observations $(x_1, \boldsymbol{Y}_1), \ldots, (x_n, \boldsymbol{Y}_n)$. Here, the $\boldsymbol{Y}_i$ might represent the vector of all outcomes of the ith cluster. The covariate values are assumed to be fixed, which could correspond either to a designed experiment or to conditioning on the values of a random covariate. The density (or probability mass function) of each $\boldsymbol{Y}_i$ is known up to the function of interest $\gamma(\cdot)$ and some k-dimensional nuisance parameter $\boldsymbol{\eta}$ ($k < \infty$). We wish to test a null hypothesis about the functional form of γ,

$$H_0 : \gamma(\cdot) \in \{\gamma(\,\cdot\,; \boldsymbol{\theta}_p) : \boldsymbol{\theta}_p = (\theta_1, \ldots, \theta_p) \in \Theta\}, \tag{9.1}$$

where Θ is a subset of a p-dimensional Euclidean space. One example would be to test for linearity of a marginal parameter. In this case $\gamma(x; \boldsymbol{\theta}_2) = \theta_1 + \theta_2 x$, where $\boldsymbol{\theta}_2 = (\theta_1, \theta_2)$ is left unspecified.

Our interest is in tests that are sensitive to essentially *any* departure from H_0. As alternative models for $\gamma(\cdot)$ we consider sequences of approximators with the property that the parametric family under the null hypothesis is nested within the alternative models, which in turn form a sequence of nested models having more and more parameters as r increases. Furthermore, we request that the approximators come closer and closer to spanning the space of all functions of interest as $r \to \infty$. For example,

$$\gamma(\cdot; \theta_1, \ldots, \theta_p) + \sum_{j=1}^{r} \theta_{p+j} u_j(\cdot),$$

where $\{u_1(\cdot), u_2(\cdot), \ldots\}$ is complete for the class of functions that are continuous on the range of the design points. If x is real-valued, possibilities for the u_j's are orthonormalized Legendre polynomials, a cosine system where $u_j(x) = \cos(Ajx)$, or linear combinations of polynomials and/or trigonometric functions that are orthogonal in some sense.

It is implicit here that the added functions in the alternative models do not repeat terms already in the null model. If this were the case, we could simply discard those u_j's from $\sum_{j=1}^{r} \theta_{p+j} u_j(x)$. For example, suppose we wish to test the hypothesis that $\gamma(x)$ has the form $\theta_1 + \theta_2 x$ and we want to use polynomial alternatives. Then we could take $u_j(x) = x^{j+1}$, $j = 1, 2, \ldots$, since it is pointless to include 1 or x in our set of alternative models.

9.2.1 A Likelihood-based Test

The basic idea of the tests is to use a data-driven method of selecting a model for $\gamma(\cdot)$, and to reject the null hypothesis (9.1) if the selected model contains more than p parameters, with p being the number of parameters in H_0. In a likelihood context, a popular method of model selection is AIC, Akaike's (1974) information criterion. Denote by $\mathcal{L}_r$ the likelihood at its maximum value in the model with r terms in addition to the p terms already in the null model, i.e., employing the approximator $\gamma(\cdot; \theta_1, \ldots, \theta_{p+r})$. $\mathcal{L}_0$ is the value corresponding to the null model. The AIC function is the penalized likelihood

$$AIC(r) = \mathcal{L}_r - (k + p + r), \quad r = 0, 1, \ldots,$$

in which k is the number of nuisance parameters. For future reference we note that the maximizer of $AIC(r)$ is equal to the maximizer of $2(\mathcal{L}_r - \mathcal{L}_0) - 2r$, and to the minimizer of another frequently used form of the criterion: $-2\mathcal{L}_r + 2r$. An estimate of $\gamma(\cdot)$ may be obtained by choosing r to maximize $AIC(r)$ over some set of the form $\{0, 1, \ldots, r_n\}$, where r_n could either be fixed or tending to infinity with n.

A possible test of H_0 against a general alternative is to reject H_0 if $\hat{r}$, the order selected by AIC, is larger than 0. Under certain regularity conditions (Aerts, Claeskens and Hart 1999), the limiting level of this test (as $n \to \infty$) is about 0.29. By most standards, a type I error probability of 0.29 is quite high. To obtain control of the test level, Aerts, Claeskens and Hart (1999) propose a modification of AIC that parallels a proposal in Eubank and Hart (1992). Define the likelihood information criterion, LIC, by

$$LIC(r; C_n) = 2(\mathcal{L}_r - \mathcal{L}_0) - C_n r, \quad r = 0, 1, \ldots, \tag{9.2}$$

where C_n is some constant larger than 1, and let $\hat{r}_{C_n}$ be the maximizer of $LIC(r; C_n)$. By appropriate choice of C_n, the asymptotic type I error probability of the test

$$\text{"reject } H_0 \text{ when } \hat{r}_{C_n} > 0\text{"} \tag{9.3}$$

can be any number between 0 and 1. For example, a test of asymptotic level 0.05 is obtained by using $C_n = 4.18$. See Hart (1997, p. 178) for values of C_n leading to other test levels. Taking C_n equal to 2 and $\log n$ in (9.2) yields the well-known AIC and the Bayesian information criterion BIC (Schwarz 1978), respectively.

The test described above in (9.3) rejects H_0 if and only if $LIC(r; C_n)$ is larger than 0 for some r in $\{1, \ldots, r_n\}$, which is equivalent to rejecting H_0 when $T_n \geq C_n$, with

$$T_n = \max_{1 \leq r \leq r_n} \frac{2(\mathcal{L}_r - \mathcal{L}_0)}{r}.$$

Hence, in addition to playing the role of penalty constant, C_n is a critical value of the statistic T_n. The test based on T_n has a nice interpretation in terms of likelihood ratio statistics. Note that $\mathcal{L}_r - \mathcal{L}_0$ is the log of the likelihood ratio that is used to test hypothesis (9.1) against the alternative with r additional parameters. Since our test of H_0 is omnibus, T_n is not a single likelihood ratio but rather the largest of a set of *weighted* log-likelihood ratio statistics. The largest weights are placed on the models with the fewest parameters. This has a similar effect to using a prior distribution that places higher probability on alternatives with fewer parameters.

Using this version of the test one may approximate the p value corresponding to an observed $T_n = t$ by using either the bootstrap or a large-sample distribution, where we approximate the p value by

$$p \text{ value} \approx 1 - \exp\left[-\sum_{j=1}^{M} \frac{P(\chi_j^2 > jt)}{j}\right],$$

for some large number M, say 80 (Hart 1997, p. 175).

9.2.2 A Score-based Test

Another method of choosing r (Aerts, Claeskens and Hart 1999) uses the score-based information criterion,

$$SIC(r; C_n) = \mathcal{S}_r - C_n r, \quad r = 0, 1, \ldots, \tag{9.4}$$

where $\mathcal{S}_0$ is defined to be 0, and the score statistic

$$\mathcal{S}_r = \boldsymbol{U}_r(\hat{\boldsymbol{\delta}}_{r0})^T \left\{J_{nr}(\hat{\boldsymbol{\delta}}_{r0})\right\}^{-1} \boldsymbol{U}_r(\hat{\boldsymbol{\delta}}_{r0}),$$

for $r = 1, 2, \ldots, r_n$, $\hat{\boldsymbol{\delta}}_{r0} = (\hat{\boldsymbol{\eta}}_0, \hat{\boldsymbol{\theta}}_{p0}, \boldsymbol{0}_r)$ and $(\hat{\boldsymbol{\eta}}_0, \hat{\boldsymbol{\theta}}_{p0})$ is the null estimate, i.e., the estimate of $(\boldsymbol{\eta}, \theta_1, \ldots, \theta_p)$ under the model specified by the null hypothesis. The vector $\boldsymbol{U}_r(\cdot)$ in $\mathcal{S}_r$ is the score vector, that is, the vector of first partial derivatives with respect to the parameter $\boldsymbol{\theta}_{p+r}$. The matrix $\boldsymbol{J}_{nr}(\cdot)$ is the observed Fisher information evaluated at the null parameter estimates. In some cases it is possible to obtain explicit expressions for the expected Fisher information, in which case one could replace $\boldsymbol{J}_{nr}(\cdot)$ by the expected information evaluated at the null estimates. There is no general consensus in the literature as to which of these two approaches is better (Efron and Hinkley 1978 and Boos 1992). Asymptotically, the two versions of $\mathcal{S}_r$ are generally equivalent to first order. An appealing aspect of using expected information is that it often leads to simpler and more readily interpretable expressions for $\mathcal{S}_r$.

The test rejects H_0 when the maximizer, $\tilde{r}_{C_n}$, of $SIC(r, C_n)$ is larger than 0. This test is equivalent to one that rejects H_0 for $\tilde{T}_n \geq C_n$, where

$$\tilde{T}_n = \max_{1 \leq r \leq r_n} \frac{\mathcal{S}_r}{r},$$

and C_n is some constant larger than 1. We may choose r to maximize criterion $SIC(r; C_n)$ over $r = 0, 1, \ldots, r_n$, with the upper bound r_n either fixed or tending to infinity with n. When the null hypothesis is true, the difference between $SIC(r; C_n)$ and $LIC(r; C_n)$ is negligible. For this reason, we shall refer to $SIC(r; 2)$ and $SIC(r; \log n)$ as score analogs of AIC and BIC.

The Wald and score test statistics are two computationally attractive and quadratic approximations of the likelihood ratio statistic. Either one could be used instead of $2(\mathcal{L}_r - \mathcal{L}_0)$. The Wald statistic (Boos 1992) needs the "unrestricted" maximum likelihood estimators, while the score statistic only requires fitting the null model. The last property is appealing in our setting where one considers a large number of alternative models. A parallel development is possible for Wald statistics, which are, however, known not to be invariant to equivalent reparameterizations of nonlinear restrictions (Phillips and Park 1988).

9.2.3 A Robust Score-based Test

Parameter estimators can be obtained by solving more general estimating equations than likelihood equations. If the parameter of interest is the (conditional) mean of the response variables, the idea of solving a set of score equations $U_r(\cdot) = 0$ in likelihood models is generalized to the construction of quasi-likelihood equations (Wedderburn 1974), and to the multivariate version, the generalized estimating equations (GEE) (Liang and Zeger 1986, for details see Chapter 5). Other frequently used estimating equations are found in the context of M-estimation (Huber 1981), resulting in robust regression models.

In the absence of a likelihood function, it is clear that an LIC-based test can no longer be constructed. Instead, a robust information criterion RIC may be defined. The basic idea of the lack-of-fit test construction is the following. Specify a nested sequence of growing models, encompassing the null model ($r = 0$). In each model, calculate the robust score statistic $\mathcal{R}_r$. We define the robust score-based information criterion (where $\mathcal{R}_0 = 0$):

$$RIC(r; C_n) = \mathcal{R}_r - C_n r, \quad r = 0, 1, \ldots, r_n.$$

We may then reject H_0 whenever the maximizer of $RIC(r; C_n)$ exceeds 0. This test is equivalent to one that rejects H_0 whenever the largest of a set of weighted robust score statistics is larger than C_n:

$$\bar{T}_n = \max_{1 \leq r \leq r_n} \frac{\mathcal{R}_r}{r} \geq C_n.$$

Under H_0 and appropriate regularity conditions, the statistic $\bar{T}_n$ will have the same limit distribution as that of T_n and $\tilde{T}_n$, based on LIC and SIC respectively. Importantly, this result does not require a correct specification of a likelihood.

More specifically, if the null model to be tested contains unknown parameters $\theta_1, \ldots, \theta_p$ and possibly a k-dimensional nuisance parameter $\boldsymbol{\eta}$, for example, consisting of components of a covariance matrix, then the estimating equations for the r-th alternative model in our model sequence is given by

$$\sum_{i=1}^{n} \boldsymbol{\psi}_r(\boldsymbol{Y}_i; \boldsymbol{\eta}, \theta_1, \ldots, \theta_{p+r}) = \boldsymbol{0}_{p+k+r},$$

where $\boldsymbol{\psi}_r$ is of dimension $p + k + r$. Let $\widehat{\boldsymbol{\delta}}_{r0} = (\widehat{\boldsymbol{\eta}}_0, \widehat{\boldsymbol{\theta}}_{p0}, \boldsymbol{0}_r)$ be the null model estimators, extended with a zero vector of length r. Define $\boldsymbol{\xi}_r$ to be the length $p + k + r$ vector equal to the estimating equations evaluated at the null estimators: $\sum_{i=1}^{n} \boldsymbol{\psi}_r(\boldsymbol{Y}_i; \widehat{\boldsymbol{\eta}}_0, \widehat{\boldsymbol{\theta}}_0, \boldsymbol{0}_r)$. The robustified score statistic, using the "sandwich" covariance estimator, is defined as

$$\mathcal{R}_r = (\boldsymbol{\xi}_r)_r^T \left\{ \left(\tilde{\boldsymbol{J}}_{nr}^{-1}(\widehat{\boldsymbol{\delta}}_{r0}) \right)_r \left[\left(\tilde{\boldsymbol{J}}_{nr}^{-1}(\widehat{\boldsymbol{\delta}}_{r0}) \tilde{\boldsymbol{K}}_{nr}(\widehat{\boldsymbol{\delta}}_{r0}) \tilde{\boldsymbol{J}}_{nr}^{-1}(\widehat{\boldsymbol{\delta}}_{r0}) \right)_r \right]^{-1} \right.$$

$$\left. \times \left(\tilde{\boldsymbol{J}}_{nr}^{-1}(\widehat{\boldsymbol{\delta}}_{r0}) \right)_r \right\} (\boldsymbol{\xi}_r)_r,$$

for $r = 1, 2, \ldots$, where $\tilde{\boldsymbol{J}}_{nr}(\cdot)$ is a $(p + k + r) \times (p + k + r)$ matrix of partial derivatives of $\boldsymbol{\psi}_r$ and

$$\tilde{\boldsymbol{K}}_{nr}(\widehat{\boldsymbol{\delta}}_{r0}) = \sum_{i=1}^{n} \boldsymbol{\psi}_r(\boldsymbol{Y}_i; \widehat{\boldsymbol{\eta}}_0, \widehat{\boldsymbol{\theta}}_{p0}, \boldsymbol{0}_r) \boldsymbol{\psi}_r(\boldsymbol{Y}_i; \widehat{\boldsymbol{\eta}}_0, \widehat{\boldsymbol{\theta}}_{p0}, \boldsymbol{0}_r)^T.$$

Further, $(\cdot)_r$ denotes the right $r \times r$ submatrix of a $(k + p + r)$-dimensional matrix or the last r components of a length $(k + p + r)$-vector. Also, when a likelihood model is constructed, but there is uncertainty about the correctness of this model, we advise using the robustified score test, since this test remains valid under model misspecification.

The score-based tests are advantageous in at least two ways. They are computationally simple, since the score criterion has the advantage of requiring MLEs of model parameters only under the null hypothesis and they can easily be robustified to protect against model misspecification, which lead to the definition of RIC.

9.2.4 The Low-iron Rat Teratology Data

Let $\pi(x)$ denote the expected proportion of dead foetuses for female rats whose hemoglobin level is x, and suppose we wish to test the following null hypothesis:

$$H_0: \text{logit}(\pi(x)) = \theta_1 + \theta_2 x, \qquad \text{for each } x. \tag{9.5}$$

We will employ the beta-binomial model which allows for correlation within a cluster. Test statistics for hypothesis (9.5) are calculated using both polynomial and cosine alternatives. First, we transform the design points to the interval $(0, 1)$. Then, we consider $u_j(x) = x^{j+1}$ and $u_j(x) = \cos(\pi j x)$, $j = 1, 2, \ldots$. In the beta-binomial model the correlation was modeled as a straight line function of x, a model that is, referring to the results of Section 8.2.2, found

Table 9.1 *Twins Data. Omnibus lack-of-fit test statistics.*

	Polynomial basis			Cosine basis		
Criterion	Value	p value	r_n	Value	p value	r_n
SIC	0.471	0.981	15	0.799	0.948	15
RIC	1.693	0.399	15	2.993	0.120	15

to be reasonable. An upper bound of $r_n = 14$ was used in each statistic, since there were computational problems when fitting models with more than 16 variables. The RIC criterion results in an observed value of 1.93 for $\bar{T}_n$ when using a cosine basis for the alternative models. The corresponding p value equals 0.31. If we use a polynomial basis instead, the observed value is 3.18 with a p value of 0.1. Neither the cosine nor the polynomial version of the robust score statistic exceeds 4.18, the asymptotic critical value for a 0.05 level test, and hence the data appear to be consistent with the hypothesis that $\pi(x)$ has the form $1/[1 + \exp\{-(\theta_1 + \theta_2 x)\}]$.

9.2.5 The POPS Twins Data

For the set of twins in the POPS study, we will test the null hypothesis of linearity in gestational age (in weeks) for the logit of the probability of mortality and morbidity within 28 days after birth,

$$H_0 : \mathrm{logit}(\pi(x)) = \theta_1 + \theta_2 x, \qquad \text{for each } x.$$

Because of the size of this data set (there are 113 twins) we model the correlation parameter, which is the nuisance parameter in a beta-binomial model, as a constant function of gestational age.

We used both a polynomial and a cosine basis to extend the null model. Observed values of test statistics and approximate p values based on the asymptotic distribution theory are given in Table 9.1. There are no results for the LIC based tests, because of convergence problems. This demonstrates clearly the advantage of using score-based tests instead of likelihood ratio-based tests. "Naive" as well as robust score-based tests do not give evidence of a more complicated structure. Based on these data, and for a level of significance smaller than 11.9%, we cannot reject the null hypothesis of linearity for the probability of mortality and morbidity within 28 days after birth.

9.2.6 Tests for Correlation in the Wisconsin Diabetes Study

We apply our proposed tests to the data set concerning the effect of the systolic blood pressure on the occurrence of macular edema at the eyes of younger onset diabetic persons. Let y_i denote the number of eyes infected (0, 1, or 2), x_i the systolic blood pressure, $\pi(x)$ the expected proportion of

Table 9.2 *Wisconsin Diabetes Data. Omnibus lack-of-fit test statistics.*

$\pi(x)$	Criterion	Value	P-value	r_n
Beta-binomial likelihood				
Linear	SIC	2.096	0.262	15
	RIC	4.350	0.045	15
	LIC	3.522	0.080	4
Quadratic	SIC	2.105	0.260	15
	RIC	4.045	0.055	15
	LIC	3.405	0.087	4
GEE2 (Bahadur)				
Linear	RIC	4.632	0.037	15
Quadratic	RIC	4.381	0.044	15

infected eyes for a diabetic person whose systolic blood pressure is x and let $\rho(x)$ denote the expected correlation between the outcomes of the left and right eye. We wish to test whether the correlation changes with systolic blood pressure, in which case the null hypothesis is

$$H_0 : \rho(x) = \theta, \qquad \text{for each } x. \qquad (9.6)$$

We will consider two models for these data, both accounting for the intra-person correlation. The first model is the full likelihood beta-binomial model (see Section 4.3.1). The second model uses generalized estimating equations (GEE2, Zhao and Prentice 1990; see also Chapter 5) based on the first four moments of the Bahadur (1961) model. Test statistics for hypothesis (9.6) are calculated using polynomial alternatives; their values are given in Table 9.2. Note that for the GEE2 model, only the RIC criterion yields relevant test statistics and that the LIC criterion is not defined in this context. Test statistics were calculated with r_n equal to 15. For the likelihood methods, however, there were convergence problems with r_n values bigger than 4. The parameters corresponding to $\pi(x)$ are, for present purposes, nuisance parameters. It turns out that for these data the test results are very similar for a linear or a quadratic form of $\pi(x)$. None of the "naive" score-based tests is able to reject the null hypothesis. In the generalized estimating equations model the p values are slightly smaller than in the full likelihood model, but in both cases, the RIC yields p values that are not larger than 5.5%. For these data, we might consider modeling the correlation as a nonconstant function of systolic blood pressure.

9.2.7 Some Other Test Statistics

Aerts, Claeskens and Hart (2000) propose and compare some other nonparametric tests of H_0 that all have two features in common: they are functions of score or likelihood statistics based on different model dimensions, and the model dimension is chosen by a data-based rule.

Define the statistics $\hat{r}_a$ and $\hat{r}_b$ as the maximizers of the SIC criteria,

$$\hat{r}_a = \text{argmax}_{0 \leq r \leq r_n} SIC(r; 2) \qquad \text{and} \qquad \hat{r}_b = \text{argmax}_{1 \leq r \leq r_n} SIC(r; \log n).$$

The first two test statistics are simply $\mathcal{S}_a \equiv \mathcal{S}_{\hat{r}_a}$ and $\mathcal{S}_b \equiv \mathcal{S}_{\hat{r}_b}$, i.e., the score statistic with number of alternative parameters chosen by the score analogs of AIC and BIC, respectively. Note that $\hat{r}_b$ maximizes $SIC(r; \log n)$ over $1, \ldots, r_n$ rather than $0, 1, \ldots, r_n$. This definition is used due to a consistency property of BIC-type order selection criteria. When $r = 0$ is included, the statistic $\hat{r}_b$ is a consistent estimator of 0 under the null hypothesis. This means that the limit distribution of $\mathcal{S}_{\hat{r}_b}$ is degenerate at 0, making the asymptotic level of a nonrandomized test either 0 or 1. Maximizing SIC over the set $\{1, \ldots, r_n\}$ (as suggested by Ledwina 1994) allows one to perform a nonrandomized (large sample) test of any level. Standardizing $\mathcal{S}_{\hat{r}_a}$ leads to the statistic (9.7)

$$T_a = \frac{\mathcal{S}_{\hat{r}_a} - \hat{r}_a}{\max(1, \sqrt{\hat{r}_a})}. \tag{9.7}$$

Under general conditions, one can show that

$$\frac{\mathcal{S}_{s_n} - s_n}{\sqrt{2 s_n}} \tag{9.8}$$

converges in distribution to a standard normal random variable as $n \to \infty$ and $s_n \to \infty$ at a sufficiently slow rate. It thus seems natural to use a test statistic of the form (9.8), and statistic (9.7) is essentially (9.8) with a data-driven choice for s_n. It turns out that standardizing $\mathcal{S}_{\hat{r}_a}$ as above greatly stabilizes the null distribution of the statistic, which has a decided effect on the power for (9.7). One could similarly standardize $\mathcal{S}_{\hat{r}_b}$, but the null distribution of this statistic is already quite stable and the standardization has a negligible effect.

Yet another test statistic is

$$T_{max} = SIC(\hat{r}_a; 2), \tag{9.9}$$

that is, the AIC-type score criterion evaluated at its maximum.

Aerts, Claeskens and Hart (2000) compare all of these statistics, including the previously defined order selection test $T_{os} \equiv \tilde{T}_n$. Their conclusion in experiments under alternative models with added higher frequency terms (for example, a high order polynomial) is that the order selection test $\tilde{T}_n$ is very good at the lowest frequency, but going to higher frequencies, its power decreases rapidly. The BIC-based test $\mathcal{S}_{\hat{r}_b}$ shows very similar behavior. It enjoys good power at low frequency alternatives, but since it has a large penalty for models with many parameters, it has almost no power at high frequencies. At low frequency alternatives, both T_a and $T_{\max}$ improve upon the unstandard-ized $S_{\hat{r}_a}$ test, but when the frequency of the alternative models exceeds the null models' frequency by at least 4, $S_{\hat{r}_a}$ has the largest power. Since $T_{\max}$ and T_a are both doing very well for the low frequency cases and still have reasonable power at high frequencies, we recommend one of these two tests. If one suspects a higher frequency alternative, $S_{\hat{r}_a}$ is the best choice. For low frequencies, the order selection test $\tilde{T}_n$ is the best one in this comparison. For combinations of low and high frequency terms, the low frequencies will be dominant.

9.3 Data-driven Tests in Multiple Regression

In this section, we consider the score-based tests in the setting of multiple re-gression. In doing so, we address the issue of choosing an appropriate sequence of nested alternative models. This issue is relatively trivial when testing the fit of a function of one variable, but increases quickly in complexity when the number of variables increases.

9.3.1 Omnibus Tests in Models With Two Covariates

To illustrate how the methods of Section 9.2 may be generalized to multiple regression, we consider the relatively simple case of two covariates. Let γ be an unknown function of the covariates x_1 and x_2. We wish to test the null hypothesis

$$H_0 : \gamma \in \{\gamma(\,\cdot\,,\,\cdot\,;\boldsymbol{\theta}) : \boldsymbol{\theta} \in \Theta\}. \tag{9.10}$$

If we use a cosine series to represent γ, an alternative model may be expressed as

$$\gamma(x_1, x_2) \;=\; \gamma(x_1, x_2; \boldsymbol{\theta})$$
$$+ \sum\sum_{(j,k)\in\Lambda} \alpha_{jk} \cos(\pi j x_1) \cos(\pi k x_2). \tag{9.11}$$

The definition of the index set Λ will, in general, depend on the specific model under the null hypothesis. For example, if we wish to test the hypothesis that $\gamma(x_1, x_2)$ has the form

$$\theta_1 + \theta_2 \cos(\pi x_1) + \theta_2 \cos(\pi x_2),$$

and we use a cosine basis, then clearly $(1, 0)$ and $(0, 1)$ should not be included in Λ. For ease of notation, we will now assume that the function $\gamma(x_1, x_2, \boldsymbol{\theta})$ is constant, but generalizations are straightforward. Under the no-effect null hypothesis, Λ is a subset of

$$\{(j, k) : 0 \leq j, k < n, \; j + k > 0\}.$$

In analogy to (9.4), we define a score statistic $\mathcal{S}_\Lambda$ and the criterion

$$SIC(\Lambda; C_n) = \mathcal{S}_\Lambda - C_n N(\Lambda),$$

respectively, where $N(\Lambda)$ denotes the number of elements in Λ.

To carry out a test we maximize $SIC(\Lambda; C_n)$ over some collection of subsets $\Lambda_1, \Lambda_2, \ldots, \Lambda_{m_n}$. It is important that this collection corresponds to nested models; otherwise, the distributions of the resultant test statistics will, in general, depend upon parameters of the null model, even when $n \to \infty$. We thus insist that $\Lambda_1 \subset \Lambda_2 \subset \cdots \subset \Lambda_{m_n}$, and we call such a collection of sets a *model sequence*. The only problem now is in deciding on how to choose a model sequence, since obviously there are many possibilities. One important consideration is whether a given sequence will lead to a consistent test. To ensure consistency against virtually any alternative to H_0, we ask that $N(\Lambda_{m_n}) \to \infty$ in such a way that, for each $(j,k) \neq (0,0)$ $(j,k \geq 0)$, (j,k) is in Λ_{m_n} for all n sufficiently large. The choice of a model sequence is further simplified if we consider only tests that place equal emphasis on the two covariates. In other words, we could insist that terms of the form $\cos(\pi j x_1) \cos(\pi k x_2)$ and $\cos(\pi k x_1) \cos(\pi j x_2)$ simultaneously enter the model.

We now give a concrete example of how to construct a model sequence. As an illustration, we use cosine basis functions; it should be clear how to follow this scheme for other bases. Below, we give explicitly the basis functions that are added to the previous model in the first four steps. For ease of notation, we assume that we want to test the null hypothesis of no effect. The same sequence of steps is graphically represented in Figure 9.1:

1. $\cos(\pi x_1)$ and $\cos(\pi x_2)$,

2. $\cos(\pi x_1) \cos(\pi x_2)$,

3. $\cos(\pi 2 x_1)$ and $\cos(\pi 2 x_2)$,

4. $\cos(\pi x_1) \cos(\pi 2 x_2)$ and $\cos(\pi 2 x_1) \cos(\pi x_2)$,

5. $\ldots$

Other choices for the model sequences are discussed in more detail in Aerts, Claeskens and Hart (2000). We also refer to this paper for more information about the large sample distribution of the test statistics as defined in Section 9.2.7.

In the case of three or more covariates, it is worthwhile to point out the price to be paid by an omnibus test as the number, d, of covariates increases. Regardless of what d is, the maximum number of parameters we should consider in a model is $O(n)$, and for simplicity let us just say n. For an omnibus test that places the same emphasis on all d covariates, this entails, roughly speaking, that r_n ought not to exceed $n^{1/d}$. Thus clearly, the ability of an omnibus test to detect higher frequency alternatives quickly wanes as the dimension of the x-space increases.

9.3.2 Tests in Additive Models

For a high-dimensional covariate vector, the omnibus tests become less attractive. However, if we can assume that an *additive* model fits the data well, the curse of dimensionality can be circumvented. Under this assumption, an

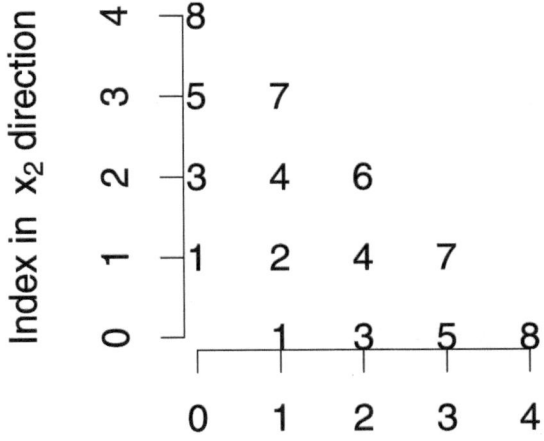

Figure 9.1 *Model sequence in two dimensions.*

alternative to the null model (9.10) can, for two covariates and the cosine basis, be written as

$$\gamma(x_1, x_2) = \gamma(x_1, x_2; \boldsymbol{\theta}) + \sum_{j \in \Lambda_a} a_j \cos(\pi j x_1) + \sum_{j \in \Lambda_b} b_j \cos(\pi j x_2),$$

which has the form of (9.11) with

$$\Lambda \subseteq \Lambda_n^* = \{(j, 0) : 1 \le j \le m_n\} \cup \{(0, k) : 1 \le k \le m_n\}.$$

As before, we define a score statistic S_Λ and the criterion

$$SIC(\Lambda; C_n) = S_\Lambda - C_n(N(\Lambda_a) + N(\Lambda_b)).$$

The test is carried out by maximizing $SIC(\Lambda; C_n)$ over a collection $\Lambda_1 \subseteq \Lambda_2 \subseteq \dots \subseteq \Lambda_{r_n}$ of subsets of Λ_n^*. Again, there are a number of ways to construct such a sequence of nested models. One possibility is to use, at step r, a series estimate of the form

$$\sum_{j=1}^{k_r} a_j \cos(\pi j x_1) + \sum_{j=1}^{\ell_r} b_j \cos(\pi j x_2),$$

where $k_r \ge k_{r-1}$ and $\ell_r \ge \ell_{r-1}$ for $r = 2, 3, \dots$. If we insist that $k_r = \ell_r$ and let k_r increase by 1 at each step, then

$$\Lambda_j = \{(1, 0), (0, 1), (2, 0), (0, 2), \dots, (j, 0), (0, j)\}.$$

We will refer to a test based on this model sequence as a diagonal test, since the "path" $\{(k_r, \ell_r) : r \ge 1\}$ corresponding to this test proceeds along the diagonal $\{(k_r, k_r) : r \ge 1\}$. At each step in a diagonal test, two terms are added to the previous model. This approach has an obvious extension to the case of more than two covariates.

Another model sequence is obtained by adding alternately a term in x_1, followed by the corresponding term in x_2, or vice versa. This may be referred to as a step-diagonal path, which has the disadvantage that the symmetry is lost.

9.3.3 The "Max" Tests in Models with Any Number of Covariates

First we explain the idea in two-covariate models. For example, consider as an alternative model

$$\gamma(x_1, x_2) = \gamma(x_1, x_2; \boldsymbol{\theta}) + \sum_{j \in \Lambda} a_j \cos(\pi j x_k), \qquad k = 1 \text{ or } 2,$$

where only one of the covariates is used to distinguish from the null model. Of course, other basis functions may be used. Unless one has a prior belief that only x_k would cause the lack of fit from the null model, this approach is not recommended. Instead, we could take as our test statistic the maximum of the values obtained by looking at each covariate "direction" separately; we refer to this as the *max test*. The level of this test can be controlled by application of Bonferronni's inequality.

This same idea can be used to extend the domain of application to models with more than two covariates. For a model with d covariates we consider, for each pair (r, s) separately, alternatives

$$\gamma(x_1, \dots, x_d) = \gamma(x_1, \dots, x_d; \boldsymbol{\theta})$$

$$+ \sum \sum_{(j,k) \in \Lambda} \alpha_{jk} \cos(\pi j x_r) \cos(\pi k x_s). \qquad (9.12)$$

For a particular choice of (r, s), we may perform any of the tests for two-covariate models, for example, one of the omnibus tests following the model sequence in Figure 9.1. Next, we take the maximum of all $d(d-1)/2$ test statistics. If the number of covariates d is large, using Bonferronni's inequality will result in a very conservative test; instead a bootstrap procedure might be applied. It seldom occurs that relevant hypotheses contain more than four or five covariates, since usually a model selection stage is passed before testing a specific hypothesis.

When both continuous and discrete covariates are present, one might want to perform the test on the continuous covariates only. In this semiparametric testing framework, we may apply the max test on the continuous variables only, while leaving the model for the discrete variables unchanged during the testing procedure.

9.3.4 Tests for More Specific Alternatives

The test in Section 9.3.2 is not necessarily consistent unless the alternative is additive. This additivity assumption can be tested by a modification of the omnibus tests. Consider the null hypothesis

$$H_0 : \gamma(x_1, x_2) = \gamma_1(x_1; \boldsymbol{\theta}_1) + \gamma_2(x_2; \boldsymbol{\theta}_2).$$

A general additive model can be estimated by a series estimator, but now the vector $\boldsymbol{\theta} = (\boldsymbol{\theta}_1, \boldsymbol{\theta}_2)$ is possibly infinite dimensional. An approach to this problem is to first estimate, by use of a model selection criterion, optimal orders k_1 and k_2 for series estimates of $\gamma_1(\cdot; \boldsymbol{\theta}_1)$ and $\gamma_2(\cdot; \boldsymbol{\theta}_2)$. Then, the null model based on these estimated orders is extended with interaction terms according to a model sequence.

Alternatively, one could perform tests of additivity for a large, fixed value of $k = k_1 = k_2$. A sensitivity analysis on k might show to what extent the choice of k_1 and k_2 influences the final conclusion.

Finally, we also mention a *goodness of link* or *single-index* test. In this case the hypothesized model (9.10) is contrasted with alternative models of the following form:

$$\gamma(x_1, x_2) = \gamma(x_1, x_2; \boldsymbol{\theta}) + \sum_{j \in \Lambda} a_j u_j [\gamma(x_1, x_2; \boldsymbol{\theta})].$$

For generalized linear models, this provides a way of testing the adequacy of the link function. It is an alternative to methods described by Collett (1991, Section 5.3) and Brown (1982).

9.3.5 The POPS Data

The Project On Preterm and Small-for-gestational age infants (POPS) has been introduced in Section 2.4 and the twins within the study have been analyzed in Section 9.2.5. We will consider information on 1338 infants born in the Netherlands in 1983 and having gestational age (x_1) less than 32 weeks and/or birthweight (x_2) less than 1500 g. The outcome of interest here concerns the situation after 2 years. The binary variable Y is 1 if an infant has died within 2 years after birth or survived with a major handicap, and 0 otherwise. After deleting observations with missing data, 1310 infants remain in the dataset.

Le Cessie and van Houwelingen (1991, 1993) examined these data to illustrate a lack-of-fit test based on a weighted sum of kernel smoothed standardized residuals. Their test failed to reject the null hypothesis of a logistic model having linear and quadratic terms in both covariates x_1 and x_2. Likewise, a likelihood ratio test showed that neither one of the third-order terms nor a first-order interaction term contributes significantly to the model. Table 9.3 shows the values of the omnibus (sequence (d)), additive (diagonal and max) and single-index test statistics. The Legendre polynomials $L_k(\cdot)$, where k denotes the polynomial order, are used to represent all models. With $L_0(x) \equiv 1$, the null hypothesis can be written as

$$H_0 : \text{logit}(E(Y)) = \sum_{k=0}^{2} \alpha_{k0} L_k(x_1) + \sum_{\ell=1}^{2} \alpha_{0\ell} L_\ell(x_2).$$

Table 9.3 *POPS Data. Results of testing H_0: "model is quadratic in x_1 and x_2".*

	S_b	S_a	T_a	T_{os}	T_{max}	$\hat{r}_a$	$\hat{r}_b$
omnibus	0.42	10.64	4.41	3.55	4.64	2	1
p value P_∞	0.811	0.073	0.039	0.036	0.034		
p value P_B	0.532	0.034	0.031	0.071	0.016		
diagonal	3.03	0.00	0.00	1.52	0.00	0	1
$\max(x_1, x_2)$	2.33	2.33	1.33	2.33	0.33	1	1
single index	3.52	8.43	0.99	3.52	2.43	6	2

We considered alternative additive models extending this null model by extra terms $L_k(x_1)$ and $L_\ell(x_2)$ with $k, \ell = 3, \ldots, 15$. For the alternative models allowing interaction terms, we included the above main effects up to the sixth order together with all interaction terms $L_k(x_1)L_\ell(x_2)$ where $2 \leq k + \ell \leq 6$. For the omnibus tests, Table 9.3 also shows p values based on the asymptotic distribution (P_∞) and on the bootstrap (P_B). Using a parametric bootstrap, 999 replications were generated under H_0 (using the null estimates). All omnibus tests except S_b indicate some evidence against H_0. The different behavior of S_b is a consequence of too large a penalty for large samples ($C_n = \log n \approx 7.18$). No additive or single index test is significant. These results suggest an extension of the null model by certain interactions. The value of $\hat{r}_a = 2$ corresponds to the interaction terms (Figure 9.1) $L_1(x_1)L_1(x_2)$, $L_1(x_1)L_2(x_2)$ and/or $L_2(x_1)L_1(x_2)$.

Inspired by these findings we investigated numerous new models extending the null model and for each we computed the classical model selection criterion (9.2) with $C_n=2$ (*AIC*) and $C_n = \log n$ (*BIC*). The *BIC* criterion selected the null model, which, again, is a consequence of the large penalty. The model selected by *AIC* is

$$\text{logit}(E(Y)) = \sum_{k=0}^{5} \alpha_{k0} L_k(x_1) + \sum_{\ell=1}^{2} \alpha_{0\ell} L_\ell(x_2)$$
$$+ \alpha_{11} L_1(x_1)L_1(x_2) + \alpha_{12} L_1(x_1)L_2(x_2), \qquad (9.13)$$

which can be rewritten in terms of $x_1, \ldots, x_1^5, x_2, x_2^2, x_1 x_2, x_1 x_2^2$. Both groups of higher order main effects x_1^3, x_1^4, x_1^5 and interactions $x_1 x_2, x_1 x_2^2$ are significant at the 5% level.

9.4 Testing Goodness of Fit

In a similar way as we test the fit of a regression model, we can perform tests for the adequacy of a likelihood function itself. Historically, this type of test is called a goodness-of-fit test. We are considering independent observations from a common density and want to test whether this density is equal to a specified f_0, against the nonparametric alternative that it is a density $f \neq f_0$.

There exist several ways to embed the null model density in a larger family. Aerts, Claeskens and Hart (1999) propose writing

$$f(x) = C_\gamma \exp[\gamma(x)] \tag{9.14}$$

and rephrase the null hypothesis in terms of $\gamma(x) = \ln[f(x)/C_\gamma]$, that is, on a logarithmic scale. More specifically, $H_0 : \gamma(x) \in \mathcal{G}$ for some parametric family of functions $\mathcal{G} = \{\gamma(x; \theta_1, \dots, \theta_p) : (\theta_1, \dots, \theta_p) \in \Theta\}$. Consider additive approximators of $\gamma(x)$ of the form

$$\gamma(x; \theta_1, \dots, \theta_p) + \sum_{j=1}^{r} \theta_j u_j(x).$$

We may then proceed to test H_0 using either LIC, SIC or RIC as described in Sections 9.2.1–9.2.3.

A second approach is to construct multiplicative approximators via a log-linear expansion:

$$f_r(x; \theta_1, \dots, \theta_{p+r}) = f_0(x; \theta_1, \dots, \theta_p) \exp\left[\sum_{j=1}^{r} \theta_{p+j} u_j(x)\right] / C_r, \tag{9.15}$$

for $r = 1, \dots, r_n$. When f_0 is a uniform density, the approaches (9.14) and (9.15) are identical; in general they are not. The basis functions $u_j(\cdot)$ are chosen to be orthogonal and normalized with respect to f_0 and also orthogonal to the function $u_0 = 1$, i.e.,

$$\int f_0(x; \boldsymbol{\theta}_p) u_j(x) u_k(x) \, dx = I\{j = k\}.$$

Because of these requirements, the basis functions will depend on the unknown parameters, i.e., $u_j(\cdot) = u_j(\cdot; \boldsymbol{\theta}_p)$. In practice, we replace the parameter by its maximum likelihood estimator. The normalizing constant C_r needs to be recalculated for each approximator since

$$C_r = \int f_0(x; \boldsymbol{\theta}_p) \exp\left[\sum_{j=1}^{r} \theta_{p+j} u_j(x)\right] \, dx.$$

To operationalize this procedure, one has first of all to decide on a practical sequence of u_j functions, which, of course, can be done in several ways, for example, by taking orthogonalized polynomials or a cosine system; see Section 9.2. Claeskens and Hjort (2001) study tests based on likelihood ratio statistics:

$$2(\mathcal{L}_r - \mathcal{L}_0) = 2 \sum_{i=1}^{n} \ln \frac{f_r(X_i; \widehat{\boldsymbol{\theta}}_{p+r})}{f_0(X_i; \widehat{\boldsymbol{\theta}}_p)} = 2 \left[\sum_{j=1}^{r} \widehat{\theta}_{p+j} \sum_{i=1}^{n} u_j(X_i) - n \ln C_r(\widehat{\boldsymbol{\theta}}_{p+r})\right],$$

for $r = 1, \dots, r_n$, where $\widehat{\boldsymbol{\theta}}_p$ maximizes the likelihood under the null model.

Tests based on the likelihood ratio statistics are compared to those based on the score statistics:

$$\mathcal{S}_r = \frac{1}{n} \sum_{j=1}^{r} \{\sum_{i=1}^{n} u_j(X_i)\}^2, \qquad r = 1, \dots, r_n.$$

The score statistic here takes a particularly simple form. This type of test has its origin in Neyman's 1937 paper on 'smooth tests'. The criteria LIC and SIC are defined as in (9.2) and (9.4). The test rejects H_0 when the model order chosen by the criterion exceeds zero. Or, equivalently, when the maximum of a set of weighted likelihood ratio or score statistics exceeds a critical value. The testing procedure in the goodness-of-fit setting is exactly as given in the lack-of-fit setting in Section 9.2.

The score test, in conjunction with BIC, has been proposed by Ledwina (1994). A comparison in Claeskens and Hjort (2001) shows that BIC applied to a sequence of nested models may have serious disadvantages. For the class of densities f where the lowest frequency term equals zero but some of the higher frequency terms do not, the asymptotic power of the deduced test under local alternatives is equal to the significance level. Since the probabilities that AIC chooses a model of dimension $m > 1$ are nonzero, although decreasing, for all dimensions $m > 1$, the AIC based test is likely to outperform the nested sequence BIC for a large class of alternatives. Therefore we recommend the use of AIC based criteria for lack-of-fit and goodness-of-fit testing.

It is important to note that the performance of tests using BIC as a model selector can be drastically improved by not restricting attention to only nested model sequences, but rather allowing all subsets within a fixed dimension m_0. The fact that m_0 is fixed is not disturbing for practical matters, since it would correspond to typical use, and since it is allowed to be arbitrarily large. For asymptotic distribution theory and more details, we refer to Claeskens and Hjort (2001).

CHAPTER 10

Quantitative Risk Assessment

Lieven Declerck

European Organization for Research and Treatment of Cancer, Brussels, Belgium

In the area of risk assessment, the focus can be on a number of issues. Interest can be placed on the characterization of the dose-response relationship, i.e., studying the dependence of a particular outcome such as the risk of a malformed foetus, on the dose which is administered to the dam. Besides investigating the dose-response relation, another issue is quantitative risk assessment. This critically important area of risk assessment is based on the relationship between dose and response, to derive a safe dose. In quantitative risk assessment, there are a number of choices that have to be made, resulting in a variety of approaches.

First, safe exposure levels can be derived from the NOAEL (No Observed Adverse Effect Level) safety factor approach. The assumption made here is that if the dose administered to a dam is below some value (the threshold), then there will be no adverse effects on the foetuses of that dam (Williams and Ryan 1996). The NOAEL is defined as the experimental dose level immediately below the lowest dose that produces a statistically or biologically significant increase in an adverse effect in comparison with the control. An "acceptably safe" daily dose for humans is then calculated by dividing the NOAEL by a safety factor (commonly 100 or 1000). In this way, sensitive subgroups of the population and extrapolation from animal experiments to human risk are taken into account. Rather than basing quantitative risk assessment on the NOAEL procedure, dose-response modeling can also be used to determine safe doses. Due to the disadvantages of the NOAEL and the benefits of dose-response models, this chapter is concerned with statistical procedures to predict safe exposure levels based on such a modeling approach.

Second, there are several ways to handle clustering. While dose-response modeling is relatively straightforward in uncorrelated settings, it is less so in the clustered context. Of course, one can ignore the clustering altogether by treating the littermates as if they were independent. Also, the litter effect issue can be avoided by modeling the probability of an affected litter. Such models are generally too simplistic but there is a multitude of models which do consider clustering. Indeed, as discussed in Chapter 4, different types of models (marginal, random effects, conditional) for clustered binary data can be formulated. In this chapter, the emphasis is on the beta-binomial model (Skellam 1948, Kleinman 1973), as well as on the conditional model of Molenberghs and Ryan (1999). Besides the choice of an appropriate dose-response model, parameters can be estimated via several inferential procedures. Estimation methods range from full likelihood to pseudo-likelihood (Chapters 6 and 7) and generalized estimating equations (Chapter 5). In this chapter, parameters are estimated using maximum likelihood methodology. Furthermore, the implications in terms of uncertainty of fitting a model based on a finite set of data are investigated. In Chapter 11, the effect of misspecifying the parametric response model on the estimation of a safe dose will be investigated.

Third, quantitative risk assessment can be based on either foetus or on litter-based risks. To perform dose-response modeling and assessment of safe doses, most authors take a foetus-based perspective, where the excess risk over background for an affected foetus is determined as a function of dose. The latter approach is straightforward for marginal models, which are expressed in terms of this marginal adverse event probability (Diggle, Liang and Zeger 1994, Pendergast et al. 1996). However, a disadvantage of this approach is that it may raise biological questions. Arguably, the entire litter is more representative of a human pregnancy than a single foetus. As a consequence, modeling litter-based excess risks is a very appealing alternative from a biological perspective. In the litter-based approach, quantitative risk assessment is based on the cluster of foetuses of a dam, i.e., the probability that at least one foetus of a dam has the adverse event under consideration is crucial. In this chapter, foetus and so-called litter-based risks are contrasted in the determination of safe doses.

Fourth, one needs to acknowledge the stochastic nature of the number of implants and the number of viable foetuses (i.e., the litter size) in a dam. Some methods (Ryan 1992) condition on the observed litter size when modeling the number of malformations. Others (e.g., Catalano et al. 1993) allow response rates to depend on litter size and then calculate a safe dose at an "average" litter size, thereby avoiding the need for direct adjustment. Krewski and Zhu (1995) use a model formulation that causes litter size to drop from the expression for excess risk. Rai and Van Ryzin (1985) compute risks by integrating over the litter size distribution. This approach will be used in this chapter.

The relatively complex data structure forces the researcher to reflect on several other questions:

1. Are linear or non-linear predictors used?

2. Are the malformation indices studied separately, collapsed into a single indicator or treated as a multivariate outcome vector per foetus within a litter?

3. Is death ignored, studied separately without taking into account malformations among the viable foetuses, combined with a collapsed malformation

indicator into a new indicator for *abnormality* (i.e., death or malformation), or studied jointly with the malformation outcomes?

4. Are continuous responses, such as birth weight, excluded from the model or not? Chen *et al.* (1991), Ryan (1992), Catalano *et al.* (1993), Krewski and Zhu (1994) discuss statistical models that allow for exposure effects on death and malformation, formulating the problem as a trinomial model with overdispersion. Catalano and Ryan (1992) and Catalano *et al.* (1993) propose models that incorporate fetal weight in addition to death and malformation. This topic is studied further in Chapter 14.

In this chapter, linear predictors are used for the parameters of the implemented models. Four approaches are considered:

1. an indicator for death,

2. a collapsed malformation indicator ignoring dead foetuses,

3. an indicator for abnormality (i.e., death or malformation) and

4. a joint model for death and malformation.

A multivariate approach for malformation and the incorporation of fetal birth weight into the model are not highlighted in this chapter.

It will be shown here how the beta-binomial and the conditional models can easily handle litter-based rates. Furthermore, it will be demonstrated how the conditional model leads to a natural formulation of the foetus-based excess risk on the number of implants in a dam, unlike marginal models such as the beta-binomial model.

In Section 10.1, foetus and litter-based risks are derived for the beta-binomial and conditional models. The collapsed outcome "abnormality" is discussed, as well as the hierarchically structured outcomes death and malformation. Section 10.2 compares the foetus and litter-based approach and contrasts a model for abnormality with a joint model for death and malformation, based on the NTP data. Section 10.3 illustrates these items based on so-called asymptotic samples. Further details can be found in Declerck, Molenberghs, Aerts and Ryan (2000).

10.1 Expressing Risks

An introduction to risk assessment is provided in Section 3.4. Suppose one wishes to estimate a safe level of exposure, based on, e.g., the beta-binomial or the conditional model. The standard approach to quantitative risk assessment requires the specification of an adverse event (e.g., abnormality), along with the risk of this event expressed as a function of dose. The risk $r(d)$ represents the probability that the adverse event occurs at dose level d.

Instead of the risk $r(d)$ itself, one might prefer to use the additive excess risk, which is the excess risk above the background rate, i.e., $r(d) - r(0)$. Assuming that at any non-zero value of dose, the chemical under investigation has more toxic effect than at dose level 0, the additive excess risk function ranges from 0 to $1 - r(0)$. This type of risk does not relate the difference in risk at dose

d and at dose 0 to the background rate. This is in contrast with the relative excess risk function $r^*(d)$. It is a "multiplicative" risk function, measuring the relative increase in risk above background and is defined as (Crump 1984)

$$r^*(d) = \frac{r(d) - r(0)}{1 - r(0)}. \tag{10.1}$$

Here, $r^*(d)$ is called the *excess risk*. Assuming again that the chemical results in more adverse effects at non-zero dose d compared to dose level 0, the excess risk ranges from 0 to 1.

From the relationship $r^*(d)$, a safe level of exposure can be determined. The terminology used to describe a "virtually safe dose" or a "benchmark dose" is not standardized and depends on the area of application (carcinogenicity, developmental toxicity) and the regulatory authorities involved (Environmental Protection Agency, Food and Drug Administration, ...). A useful overview is given in Williams and Ryan (1997). Here, the benchmark dose (BMD), sometimes called the effective dose (ED), is the dose at which the excess risk over the background rate is small, say 10^{-4} and the BD or VSD is the lower confidence limit of the effective dose.

Using $r(d) \equiv r(d; \boldsymbol{\beta})$, i.e., the risk at dose level d corresponding to the parameter vector $\boldsymbol{\beta}$ in the model considered, the BMD can be defined as the value d that solves $r^*(d; \boldsymbol{\beta}) = 10^{-4}$. The ML estimate of the effective dose is the solution to $\hat{r}^*(d) = r^*(d; \hat{\boldsymbol{\beta}}) = 10^{-4}$ where $\hat{\boldsymbol{\beta}}$ is the ML estimate of $\boldsymbol{\beta}$.

For setting confidence limits in low dose extrapolation, i.e., to determine the BMDL, several approaches can be considered. For example, Crump and Howe (1985) recommend to use the asymptotic distribution of the likelihood ratio. According to this method, an approximate $100(1-\alpha)\%$ lower confidence limit for the BMD corresponding to an excess risk of 10^{-4} is defined as

$$\min\{d(\boldsymbol{\beta}) : r^*(d; \boldsymbol{\beta}) = 10^{-4} \text{ over all } \boldsymbol{\beta}$$
$$\text{such that } 2(\ell(\hat{\boldsymbol{\beta}}) - \ell(\boldsymbol{\beta})) \le \chi_p^2(1 - 2\alpha)\}, \tag{10.2}$$

with p the number of regression parameters. This might imply that a dose-response model with more regression parameters (and thus more uncertainty) leads to a larger confidence region and thus to a smaller BMDL. Of course, it is important that a likelihood ratio test be available, making the method less straightforward to use in non-likelihood settings. For pseudo-likelihood, the proposal of Geys, Molenberghs and Ryan (1999) for pseudo-likelihood ratio tests could be followed. As another appealing alternative, Section 11.4 studies a profile score based approach.

A variation to this theme, suggested by many authors (Chen and Kodell 1989, Gaylor 1989, Ryan 1992), first determines a lower confidence limit, e.g., corresponding to an excess risk of 1% and then linearly extrapolates it to a BMDL. The main advantage quoted for this procedure is that the determination of a BMDL is less model dependent.

Several other methods have been proposed. A lower confidence bound for the benchmark dose can be computed based on the profile likelihood method

(Morgan 1992). This idea will be pursued in Section 11.3. Also, a Wald-based version can be obtained using the delta method. Several authors have indicated that the latter method suffers from drawbacks, especially with low dose extrapolation (Krewski and Van Ryzin 1981, Crump 1984, Crump and Howe 1985). One of the disadvantages of a Wald-based confidence interval for the effective dose is that its lower limit may fail to be positive. The NOAEL provides another alternative.

10.1.1 Foetus and Litter-Based Risks

In this section, different risks and corresponding excess risks are presented for the beta-binomial and conditional models. They can be foetus or litter based and they can be defined for a single adverse event like "death" or "malformation" as well as for both events jointly. In the next sections, these different approaches to risk and BMD estimation are compared for the NTP data and for asymptotic samples.

The main issue deals with the choice between foetus and litter-based risks. Here, for simplicity, the presentation is restricted to the adverse event "abnormality" in a litter with m implants. A foetus-based approach focuses on the risk of a foetus as a function of the level of exposure d given to the dam. Let $q_F(m; d)$ be the probability that a foetus is abnormal, given that the foetus is selected from a litter with m implants. Consider all values of the number of implants m with non-zero probability $P(m)$. Administering some specified dose d to M dams, the foetus-based risk is:

$$r_F(d) = \frac{\sum_m M P(m) m q_F(m; d)}{\sum_m M P(m) m} = \frac{\sum_m P(m) m q_F(m; d)}{\sum_m P(m) m}. \tag{10.3}$$

Hence, the foetus-based risk at some specified dose is an average of conditional probabilities $q_F(m; d)$ with weights $M P(m) m$, i.e., the expected number of foetuses in litters with m implants resulting from the M dams.

In marginal models such as, for example, the beta-binomial model, the probability $q_F(m; d)$ does not depend on the number of implants m (except when it is explicitly incorporated in the model as a covariate) and, hence, $r_F(d) = q_F(d)$. It will be shown that this is in contrast with the conditional model of Molenberghs and Ryan (1999), where q_F is related to the number of implants m in a natural way.

In a litter-based approach, the event of interest is whether at least one foetus in a litter is abnormal. Let $q_L(m; d)$ be the probability that at least one foetus in a litter of size m has the adverse event. The litter-based risk is

$$r_L(d) = \sum_m P(m) q_L(m; d), \tag{10.4}$$

which is an average of conditional probabilities $q_L(m; d)$ with weights $P(m)$.

Since a particular adverse effect in one or more foetuses of a litter is at least as probable as the occurrence of this adverse event in a specific foetus, it follows that $q_F(m; d) \leq q_L(m; d)$. Considering this inequality for any number of implants m with non-zero probability $P(m)$, it follows that

$$\sum_m P(m)q_F(m; d) \leq r_L(d) = \sum_m P(m)q_L(m; d).$$

For a single adverse event in a marginal model, the conditional probability $q_F(m; d) = q_F(d)$ and, hence, the first sum equals $r_F(d)$. In this case, the foetus-based risk is smaller than or equal to the litter-based risk. Note however that in general the first sum is different from $r_F(d)$. One can easily find examples in which $r_F(d)$ is smaller than, equal to or greater than $r_L(d)$. Indeed, consider the case of two litters, litter 1 with one foetus being abnormal and litter 2 with two foetuses being healthy. Then, $r_F(d)$ for the adverse event "abnormality" is $1/3$, while $r_L(d) = 1/2$. If litter 1 would have had two abnormal foetuses, then $r_F(d) = r_L(d) = 1/2$. Finally, if litter 1 consisted of three abnormal foetuses, then $r_F(d) = 3/5 > r_L(d) = 1/2$. Similarly, there are cases where the foetus-based excess risk $r_F^*(d)$ is smaller than, equal to or greater than the litter-based excess risk $r_L^*(d)$.

For a specific model, these risks can be estimated by replacing all parameters by their maximum likelihood estimates. Also, the values $P(m)$, i.e., the distribution of the number of implants in a litter, have to be estimated. This is discussed in more detail in Declerck $et\ al.$ (2000).

In what follows, foetus and litter-based risks will be discussed for the beta-binomial model and the conditional model of Molenberghs and Ryan. For both models, the approach of a single adverse event (focusing on "abnormality") will be given, as well as the approach where the adverse events "death" and "malformation" are studied jointly.

10.1.2 Risks for a Beta-binomial Model for Abnormality

The probability q_F that a foetus is abnormal, given that the foetus is selected from a litter with m implants, is π. Based on (4.3) and (4.4),

$$q_F = \frac{1}{1 + \exp(-\beta_0 - \beta_d d)}.$$

As mentioned before, the probability q_F does not depend on the number of implants m and hence the foetus-based excess risk equals

$$r_F^* = \frac{q_F(d) - q_F(0)}{1 - q_F(0)} = \frac{1 - \exp(-\beta_d d)}{1 + \exp(-\beta_0 - \beta_d d)}.$$

Since r_F^* does not depend on the correlation parameter ρ, the above expression is also valid for the ordinary logistic regression model.

The probability that at least one foetus of a litter is abnormal is

$$q_L = 1 - \frac{B(\pi(\rho^{-1} - 1), (1 - \pi)(\rho^{-1} - 1) + m)}{B(\pi(\rho^{-1} - 1), (1 - \pi)(\rho^{-1} - 1))}.$$

This expression can be rewritten as

$$q_L = 1 - \prod_{k=0}^{m-1} \left(1 - \pi + \frac{k\pi\rho}{1 + (k-1)\rho}\right).$$

Note that, in cases of overdispersion, the litter-based probability of an adverse event q_L is smaller than the probability $1 - (1 - \pi)^m$, corresponding to $\rho = 0$ (no clustering). From (10.1) and (10.4), the litter-based excess risk can be computed as

$$r_L^* = 1 - \frac{\displaystyle\sum_m P(m) \prod_{k=0}^{m-1} (1 - \pi(d) + k\pi(d)\rho/(1 + (k-1)\rho))}{\displaystyle\sum_m P(m) \prod_{k=0}^{m-1} (1 - \pi(0) + k\pi(0)\rho/(1 + (k-1)\rho))}.$$

In case of no clustering, this expression reduces to

$$r_L^* = 1 - \frac{G(1 - \pi(d))}{G(1 - \pi(0))},$$

where $G(\cdot)$ is the probability generating function of the number of implants. For $m = 1$, $G(z) = z$, such that $r_L^* = r_F^*$.

10.1.3 Risks for a Beta-binomial Model For Death and Malformation Jointly

Here, it is proposed to model both components of

$$
\begin{aligned}
f(r_i, z_i | m_i, d_i) &= f(r_i | m_i, d_i) f(z_i | r_i, m_i, d_i) \\
&= f(r_i | m_i, d_i) f(z_i | n_i, m_i, d_i), \quad (10.5)
\end{aligned}
$$

with a model similar to (4.32):

$$f(r \mid m, d) =$$
$$\binom{m}{r} \frac{B(\pi_{\text{dth}}(\rho_{\text{dth}}^{-1} - 1) + r, (1 - \pi_{\text{dth}})(\rho_{\text{dth}}^{-1} - 1) + m - r)}{B(\pi_{\text{dth}}(\rho_{\text{dth}}^{-1} - 1), (1 - \pi_{\text{dth}})(\rho_{\text{dth}}^{-1} - 1))}, \quad (10.6)$$

$$f(z \mid n, d) =$$
$$\binom{n}{z} \frac{B(\pi_{\text{mal}}(\rho_{\text{mal}}^{-1} - 1) + z, (1 - \pi_{\text{mal}})(\rho_{\text{mal}}^{-1} - 1) + n - z)}{B(\pi_{\text{mal}}(\rho_{\text{mal}}^{-1} - 1), (1 - \pi_{\text{mal}})(\rho_{\text{mal}}^{-1} - 1))}. \quad (10.7)$$

Again, one can distinguish between risk assessment at foetus level and at litter level.

The probability that foetus j is dead or malformed, given that the number of implants equals m, is

$$
\begin{aligned}
q_F &= P(\text{foetus } j \text{ is dead} \mid m \text{ implants}) \\
&\quad + P(\text{foetus } j \text{ is malformed} \mid m \text{ implants}) \\
&= \pi_{\text{dth}} + \sum_{r=0}^{m-1} (P(\text{foetus } j \text{ is alive and } R = r \mid m \text{ implants}) \\
&\quad \times P(\text{foetus } j \text{ is malformed} \mid \\
&\qquad \text{foetus } j \text{ alive \& } r \text{ deaths out of } m \text{ implants}) \quad (10.8)
\end{aligned}
$$

where R denotes the number of deaths in a litter. This can be re-expressed as

$$
q_F = \pi_{\text{dth}} + \frac{\pi_{\text{mal}}}{B(\pi_{\text{dth}}(\rho_{\text{dth}}^{-1} - 1), (1 - \pi_{\text{dth}})(\rho_{\text{dth}}^{-1} - 1))} \sum_{r=0}^{m-1} \binom{m-1}{r}
$$

$$
\times B(\pi_{\text{dth}}(\rho_{\text{dth}}^{-1} - 1) + r, (1 - \pi_{\text{dth}})(\rho_{\text{dth}}^{-1} - 1) + m - r). \quad (10.9)
$$

Expressions (10.1), (10.3) and (10.9) enable the calculation of the foetus-based excess risk.

The probability that at least one foetus is dead or malformed, given m, is based upon (10.5) and reduces to

$$
q_L = 1 - P(R = 0, Z = 0 \mid m, d) = 1 - P(R = 0 \mid m, d)P(Z = 0 \mid n, d).
$$

Explicitly, in terms of (10.6) and (10.7),

$$
q_L = 1 - \prod_{k,\ell=0}^{m-1} \left(1 - \pi_{\text{dth}} + \frac{k\pi_{\text{dth}}\rho_{\text{dth}}}{1 + (k-1)\rho_{\text{dth}}}\right)
$$

$$
\times \left(1 - \pi_{\text{mal}} + \frac{\ell\pi_{\text{mal}}\rho_{\text{mal}}}{1 + (\ell-1)\rho_{\text{mal}}}\right). \quad (10.10)
$$

Formulas (10.1), (10.4) and (10.10) allows one to compute the litter-based excess risk.

10.1.4 Risks for a Conditional Model for Abnormality

For the conditional model, the probability q_F that foetus j is abnormal, given implant size m, can be expressed in several ways. The probability q_F can be written as in (4.29) in terms of S, the number of abnormals. It can also be computed based on (4.26), by summing over the distribution of the outcomes of the other littermates:

$$
\begin{aligned}
q_F &= \sum_{s=1}^{m} \binom{m-1}{s-1} \exp\left\{\psi s - \phi s(m-s) - A(\psi, \phi, m)\right\} \\
&= \exp\left\{-A(\psi, \phi, m) + A(\psi + \phi, \phi, m-1) + \psi - \phi(m-1)\right\},
\end{aligned}
$$

with s the number of abnormal foetuses in a litter. Finally, one can derive an expression for q_F by calculating the probability that foetus j is healthy, given

m implants:

$$q_F = 1 - \sum_{s=0}^{m-1} \binom{m-1}{s} \exp\{\psi s - \phi s(m-s) - A(\psi, \phi, m)\}$$

$$= 1 - \exp\{-A(\psi, \phi, m) + A(\psi - \phi, \phi, m-1)\}. \tag{10.11}$$

Using the expression for a healthy foetus is slightly more convenient. These formulas show how, for the conditional model, the probabilities q_F depend on the number of implants. Based on (10.1), (10.3) and (10.11), the foetus-based excess risk follows.

Now, the probability that at least one foetus is abnormal, given m, is $q_L = 1 - P(S = 0)$, which, based on (4.26), is given by

$$q_L = 1 - \exp\{-A(\psi, \phi, m)\}. \tag{10.12}$$

This result is an appealing counterpart to (10.11). It differs from (10.11) by the deletion of one normalizing constant. Expression (10.1), (10.4), and (10.12) can be used to calculate the litter-based excess risk.

10.1.5 Risks for a Conditional Model for Death and Malformation Jointly

The conditional model for death and malformation jointly is the product of

$$f(r \mid m, d) = \binom{m}{r}$$

$$\times \exp\{\psi_{\text{dth}} r - \phi_{\text{dth}} r(m-r) - A(\psi_{\text{dth}}, \phi_{\text{dth}}, m)\}, \tag{10.13}$$

$$f(z \mid n, d) = \binom{n}{z}$$

$$\times \exp\{\psi_{\text{mal}} z - \phi_{\text{mal}} z(n-z) - A(\psi_{\text{mal}}, \phi_{\text{mal}}, n)\}. \tag{10.14}$$

Using (10.8), the conditional probability that foetus j exhibits an adverse event, given that a litter contains m implants, can be rewritten as:

$$q_{\bar{r}} = \pi_{\text{dth}} + \sum_{r=0}^{m-1} \binom{m-1}{r} \exp\{\psi_{\text{dth}} r - \phi_{\text{dth}} r(m-r) - A(\psi_{\text{dth}}, \phi_{\text{dth}}, m)\}$$

$$\times \{1 - \exp\{-A(\psi_{\text{mal}}, \phi_{\text{mal}}, m-r)$$

$$+A(\psi_{\text{mal}} - \phi_{\text{mal}}, \phi_{\text{mal}}, m-r-1)\}\}.$$

Based on (10.11), this expression can be simplified to

$$q_F = 1 - \exp\{-A(\psi_{\text{dth}}, \phi_{\text{dth}}, m)\}$$

$$\times sum_{r=0}^{m-1} \binom{m-1}{r} \exp(B(r)), \tag{10.15}$$

with

$$B(r) = \psi_{\mathrm{dth}} r - \phi_{\mathrm{dth}} r (m-r) - A(\psi_{\mathrm{mal}}, \phi_{\mathrm{mal}}, m-r) + A(\psi_{\mathrm{mal}} - \phi_{\mathrm{mal}}, \phi_{\mathrm{mal}}, m-r-1).$$

Again, by means of (10.1), (10.3) and (10.15), the foetus-based excess risk can be computed.

Considering the conditional model for death and malformation jointly, the probability that a litter has at least one adverse event, given m, can be based on (10.5), (10.13) and (10.14):

$$q_L \quad = \quad 1 - \exp\left\{-A(\psi_{\mathrm{dth}}, \phi_{\mathrm{dth}}, m) - A(\psi_{\mathrm{mal}}, \phi_{\mathrm{mal}}, m)\right\}. \qquad (10.16)$$

The rather complicated sum in (10.15) is replaced by a normalizing constant. Expressions (10.1), (10.4) and (10.16) enable the calculation of the litter-based excess risk.

10.2 Analysis of NTP Data

The different risk and corresponding BMD estimators of Section 10.1 are compared for the NTP data, introduced in Section 2.1. Excess risk functions are estimated by maximum likelihood for a grid of dose values, based on the beta-binomial model and the conditional model. These results are also compared to those of the logistic model for which all information of a litter is collapsed into a single, binary variable indicating whether there is at least one abnormal foetus. For the toxic agent DEHP, the resulting curves for the adverse events "abnormality" and "death and malformation jointly" are shown in Figure 10.1. As expected intuitively, litter-based excess risks are clearly larger than foetus-based risks at the same dose level. Also, the risk curve for the adverse event "abnormality" and the corresponding curve for "death and malformation jointly" are rather close to each other. This holds for both models (beta-binomial and conditional) and for both foetus and litter-based approaches. The litter-based excess risks for the beta-binomial model are clearly larger than the corresponding risks for the conditional model. This is also true for the foetus-based risks except for large dose values. The risks for the logistic regression model are comparable to the beta-binomial litter-based risks for all dose levels below 0.25.

Besides the adverse effects considered in Figure 10.1, also "death" and "malformation among the viable foetuses" are investigated for DEHP, DYME, EG, TGDM and THEO. Benchmark doses are calculated for several adverse events. These are shown in Table 10.1. The BMD of a foetus-based risk curve is in general about 5 to 10 times larger than the corresponding litter-based BMD. This is in line with the excess risk curves of Figure 10.1. The effective doses of "abnormal" and of "joint" are well comparable, except for the chemical TGDM. Comparing the three models under investigation, the ED of the conditional model is most often larger than the BMD of the beta-binomial model. The logistic model results in the smallest BMD. Since the models considered come from fundamentally different modeling families (conditional and marginal), a somewhat different behavior in key aspects is not unexpected. Aerts, Declerck

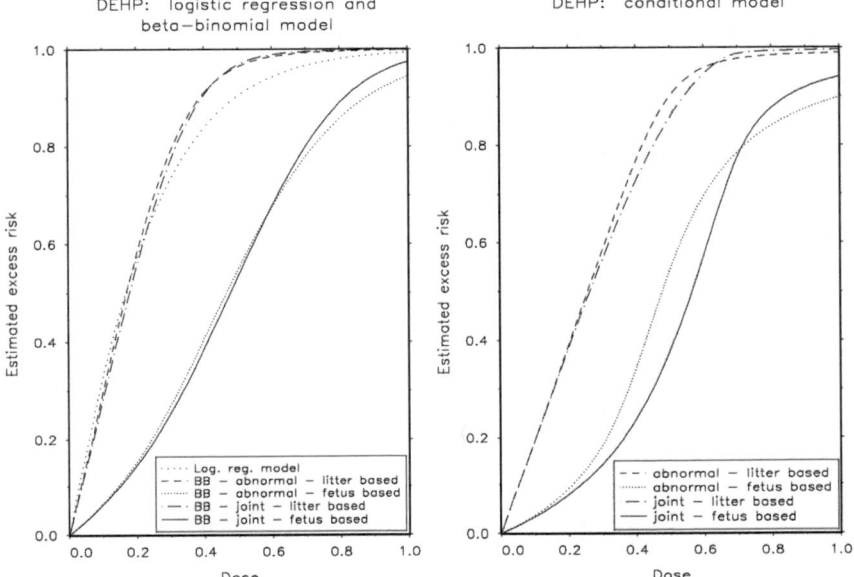

Figure 10.1 *NTP Data. Excess risk curves for DEHP based on the logistic regression, beta-binomial and conditional models.*

and Molenberghs (1997) addressed BMD determination in the foetus-based setting. They also concluded that BMDs tend to be somewhat higher in the conditional model, as opposed to the beta-binomial, the Bahadur and the George-Bowman models. The next section examines whether these findings are confirmed by a large sample simulation study.

10.3 Asymptotic Study

To compare the asymptotic effect of a foetus-based versus a litter-based approach on the benchmark dose, the ideas of Rotnitzky and Wypij (1994) are followed here. This procedure was employed in Section 7.2 as well. Recall that an artificial (asymptotic or "large") sample is constructed where each possible realization of dose d, number of implants m, number of deaths r and number of malformations z is weighted according to the probability in the underlying model. Precisely, all realizations of the form (d, m, r, z) are included and are assigned the weight $f(d, m, r, z)$ where f denotes a probability mass function. Hence, one has to specify: (1) $f(d)$, the relative frequency of the dose group as prescribed by the design; (2) $f(m|d)$, which equals $f(m)$ since a dam is randomly assigned to a dose group and exposure occurs after mating; (3) $f(r|m, d)$, the actual model probability for the occurrence of r deaths and (4) $f(z|r, m, d) = f(z|n, m, d)$, which is assumed here to be $f(z|n, d)$, the actual model probability for z malformations. The doses 0, 0.25, 0.5 and 1 are

Table 10.1 *NTP Data. Effective doses of DEHP, DYME, EG, TGDM and THEO corresponding to an excess risk of* 10^{-4}. *All quantities shown should be divided by* 10^4.

Model	Unit	Adv. event	DEHP	DYME	EG	TGDM	THEO
Logist. regr.	Litter	Abnormal	0.3	0.4	0.3	1.7	1.4
Beta-bin.	Foetus	Dead	2.5	7.8	12.3	40.4	8.1
		Malformed	7.7	15.6	5.1	81.3	150.6
		Abnormal	1.9	4.6	2.5	10.0	7.5
		Joint	1.9	5.2	3.6	27.1	7.7
	Litter	Dead	0.5	0.9	1.5	7.7	1.9
		Malformed	1.4	2.6	1.1	12.1	17.6
		Abnormal	0.3	0.6	0.4	1.7	1.7
		Joint	0.4	0.6	0.6	4.7	1.7
Condit.	Foetus	Dead	5.2	10.7	15.9	31.7	14.9
		Malformed	9.8	15.4	8.2	93.6	182.4
		Abnormal	3.4	6.6	4.0	15.2	13.7
		Joint	3.5	6.7	5.3	23.7	13.8
	Litter	Dead	0.8	1.0	1.8	5.9	2.3
		Malformed	1.1	1.5	1.2	8.6	16.8
		Abnormal	0.5	0.6	0.6	2.0	2.0
		Joint	0.5	0.6	0.8	3.6	2.0

considered when generating asymptotic samples and each dose is assigned a relative frequency of $1/4$. The distribution of the number of implants, $f(m)$, is based on the NTP data. The relative frequencies of m for all NTP datasets under investigation are smoothed via a local linear smoothing technique. Least squares cross-validation has been used to choose the bandwidth. The absolute and relative frequency distribution resulting from the NTP data, as well as the smoothed relative frequencies, are presented in Table 10.2. The conditional model is used for generating the number of deaths and the number of malformations as in (10.13) and (10.14). The parameters are modeled as

$$\psi_{\text{dth}} = \beta_{0,\text{dth}} + \beta_{d,\text{dth}}d, \qquad \phi_{\text{dth}} = \beta_{2,\text{dth}},$$
$$\psi_{\text{mal}} = \beta_{0,\text{mal}} + \beta_{d,\text{mal}}d, \qquad \phi_{\text{mal}} = \beta_{2,\text{mal}}.$$

Based on the parameter estimates from the conditional model for each NTP dataset, 60 parameter combinations are selected (Table 10.3). Next, for each parameter vector, an asymptotic sample is generated based on a conditional

Table 10.2 *NTP Data. Absolute and relative frequencies of the number of implants.*

Number of implants	absolute frequency	relative frequency	smoothed relative frequency
1	4	0.0073	0.0073
2	3	0.0054	0.0063
3	4	0.0073	0.0081
4	7	0.0127	0.0094
5	2	0.0036	0.0074
6	6	0.0109	0.0092
7	6	0.0109	0.0113
8	7	0.0127	0.0189
9	23	0.0417	0.0376
10	29	0.0526	0.0676
11	71	0.1289	0.1226
12	98	0.1779	0.1712
13	109	0.1978	0.1812
14	80	0.1452	0.1469
15	55	0.0998	0.1002
16	31	0.0563	0.0579
17	11	0.0200	0.0249
18	3	0.0054	0.0084
19	2	0.0036	0.0036
	551	1	1

model for death and malformation jointly. Foetus and litter-based excess risk curves are computed for death and malformation jointly as well as for abnormality. Figure 10.2 shows a selection of curves for six parameter combinations. Again, foetus-based excess risks are markedly smaller than litter-based excess risks. For $\beta_{0,dth} = \beta_{0,mal} = 0$, the difference is less pronounced. In general, foetus and litter-based curves are relatively close to each other for large background rates for death and malformation. The plots of Figure 10.2 also show that the curve for "abnormality" and the corresponding curve for "death and malformation jointly" are relatively close to each other. This is true for the foetus-based as well as for the litter-based approach. The 54 other parameter combinations considered here result in excess risk functions for "abnormal" and for "joint", which are often comparable. However, there are a number of parameter combinations for which the curve for "abnormal" is strikingly

Table 10.3 *Asymptotic Study. Parameter settings.*

Parameter	values			
$\beta_{0,\mathrm{mal}}$	-6	-4	-2	0
$\beta_{0,\mathrm{dth}} = f\beta_{0,\mathrm{mal}}$ with f	0.25	0.5	1	
$\beta_{d,\mathrm{mal}} = \beta_{d,\mathrm{dth}}$	2	4	6	
$\beta_{2,\mathrm{mal}} = \beta_{2,\mathrm{dth}}$	0.0	0.2		

larger than for "joint", i.e., the overly simplistic model leads to higher excess risks than the correctly specified joint model.

In general, for an increasing value of the ratio $f = \beta_{0,\mathrm{dth}}/\beta_{0,\mathrm{mal}}$, the risk curves seem to get closer to each other. The same holds for increasing values of $\beta_{0,\mathrm{mal}}$ and of $\beta_{d,\mathrm{mal}} = \beta_{d,\mathrm{dth}}$. Furthermore, in case of association ($\beta_{2,\mathrm{mal}} = \beta_{2,\mathrm{dth}} = 0.2$), the excess risk at a particular dose is in general smaller than in the case of independence.

10.4 Concluding Remarks

Developmental toxicity studies are complicated by the hierarchical (death, malformation, healthy foetus), clustered (foetuses within litters) and multivariate (several malformation indicators and low birth weight) nature of the data. As a consequence, a multitude of modeling strategies, with varying degrees of simplification, have been proposed in the literature. Such choices are often subjective and can affect the quantitative risk assessment based on the fitted models.

The emphasis was on the choice between (1) the beta-binomial model versus the conditional model proposed by Molenberghs and Ryan (1999), (2) modeling death only, modeling malformation only, modeling a collapsed outcome indicating death or malformation (termed "abnormal") or a joint model for death and malformation. The main emphasis has been put on (3) the distinction between foetus-based and litter-based risk assessment.

It has been argued that benchmark doses calculated from the litter-based approach are between 5 and 10 times smaller than those obtained from the foetus-based approach. Thus, while the latter seems to be the standard in practice, it is deduced that a litter-based approach should be considered far more often. Furthermore, from a biological perspective, one could argue that litter-based inference makes sense since a litter represents the typical pregnancy outcome in a rodent, compared with a single birth in humans. However, in general, litter-based risk assessment has not been widely studied, nor compared with foetus-based risk assessment in a systematic way. While this chapter does not resolve the issue of whether to use foetus or litter-based risk

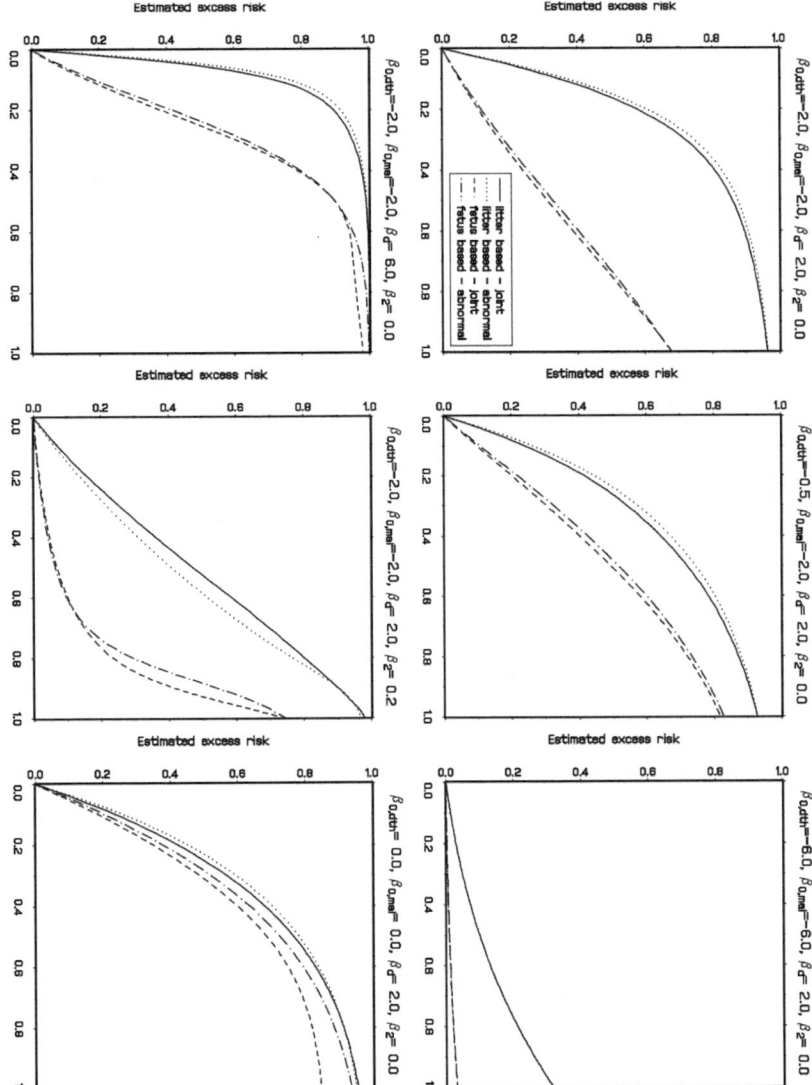

Figure 10.2 *NTP Data. Excess risk versus dose for asymptotic samples based on the conditional model.*

assessment procedures, it raises the question in a new way and provides a convincing argument that further work, statistical and biological, is needed on this topic.

In most cases, the beta-binomial model yields somewhat smaller BMDs than the conditional approach, but the differences are less pronounced. A joint model for death and malformation yields in some cases approximately the same risk as a collapsed indicator for abnormality, but there are regions in the parameter space where the former yields considerably larger BMDs. As a result, a joint modeling strategy is recommended.

CHAPTER 11

Model Misspecification

Gerda Claeskens

Texas A & M University, College Station, TX

Marc Aerts

transnationale Universiteit Limburg, Diepenbeek, Belgium

Lieven Declerck

European Organization for Research and Treatment of Cancer, Brussels, Belgium

In this chapter, the effect of misspecifying the parametric response model on dose effect estimation, hypothesis testing and safe dose determination for the clustered binary data models is investigated. In Section 11.1 the impact on testing for dose effect under a possibly incorrectly specified model is investigated. A robust alternative to the classical parametric approaches is proposed in Section 11.2. The impact of misspecifying a model on the determination of a safe dose is studied in Section 11.3, while profile score-based alternatives are introduced in Section 11.4.

11.1 Implications of Misspecification on Dose Effect Assessment

If the correct model is fit to a dataset of finite sample size, it is well known that score, Wald and likelihood ratio test statistics have the same asymptotic χ^2 distribution under the null model as well as under contiguous alternatives (Serfling 1980). For fixed alternatives, the picture is less clear and when fitting the incorrect model, asymptotic theory is not always available. In Molenberghs, Declerck and Aerts (1998), ideas of Rotnitzky and Wypij (1994) are adapted to get asymptotic information on the effect of model misspecification on dose effect assessment. These have been described in Sections 7.2 and 10.3 An artificial sample is constructed, where each combination of dose d_i, number of viable foetuses n_i and number of malformations z_i is assigned a weight equal to the probability $f(d_i, n_i, z_i)$ that this combination occurs in the underlying model. For a specified probability model the value of the test statistic is computed for this artificial sample. These are the population test statistics and have the following interpretation. Suppose a very large sample of size N is obtained, in which a given cluster occurs exactly with the multiplicity predicted by the model; then the corresponding test statistics are N times the population values. Therefore, this method is useful to investigate the large sample effect of choosing an incorrect model.

A comparison of the likelihood ratio (LR) and Wald (W) statistics for the Bahadur (Bah), beta-binomial (BB) and conditional (MR) models in Molenberghs, Declerck and Aerts (1998) shows that all population test statistics are very close when the true dose effect is small. For higher dose effects, one observes that

$$LR(Bah) \gg LR(Cond) > LR(BB)$$

and that

$$W(Bah) \geq W(BB) \gg W(Cond).$$

For the conditional model, one should bear in mind that all parameters, including the dose effect parameter, are conditional in nature. A marginal dose effect is likely to depend in a complex way on the model parameters. Since the Wald test is known to depend on the particular parameterization (in contrast to likelihood ratio and score tests), it might be a less relevant measure, in particular for conditional models. Finally, all Wald tests show a non-monotone trend, an aberrant behavior in agreement with Hauck and Donner (1977). Likelihood ratio test statistics all increase with increasing dose.

For more details on the comparison of the test statistics for a set of models for clustered binary data, we refer to Molenberghs, Declerck and Aerts (1998). After illustrating the effect of the model on two NTP datasets in Section 11.1.1, we explain how robust test statistics in combination with bootstrap methods can be used to largely circumvent model misspecification problems.

11.1.1 Analysis of NTP Data

Apart from the external, skeletal, and visceral malformation outcomes, a collapsed malformation outcome is considered, which is one if a foetus exhibits at least one type of malformation and zero otherwise. Tables 11.1 and 11.2 contain maximum likelihood estimates (MLE) and standard errors for the Bahadur, beta-binomial and conditional models. Estimates of the Bahadur model parameters obtained by a GEE2 (Chapter 5) method are also shown.

Bahadur (MLE and GEE2) and beta-binomial parameters have the same interpretation, but they are not directly comparable with the parameters of the conditional model. For the first three models, the intercepts β_0 and dose effect parameters β_d have similar numerical values but the situation is slightly different for the association parameter β_a, where in 6 out of 8 cases, $\beta_a(Bah) < \beta_a(GEE2) < \beta_a(BB)$. The only exceptions are EG (visceral), where

Table 11.1 *NTP Data. Parameter estimates (standard errors) for the DEHP study.*

Outcome	Par.	Bah	GEE2(Bah)	BB	Cond
External	β_0	-4.93(0.39)	-4.98(0.37)	-4.91(0.42)	-2.81(0.58)
	β_d	5.15(0.56)	5.29(0.55)	5.20(0.59)	3.07(0.65)
	β_a	0.11(0.03)	0.15(0.05)	0.21(0.09)	0.18(0.04)
Visceral	β_0	-4.42(0.33)	-4.49(0.36)	-4.38(0.36)	-2.39(0.50)
	β_d	4.38(0.49)	4.52(0.59)	4.42(0.54)	2.45(0.55)
	β_a	0.11(0.02)	0.15(0.06)	0.22(0.09)	0.18(0.04)
Skeletal	β_0	-4.67(0.39)	-5.23(0.40)	-4.88(0.44)	-2.79(0.58)
	β_d	4.68(0.56)	5.35(0.60)	4.92(0.63)	2.91(0.63)
	β_a	0.13(0.03)	0.18(0.02)	0.27(0.11)	0.17(0.04)
Collapsed	β_0	-3.83(0.27)	-5.23(0.40)	-3.83(0.31)	-2.04(0.35)
	β_d	5.38(0.47)	5.35(0.60)	5.59(0.56)	2.98(0.51)
	β_a	0.12(0.03)	0.18(0.02)	0.32(0.10)	0.16(0.03)

the association is not statistically significant and EG (collapsed), where the three estimates are very close. In the other cases, the beta-binomial MLE for β_a is typically about double the corresponding Bahadur MLE. This is due to range restrictions on β_a in the Bahadur model. For instance, the allowable range of β_a for the external outcome in the DEHP data is $(-0.0164; 0.1610)$ when β_0 and β_d are fixed at their MLE. This range excludes the MLE under a beta-binomial model. It translates to $(-0.0082; 0.0803)$ on the correlation scale (see also Appendix A). A GEE2 estimate is valid as soon as the second, third and fourth order joint probabilities are non-negative, whereas the likelihood analysis requires all joint probabilities to be non-negative. Thus, a correlation valid for GEE2 estimation is allowed to violate the full likelihood range restrictions. The standard errors, obtained by Bahadur and GEE2, are very similar, except for EG(skeletal) and EG(collapsed). It is no coincidence that exactly in these cases, β_a attains very high values, probably very close to the boundary of the admissible range, implying that boundary effects might distort large sample approximations to the null distribution. The beta-binomial model features all positive correlations. Hence, the dominant ordering of the estimated β_a parameters reflects the severity of the parameter restrictions.

Since the conditional model has no restrictions on the parameters, it is easier to fit than the others. In all 8 examples, standard starting values (all parameters equal to zero) led to convergence.

After having discussed the parameter estimates of the fitted models, the focus is now on the problem of testing the null hypothesis of no dose effect.

Table 11.2 *NTP Data. Parameter estimates (standard errors) for the EG study.*

Outcome	Par.	Bah	GEE2(Bah)	BB	Cond
External	β_0	-5.25(0.66)	-5.63(0.67)	-5.32(0.71)	-3.01(0.79)
	β_d	2.63(0.76)	3.10(0.81)	2.78(0.81)	2.25(0.68)
	β_a	0.12(0.03)	0.15(0.05)	0.28(0.14)	0.25(0.05)
Visceral	β_0	-7.38(1.30)	-7.50(1.05)	-7.45(1.17)	-5.09(1.55)
	β_d	4.25(1.39)	4.37(1.14)	4.33(1.26)	3.76(1.34)
	β_a	0.05(0.08)	0.02(0.02)	0.04(0.09)	0.23(0.09)
Skeletal	β_0	-2.49(0.11)	-4.05(0.33)	-2.89(0.27)	-0.84(0.17)
	β_d	2.96(0.18)	4.77(0.43)	3.42(0.40)	0.98(0.20)
	β_a	0.27(0.02)	0.30(0.03)	0.54(0.09)	0.20(0.02)
Collapsed	β_0	-2.51(0.09)	-4.07(0.71)	-2.51(0.09)	-0.81(0.16)
	β_d	3.05(0.17)	4.89(0.90)	3.05(0.17)	0.97(0.20)
	β_a	0.28(0.02)	0.26(0.14)	0.28(0.02)	0.20(0.02)

Results are summarized in Table 11.3. They are in agreement with previous findings.

For the LR tests, one observes that LR(Bah) dominates the others. LR(BB) is considerably smaller and the smallest values are found with LR(Cond). This picture is seen in 7 out of 8 cases. A slightly different picture is seen for EG (visceral and external outcomes), where all three statistics are in fact very close to each other. However, although there are discrepancies between the magnitudes of the LR statistics, they all reject the null hypothesis.

Comparing LR to Wald tests, the former ones are seen to dominate the latter in most cases: LR(Cond)>W(Cond) in all 8 cases and LR(Bah)>W(Bah) in 6 cases. However, LR(BB)>W(BB) in only two cases and, more importantly, agreement between both test statistics is very close, providing evidence for approximate equivalence of both tests under a range of alternatives. This feature is in agreement with the asymptotic findings. Recall that both Bahadur test statistics might differ due to a misspecified higher order correlation structure, whereas for the conditional model, the Wald statistic could be low due to sensitivity of the test to the particular parameterization adopted. For example, the correlations between $\hat{\beta}_d$ and the other parameter estimates for the external outcome in the DEHP study are corr$(\hat{\beta}_0, \hat{\beta}_d) = -0.96$ and corr$(\hat{\beta}_a, \hat{\beta}_d) = -0.79$, as opposed to -0.91 and 0.27 for Bahadur and -0.90 and 0.23 for beta-binomial.

Among the Wald tests, W(Bah) and W(BB) are reasonably close to each other, apart from two aberrant cases (EG skeletal and EG collapsed). Misspecification might be one of the sources for the observed discrepancies. When

Table 11.3 *NTP Data. Wald and likelihood ratio test statistics for the DEHP and EG data. H_0: no dose effect, in a constant association model. W_r denotes the robustified Wald statistic.*

		External		Visceral		Skeletal		Collapsed	
		DEHP	EG	DEHP	EG	DEHP	EG	DEHP	EG
Bah	LR	96.48	15.05	81.28	16.00	76.40	182.45	164.75	189.99
	W	85.94	11.89	78.82	9.40	71.02	261.22	130.31	314.40
BB	LR	71.58	13.18	59.78	17.37	58.51	63.88	91.66	65.36
	W	76.78	11.61	66.80	11.82	61.39	74.89	98.46	71.58
Cond	LR	43.20	14.43	33.72	13.98	38.95	49.95	74.48	50.74
	W	22.30	10.78	19.91	7.81	21.46	23.15	33.71	23.39
GEE2	W	79.40	12.50	71.45	10.32	70.72	120.99	113.75	121.87
	W_r	92.41	14.70	58.23	14.64	78.87	58.53	92.27	29.69

GEE2 based tests are believed to correct for (at least part) of aforementioned misspecification, their values are smaller than W(Bah) and much closer to W(BB). The most striking phenomenon is that the two aberrant W(Bah) values in the EG data are indeed largely corrected downwards by the GEE2 versions.

Liang and Hanfelt (1994) have shown that assuming a constant intraclass correlation in the beta-binomial model might substantially bias mean parameter estimation and testing. Therefore, it is useful to study at least one possible departure from the constant association model. In the Bahadur, beta-binomial and conditional models, the association parameter β_a was allowed to vary linearly with dose level, $\beta_{2i} = \beta_{20} + \beta_{2d}d_i$, extending the three parameter families $(\beta_0, \beta_d, \beta_a)$ to four parameter versions $(\beta_0, \beta_d, \beta_{20}, \beta_{2d})$. Reconsider now the problem of testing the null hypothesis of no dose effect, neither on the malformation rate nor on the association. The corresponding test statistics are shown in Table 11.4. Values in bold correspond to those cases where the null hypothesis of a constant association parameter $H_0 : \beta_{2d} = 0$ was rejected on the basis of a one degree of freedom likelihood ratio test. Clearly, a non-constant association in one model (e.g., the conditional model) does not necessarily imply the same for the other models (e.g., the Bahadur and beta-binomial models). Next, the test statistics for dose effect are considered, which in the four parameter model becomes $H_0 : \beta_d = \beta_{2d} = 0$. In most cases, the statistics vary only mildly, although W(Bah) tends to increase somewhat more. The discrepancy is larger when the null hypothesis of a constant association is rejected. Of course, one has to bear in mind that these test statistics should be compared to a null χ^2 distribution with *two* degrees of freedom, diluting power when there is no evidence for non-constant association. Finally, failure to detect a linear trend on the association does not imply that the

Table 11.4 *NTP Data. Wald and likelihood ratio test statistics, H_0: no dose effect in a linear model with linear dose effect on the association parameter. Bold figures refer to cases where the dose effect on the association is significant at the 5% nominal level.*

		External		Visceral		Skeletal		Collapsed	
		DEHP	EG	DEHP	EG	DEHP	EG	DEHP	EG
Bah	LR	99.09	**19.99**	**86.13**		76.59	**192.77**	**173.07**	**196.06**
	W	93.04	**18.07**	**90.00**		69.60	**207.93**	**157.37**	**211.28**
BB	LR	71.90	13.57	61.56	17.55	58.65	65.74	**97.98**	67.45
	W	73.70	11.20	67.17	9.48	63.07	63.50	**107.17**	63.88
Cond	LR	44.99	16.19	33.72	14.89	40.70	**58.90**	75.61	**60.01**
	W	25.98	12.39	19.99	7.44	24.98	**30.72**	33.00	**30.35**

association is constant, since the association function might have an entirely different shape (e.g., quadratic). It is advisable to explore these functions in a bit more detail (Molenberghs and Ryan 1999). Also smoothing methods are an excellent way to get some idea about the functional form; see Chapter 8 for a discussion.

11.2 A Robust Bootstrap Procedure

When the data do not come from the assumed parametric model, the usual asymptotic chi-squared distribution under the null hypothesis remains valid for "robustified" Wald and score test statistics. For full likelihood models, robust Wald and score tests have been described by Kent (1982), Viraswami and Reid (1996), amongst others. The modified tests again have an asymptotic chi-squared distribution, even when the assumed model is not correct. We are not aware of a modified likelihood ratio test with asymptotic chi-squared distribution. Robust test statistics are also used in the context of generalized estimating equations (Rotnitzky and Jewell 1990) and in the pseudo-likelihood approach (Geys, Molenberghs and Ryan 1999).

Aerts and Claeskens (2001) compare the performance of this chi-squared approximation to that of a semiparametric bootstrap method. The bootstrap is a well-established statistical methodology nowadays. There are several papers and books showing a multitude of examples where the bootstrap can be implemented and applied successfully (e.g., Davison and Hinkley 1997 and references therein). The main difficulty in bootstrap hypothesis testing is the generation of bootstrap data reflecting the null hypothesis. Assuming the true likelihood of the data to be known, this can be achieved by the parametric bootstrap based on the null estimates. Such an approach has been generalized to pseudo-likelihood models and applied to clustered binary data in Aerts and Claeskens (1999). There it is shown that the parametric bootstrap test leads to a substantial improvement. In practice, however, the assumed probability model can be wrong, in which case the parametric bootstrap leads to incorrect results. The semiparametric bootstrap method of Aerts and Claeskens (2001), which will be explained in this section, remains valid when the assumed model is incorrect.

11.2.1 A One-Step Bootstrap Procedure

We formulate the procedure in general, such that it can be applied to essentially any type of estimating equations, whether maximum likelihood, quasi-likelihood, pseudo-likelihood or generalized estimating equations. Let $\boldsymbol{Y}_i$ ($i = 1, \ldots, n$) be independent response variables of length m with unknown joint density or discrete probability function (abbreviated: pdf) $g(\boldsymbol{y}_i; \boldsymbol{x}_i)$ where $\boldsymbol{y}_i = (y_{i1}, \ldots, y_{im})$ and $\boldsymbol{x}_i = (x_{i1}, \ldots, x_{ip})$, the latter representing a vector of p explanatory variables. In the context of clustered binary data, m corresponds to the size of the cluster.

In general, parametric inference is based on an r dimensional score or estimating function $\boldsymbol{\psi}(\boldsymbol{y}; \boldsymbol{x}, \boldsymbol{t})$, where the "true" parameter $\boldsymbol{\theta} = (\theta_1, \ldots, \theta_r)$ is defined as the solution $\boldsymbol{t}$ to the equations

$$\sum_{i=1}^{n} E\{\boldsymbol{\psi}(\boldsymbol{Y}_i; \boldsymbol{x}_i, \boldsymbol{t})\} = \boldsymbol{0}$$

where all expectations are with respect to the true pdf $g(\boldsymbol{y}_i; \boldsymbol{x}_i)$. Solving the system of equations

$$\sum_{i=1}^{n} \boldsymbol{\psi}(\boldsymbol{Y}_i; \boldsymbol{x}_i, \boldsymbol{t}) = \boldsymbol{0}$$

leads to the estimator $\hat{\boldsymbol{\theta}}_n$ for $\boldsymbol{\theta}$. Within classical maximum likelihood

$$\boldsymbol{\psi}(\boldsymbol{y}; \boldsymbol{x}, \boldsymbol{t}) = (\partial/\partial \boldsymbol{t}) \log f(\boldsymbol{y}; \boldsymbol{x}, \boldsymbol{t}).$$

For example, for clustered binary data, $f(\boldsymbol{y}; \boldsymbol{x}, \boldsymbol{t})$ might represent the beta-binomial distribution or the conditional model proposed by Molenberghs and Ryan (1999). Note that, in this setting, the assumed pdf $f(\boldsymbol{y}; \boldsymbol{x}, \boldsymbol{t})$, the model used for fitting the data, might not contain the true data generating structure $g(\boldsymbol{y}; \boldsymbol{x})$.

The proposal of Aerts and Claeskens (2001) is to resample the score and the differentiated score values. Based on a linear approximation, a bootstrap replicate of $\hat{\boldsymbol{\theta}}_n$ is constructed as

$$\hat{\boldsymbol{\theta}}_n^* = \hat{\boldsymbol{\theta}}_n - \left(\sum_{i=1}^{n} \dot{\boldsymbol{\psi}}_i^*(\hat{\boldsymbol{\theta}}_n)\right)^{-1} \sum_{i=1}^{n} \boldsymbol{\psi}_i^*(\hat{\boldsymbol{\theta}}_n) \tag{11.1}$$

where $(\boldsymbol{\psi}_i^*(\hat{\boldsymbol{\theta}}_n), \dot{\boldsymbol{\psi}}_i^*(\hat{\boldsymbol{\theta}}_n))$, $i = 1, \ldots, n$ is a sample with replacement from the set

$$\left\{ \left(\boldsymbol{\psi}(\boldsymbol{Y}_i; \boldsymbol{x}_i, \hat{\boldsymbol{\theta}}_n), (\partial/\partial\boldsymbol{\theta})\boldsymbol{\psi}(\boldsymbol{Y}_i; \boldsymbol{x}_i, \hat{\boldsymbol{\theta}}_n) \right), i = 1, \dots, n \right\}.$$

Consider the hypothesis $H_0 : \boldsymbol{\theta} \in \boldsymbol{\Theta}_0$ versus $H_1 : \boldsymbol{\theta} \in \boldsymbol{\Theta} \backslash \boldsymbol{\Theta}_0$ where $\boldsymbol{\Theta}_0$ is a $(r-t)$ dimensional subspace of the parameter space $\boldsymbol{\Theta}$ such that the parameter of interest $\boldsymbol{\theta} = (\theta_1, \dots, \theta_r)$ belongs to $\boldsymbol{\Theta}_0$ if and only if $\theta_1 = \dots = \theta_t = 0$, $1 \leq t \leq r$. More general situations, in which H_0 is of the form $H_0 : h_1(\boldsymbol{\theta}) = \dots = h_t(\boldsymbol{\theta}) = 0$ for some smooth real-valued functions $h_1, \dots, h_t$, can be put into this form by a reparametrization.

The bootstrap Wald and score test statistics based on $\hat{\boldsymbol{\theta}}_n^*$ as defined in (11.1) coincide and are given by

$$W_n^* = S_n^* = n(\hat{\boldsymbol{\theta}}_n^* - \hat{\boldsymbol{\theta}}_n^{(0)})_L^T \left(\boldsymbol{J}_n^*(\hat{\boldsymbol{\theta}}_n)^{-1} \boldsymbol{K}_n^*(\hat{\boldsymbol{\theta}}_n) \boldsymbol{J}_n^*(\hat{\boldsymbol{\theta}}_n)^{-1} \right)_{LL}^{-1} (\hat{\boldsymbol{\theta}}^* - \hat{\boldsymbol{\theta}}_n^{(0)})_L,$$

where $\boldsymbol{A}_{LL}$ denotes the left upper $t \times t$ submatrix of A and $\boldsymbol{m}v_L$ the subvector of the first t elements. Further,

$$\boldsymbol{J}_n(\boldsymbol{\theta}) = -\frac{1}{n} \sum_{i=1}^{p} \sum_{j=1}^{n_i} \frac{\partial}{\partial\boldsymbol{\theta}} \boldsymbol{\psi}_i(\boldsymbol{Y}_{ij}, \boldsymbol{\theta}),$$

$$\boldsymbol{K}_n(\boldsymbol{\theta}) = \frac{1}{n} \sum_{i=1}^{p} \sum_{j=1}^{n_i} \boldsymbol{\psi}_i(\boldsymbol{Y}_{ij}, \boldsymbol{\theta}) \boldsymbol{\psi}_i(\boldsymbol{Y}_{ij}, \boldsymbol{\theta})^T.$$

Finally, $\boldsymbol{J}_n^*(\boldsymbol{\theta})$ and $\boldsymbol{K}_n^*(\boldsymbol{\theta})$ are defined the same way as $\boldsymbol{J}_n(\boldsymbol{\theta})$ and $\boldsymbol{K}_n(\boldsymbol{\theta})$, but using the bootstrap scores and derivatives of the scores $\{\boldsymbol{\psi}_{ij}^*(\hat{\boldsymbol{\theta}}_n), \dot{\boldsymbol{\psi}}_{ij}^*(\hat{\boldsymbol{\theta}}_n)\}$, $(j = 1, \dots, n_i)$ instead of $\{\boldsymbol{\psi}_i(\boldsymbol{Y}_{ij}, \hat{\boldsymbol{\theta}}_n), (\partial/\partial\boldsymbol{\theta})\boldsymbol{\psi}_i(\boldsymbol{Y}_{ij}, \hat{\boldsymbol{\theta}}_n)\}$, $(j = 1, \dots, n_i)$. Further, $\hat{\boldsymbol{\theta}}^{(0)}$ is a $\sqrt{n}$-consistent estimator of $\boldsymbol{\theta}$ under the null hypothesis.

The motivation for defining S_n^* equal to W_n^* follows from the classical arguments in proving the asymptotic normality of the score test statistic. It is well known that both test statistics are first order equivalent. A typical way of obtaining the asymptotic distribution of the score test statistic is by substituting $(\hat{\boldsymbol{\theta}}_n^* - \hat{\boldsymbol{\theta}}_n^{(0)})_L$ in the Wald statistic by the first t components of the second term in (11.1). By definition, (11.1) of the one-step linear estimator, this substitution is exact; for estimators in general, this is only approximate.

Definition (11.1) follows from a linear approximation of the score equations. One might improve on this by including quadratic and higher order terms. We focus attention to the following second order approximation,

$$\begin{aligned}
\boldsymbol{0} = & \sum_{j=1}^{n} \boldsymbol{\psi}(\boldsymbol{Y}_j, \boldsymbol{\theta}) + \sum_{k=1}^{r} \sum_{j=1}^{n} \frac{\partial}{\partial\theta_k} \boldsymbol{\psi}(\boldsymbol{Y}_j, \boldsymbol{\theta})(\hat{\theta}_{nk} - \theta_k) \\
& + \sum_{k=1}^{r} \sum_{\ell=1}^{r} \sum_{j=1}^{n} \frac{\partial^2}{\partial\theta_k \partial\theta_\ell} \boldsymbol{\psi}(\boldsymbol{Y}_j, \boldsymbol{\theta})(\hat{\theta}_{nk} - \theta_k)(\hat{\theta}_{n\ell} - \theta_\ell) \\
& + O_P(n^{-1/2}).
\end{aligned} \qquad (11.2)$$

By calculations similar to those of Ghosh (1994), a Taylor expansion suggests the following one-step quadratic estimator

$$\hat{\boldsymbol{\theta}}_n^* = \hat{\boldsymbol{\theta}}_n^{(0)} + \boldsymbol{U}_n^*$$

$$-\frac{1}{2}\left(\sum_{j=1}^n \dot{\boldsymbol{\psi}}_j^*(\hat{\boldsymbol{\theta}}_n)\right)^{-1}\sum_{k=1}^r\sum_{\ell=1}^r\sum_{j=1}^n \ddot{\boldsymbol{\psi}}_j^*(\hat{\boldsymbol{\theta}}_n)_{k,\ell}U_{nk}^*U_{n\ell}^*, \quad (11.3)$$

with

$$\boldsymbol{U}_n^* = -\{\sum_{i=1}^p\sum_{j=1}^{n_i}\dot{\boldsymbol{\psi}}_{ij}^*(\hat{\boldsymbol{\theta}}_n)\}^{-1}\sum_{i=1}^p\sum_{j=1}^{n_i}\boldsymbol{\psi}_{ij}^*(\hat{\boldsymbol{\theta}}_n).$$

This bootstrap estimator is based on the values

$$(\boldsymbol{\psi}_j^*(\hat{\boldsymbol{\theta}}_n), \dot{\boldsymbol{\psi}}_j^*(\hat{\boldsymbol{\theta}}_n), \ddot{\boldsymbol{\psi}}_j^*(\hat{\boldsymbol{\theta}}_n)),$$

$(j = 1, \ldots, n)$ taken with replacement from the set

$$\left\{\left(\boldsymbol{\psi}(\boldsymbol{Y}_j, \hat{\boldsymbol{\theta}}_n), (\partial/\partial\boldsymbol{\theta})\boldsymbol{\psi}(\boldsymbol{Y}_j, \hat{\boldsymbol{\theta}}_n), (\partial^2/\partial\boldsymbol{\theta}\partial\boldsymbol{\theta}^T)\boldsymbol{\psi}(\boldsymbol{Y}_j, \hat{\boldsymbol{\theta}}_n)\right), j = 1, \ldots, n\right\}.$$

It is expected that the last term at the right-hand side of (11.3) improves the representation of the random variation about the null estimate $\hat{\boldsymbol{\theta}}_n^{(0)}$. This is confirmed in a simulation study in Aerts and Claeskens (2001). For the no dose effect null hypothesis we might consider two different resampling schemes. In a first method scores and differentiated scores are resampled for each dose level separately. We denote this by B_1/D for the linear one-step approximation, or by B_2/D when the quadratic approximation is being constructed. For the no dose effect null hypothesis, an alternative valid resampling scheme is to ignore the presence of the dose and to resample from the complete set of scores and differentiated scores. This resampling scheme is denoted by B_i/A $(i = 1, 2)$.

Both bootstrap procedures (linear and quadratic) lead to consistent estimators for the null distribution of the robust Wald and score test statistics.

11.2.2 The Theophylline Data

Lindström et al. (1990) investigated the effect in mice of the chemical theophylline. In each of the three models, the beta-binomial (BB) model (Section 4.3.1), the conditional model (Section 4.2) of Molenberghs and Ryan (1999; MR) and the pseudo-likelihood approach (PL) we used a linear/constant parameterization, i.e., a linear function for the mean function and a constant association function. We are interested in testing the no dose effect hypothesis on the main effect parameter ($H_0 : \theta_{11} = 0$). The results are shown in Table 11.5. The table shows the p values (as %) of the different test statistics discussed before. We also included the results from a GEE2 estimation method based on the first four order moments of the Bahadur model.

For *external* malformations, the results of the different tests are almost the same for the beta-binomial model and close to each other for the other

Table 11.5 *NTP Data. Analysis of theophylline with* $H_0 : \beta_d = 0$. *Robust Wald statistic* W_r *and robust score statistic* R; *p values are shown as* %.

		External		Visceral		Skeletal		Collapsed	
		W_r	R	W_r	R	W_r	R	W_r	R
BB	χ^2	22.85	23.54	4.50*	—	0.00*	31.98	4.58*	5.33
	B_1/A	22.78	23.25	0.50*	—	0.00*	13.10	6.16	6.96
	B_1/D	23.98	24.66	0.90*	—	0.00*	13.70	5.32	6.00
GEE2	χ^2	14.60	13.90	3.61*◇	17.27◇	0.00*	36.86	3.45*	4.20*
	B_1/A	15.70	14.40	1.40*◇	18.80◇	0.00*	19.43	4.05*	4.80*
	B_1/D	14.20	13.30	1.00*◇	20.20◇	0.00*	22.86	1.70*	2.20*
MR	χ^2	9.76	16.98	3.55*	18.26	2.19*	18.82	3.40*	7.54
	B_1/A	9.30	18.80	0.60*	21.60	0.30*	16.60	2.30*	7.20
	B_1/D	7.80	16.30	0.50*	19.70	0.20*	14.50	1.50*	5.20
	B_2/A	8.70	—	5.20	—	28.10	—	2.50*	—
	B_2/D	7.70	—	3.40*	—	27.40	—	1.70*	—
PL	χ^2	12.26	17.06	3.47*	18.56	0.64*	16.08	4.11*	6.83
	B_1/A	12.30	19.10	0.50*	21.70	0.00*	7.30	3.90*	6.70
	B_1/D	11.10	16.10	0.50*	19.70	0.00*	6.70	2.20*	5.20
	B_2/A	17.60	—	10.30	—	1.90*	—	8.20	—
	B_2/D	13.90	—	8.60	—	1.10*	—	5.40	—

A $*$ denotes rejection at the 5% level and a ◇ indicates that a Moore-Penrose generalized inverse is used to obtain the results.

models. There seems to be no significant effect of theophylline on the external malformation probability.

For *visceral* and *skeletal* malformations, there is a striking discrepancy between the Wald and the score test. The significance of the Wald test should be interpreted with care because of inflated type I errors for model misspecification settings. The quadratic bootstrap seems to correct the Wald test in the direction of the score test. Compared with the chi-squared tests, the bootstrap score test has higher p values for visceral malformation and lower p values for skeletal malformation. For the beta-binomial model there were convergence problems when fitting the null model. For this reason, the score statistics could not be obtained. Also, for the GEE2 model some problems arose, but these could be avoided by using a Moore-Penrose generalized inverse of the matrix J. For visceral malformations, the null hypothesis cannot be rejected. The picture is less clear for skeletal malformations. Except for the quadratic bootstrap for the MR model, all Wald tests indicate a significant dose effect, while all score tests indicate no effect (although the p value of the bootstrap score PL test is nearing 5%). Since the score tests showed also a good behavior in a simulation study, we tend to better believe the results of these tests, though further investigation might be necessary to come to a definite conclusion. Finally, also note that, for all three types of malformation, the different Wald tests lead to highly variable p values whereas the score tests are much stabler.

The *collapsed* version seems to indicate that theophylline might have an effect on the development of foetuses. Here, all score p values are between 0.0220 and 0.0754 and the Wald p values are between 0.015 and 0.082. This indicates that a separate analysis of each type of malformation can lead to misleading conclusions and that for this type of problems one has to consider all types jointly as a multivariate response or at least, as we did here, a collapsed malformation indicator.

11.2.3 Bias Correction and Double Bootstrap

Although the maximum likelihood estimator $\hat{\boldsymbol{\theta}}_n$ is asymptotically unbiased, the quadratic one-step bootstrap procedure can be used for finite sample bias correction. In practical applications, a large number, B say, resamples are taken, resulting in a set of B bootstrap estimators $\hat{\boldsymbol{\theta}}_n^{*1}, \ldots, \hat{\boldsymbol{\theta}}_n^{*B}$. From this set a bias-corrected estimator is defined as

$$\boldsymbol{\theta}_n^{bc} = 2\hat{\boldsymbol{\theta}}_n - \frac{1}{B}\sum_{i=1}^{B} \hat{\boldsymbol{\theta}}_n^{*i}.$$

The intuition behind this definition is clear. We subtract from the estimator $\boldsymbol{\theta}_n$, the estimated bias based on the B bootstrap replicates. For bias estimation, the second order approximation turns out to be very useful, which is not completely unexpected since the bias is a second order aspect of the asymptotic properties of the estimator. Simulations show that this bias correction might even decrease the variance. Using a double bootstrap procedure, Aerts, Claeskens and Molenberghs (1999) study the distribution of $\hat{\boldsymbol{\theta}}_n^{bc}$ and define a bootstrap based variance estimator for $\hat{\boldsymbol{\theta}}_n^{bc}$. An important observation in their simulation study is that the bias correction even decreases the variance, as the simulated standard deviation $\sigma(\hat{\beta}_0^{bc})$ and $\sigma(\hat{\beta}_1^{bc})$ are, for all settings in their study, smaller than the corresponding simulated values of $\sigma(\hat{\beta}_0)$ and $\sigma(\hat{\beta}_1)$, respectively.

To study the distribution of the bias corrected estimator, which is already based on a bootstrap resampling scheme, we will need a second bootstrap stage. The double bootstrap procedure reads as follows. Using a non-parametric resampling scheme (i.e., by resampling the data directly), we construct a set of bootstrap estimators. In the same way as described before (i.e., we construct the one-step quadratic bootstrap), perform a bias correction, and next, construct the bias-corrected estimator. To study the distribution of the resulting estimator, we construct from each of the resampled data new bootstrap estimators via the one-step quadratic bootstrap. These values can now be used to compute the estimator's variance, quantiles, etc.

11.2.4 Bootstrap Confidence Intervals

Confidence intervals for the parameter $\boldsymbol{\theta}$ can be derived from the asymptotic normality result, by using the Wald statistic as a pivot. Next to this classical approach, the appropriate quantiles can be selected from the bootstrap approximation to the asymptotic distribution. Aerts, Claeskens and Molenberghs (1999) construct bootstrap confidence intervals from the so-called hybrid bootstrap (Shao and Tu 1995, Sections 4.1 and 4.2). A $100(1-\alpha)\%$ confidence interval for the parameter $\boldsymbol{\theta}$ is defined as

$$\{\boldsymbol{\theta} : z_L^* \leq \sqrt{n}(\hat{\boldsymbol{\theta}}_n - \boldsymbol{\theta}) \leq z_R^*\},$$

where z_L^* is the $100\alpha/2\%$ and z_R^* is the $100(1-\alpha/2)\%$ quantile of the distribution of $\sqrt{n}(\hat{\boldsymbol{\theta}}_n^* - \hat{\boldsymbol{\theta}}_n)$.

Simulations show that the quadratic bootstrap everywhere reduces the length of the confidence intervals. This reduction in length is usually more pronounced for the highest confidence levels. For more information we refer to Aerts, Claeskens and Molenberghs (1999).

Although the one-step linear method is asymptotically equivalent to the fully iterative one, our experience in the setting of binary response data showed to interpret its results with care. Compared to the normal based confidence intervals, the linear one-step bootstrap tends to produce shorter confidence intervals but simulations show an equivalent decrease in coverage probability. Also, by definition, the one-step approach is not able to detect the bias of the estimator. These shortcomings seem to be greatly eliminated by the one-step *quadratic* bootstrap, which allows the construction of a bias corrected estimator and improved confidence intervals.

11.3 Implications of Misspecification on Safe Dose Determination

Quantitative risk assessment is based on the relationship between dose and response to derive a safe dose. Whereas in Chapter 10, we contrasted foetus and litter-based risks, here we focus on effects of model misspecification on the determination of a safe dose. As explained in Section 10.1, the benchmark dose BMD is this dose level d such that the excess risk function $r(d; \beta) = q$, where q is a user-specified value. Popular values of q are $0.05, 0.01$, or 10^{-4} depending on the specific area of application.

For example, for a linear dose-response effect in a Bahadur or a beta-binomial model:

$$BMD = \ln\left(\frac{1 + q\exp(-\beta_0)}{1 - q}\right)\beta_1^{-1}. \tag{11.4}$$

The maximum likelihood (ML) estimate of the effective dose is the solution to $r(d; \hat{\boldsymbol{\beta}}) = q$ where $\hat{\boldsymbol{\beta}}$ is the ML estimate of $\boldsymbol{\beta}$. Following Crump and Howe (1985), in the next section we will use a full likelihood method to determine a safe dose.

11.3.1 Likelihood Determination of a Safe Dose

A full likelihood approach to safe dose determination defines the virtually safe dose corresponding to an excess risk value q, as an approximate $100(1 - \alpha)\%$ lower confidence limit for the BMD; see also Section 10.1:

$$\mathrm{BMDL}_{m\ell} = \min\{d(\boldsymbol{\beta}) : r(d; \boldsymbol{\beta}) = q \text{ over all } \boldsymbol{\beta} \text{ such that}$$

$$2(\ell(\hat{\boldsymbol{\beta}}) - \ell(\boldsymbol{\beta})) \leq \chi_p^2(1 - 2\alpha)\}, \tag{11.5}$$

with $\ell(\boldsymbol{\beta})$ the value of the log-likelihood at $\boldsymbol{\beta}$, p the total number of parameters in the model and $\chi_p^2(1 - 2\alpha)$ the $(1 - 2\alpha)$-quantile of a χ_p^2 distributed random variable. The dependence of the log-likelihood function on the data is, for simplicity, omitted in the notation. For the Bahadur and beta-binomial models, the $(1 - \alpha)100\%$ lower limit can be rewritten as

$$\min\left\{\ln\left(\frac{1 + qe^{-\beta_0}}{1 - q}\right)\beta_1^{-1} \text{ over all } \beta_0, \beta_1\right.$$

$$\left. \text{s.t. } 2\{\ell(\hat{\beta}_0, \hat{\beta}_1, \hat{\beta}_a) - \ell(\beta_0, \beta_1, \beta_a)\} \leq \chi_3^2(1 - \alpha)\right\}. \tag{11.6}$$

The procedure is somewhat more involved for the conditional model, where (11.5) is solved numerically.

In Section 11.4, we will give an alternative definition which does not require a likelihood specification of the model.

11.3.2 A One-Dimensional Profile Likelihood Determination of a Safe Dose

Since expression (11.4) contains only the *two* coefficients β_0 and β_1 but the critical point is calculated from a chi-squared distribution with *three* degrees of freedom, this full likelihood approach is expected to be too conservative. Indeed, the parameter β_a can be considered as a nuisance parameter. A technique which treats (nuisance) parameters in a more parsimonious way is the profile likelihood method (Morgan 1992).

The partially maximized log-likelihood function

$$\ell_P(\beta_1) = \max_{\beta_0, \beta_a} \ell(\beta_0, \beta_1, \beta_a)$$

is the profile likelihood for β_1. A one-dimensional profile likelihood approach for BMDL calculation can be defined as follows:

$$\mathrm{BMDL}_{p\ell 1} = \min\left\{d(\beta_1|\hat{\beta}_0, \hat{\beta}_a) : r(d; \beta_1, \hat{\boldsymbol{\beta}}_n(\beta_1)) = q \text{ over all } \beta_1 \text{ such}\right.$$

$$\left. \text{that } 2\{\ell(\hat{\beta}_d, \hat{\boldsymbol{\beta}}_n(\hat{\beta}_1)) - \ell(\beta_1, \hat{\boldsymbol{\beta}}_n(\beta_1))\} \leq \chi_1^2(1 - 2\alpha)\right\}. \tag{11.7}$$

In Aerts, Declerck and Molenberghs (1997), an asymptotic study is performed, comparing both procedures $BMDL_{ml}$ and $BMDL_{pl1}$. Also, a comparison between an extrapolated version and a model based one is made. An extrapolated version, as suggested by many authors (Chen and Kodell 1989, Gaylor 1989, Ryan 1992), first determines a lower confidence limit, e.g., corresponding to an excess risk of 1% and then linearly extrapolates it to a BMD. The main advantage quoted for this procedure is that the determination of a BMD is less model dependent. In our case, the effective dose values are found to be too low under the extrapolated version. Morgan (1992, p. 175) and the Scientific Committee of the Food Safety Council (1980) point out that blind adherence to a conservative procedure is to be regarded as scientifically indefensible.

11.3.3 Analysis of NTP Data

Benchmark (effective) doses and safe doses are obtained for the NTP studies which investigate the effects in mice of the toxic agents DEHP, DYME, and EG. Details on these data are provided in Section 2.1.

Two estimates of the benchmark dose corresponding to an excess risk of $q = 10^{-4}$ are obtained. Besides an entirely model based (MB) effective dose, a linear extrapolation (EP) version is computed. Furthermore, four quantities for the lower confidence limit of the BMD are determined. In addition to the determination of the confidence region based $BMDL(ml)$, the profile likelihood version $BMDL_{pl1}$ is calculated. Table 11.6 shows model based BMDL's and the corresponding extrapolated versions. The conditional model in general yields the highest values for both BMD and $BMDL_{ml}$, in both the extrapolation and the model based methods, but the effect is somewhat clearer in the extrapolation procedure. To some extent, this result differs for the three chemicals under consideration.

The number of cases in which the Bahadur model results in higher values relative to the beta-binomial model is comparable to the number of cases with lower values for the Bahadur model.

One also notices that the extrapolation method yields much smaller values in all cases. This is in line with an asymptotic study, which has shown that effective doses computed by means of the extrapolation procedure result in lower values as compared to the model based estimates.

Next, the focus is on the various procedures to calculate BMDL's. First, one observes that $BMDL(pl1)$ is virtually always higher than $BMDL_{ml}$. This is to be expected, since it is based on a one degree of freedom procedure, whereas for $BMDL_{ml}$, three degrees of freedom are spent in case of the Bahadur and beta-binomial models. Of course, as pointed out in the previous section, a lower BMDL is "safer", but one should be careful not to be overly cautious (Morgan 1992). Second, the linearly extrapolated versions of Table 11.6 are smaller than their purely model based counterparts. These two observations

Table 11.6 *NTP Data. Estimates of effective doses and lower confidence limits. All quantities shown should be divided by* 10^4.

Outc.	Model		Ent. model based DEHP	EG	DYME	Linear extrap. DEHP	EG	DYME
Ext.	Bah	BMD	27	72	165	17	41	34
		$BMDL_{ml}$	15	47	48	12	38	23
		$BMDL_{pl1}$	18	55	63	14	39	25
	BB	BMD	26	73	168	17	41	34
		$BMDL_{ml}$	14	45	47	12	37	22
		$BMDL_{pl1}$	17	56	62	13	39	25
	Cond	BMD	36	124	141	24	57	36
		$BMDL_{ml}$	22	66	55	17	42	25
Visc.	Bah	BMD	19	350	171	14	67	44
		$BMDL_{ml}$	13	189	48	11	62	28
		$BMDL_{pl1}$	15	126	72	12	62	34
	BB	BMD	18	367	98	14	67	36
		$BMDL_{ml}$	11	199	63	10	62	31
		$BMDL_{pl1}$	14	131	40	11	62	26
	Cond	BMD	28	504	202	21	78	64
		$BMDL_{ml}$	18	134	95	15	55	47
Skel.	Bah	BMD	23	4	13	16	4	10
		$BMDL_{ml}$	14	4	9	11	4	8
		$BMDL_{pl1}$	16	4	7	13	4	6
	BB	BMD	27	6	25	17	5	14
		$BMDL_{ml}$	14	4	11	12	4	9
		$BMDL_{pl1}$	18	5	14	14	5	10
	Cond	BMD	34	11	25	23	10	18
		$BMDL_{ml}$	20	9	17	16	8	14
Coll.	Bah	BMD	9	4	25	7	4	14
		$BMDL_{ml}$	6	4	13	5	4	9
		$BMDL_{pl1}$	7	4	15	6	4	10
	BB	BMD	8	5	27	7	5	14
		$BMDL_{ml}$	6	4	13	5	4	9
		$BMDL_{pl1}$	6	4	15	6	4	11
	Cond	BMD	14	11	27	11	10	17
		$BMDL_{ml}$	9	8	18	8	8	13

yield the following ordering:

$$BMDL_{ml}(EP) \leq BMDL_{pl1}(EP) \leq BMDL_{ml}(MB) \leq BMDL_{pl1}(MB).$$

This ordering is found in 30 out of 36 cases. In addition, the discrepancies between the BMDL's from different models and between $\text{BMDL}_{p\ell1}$ and BMDL_{ml} are smaller with the linear extrapolation method.

The exception to the rule seems to be visceral malformation in the EG study. At the same time, it is the only outcome for which virtually no intra-litter clustering has been found.

11.3.4 Profile Likelihood After Reparametrization for the Determination of a Safe Dose

Simulations reported in Claeskens, Aerts, Molenberghs and Ryan (2001) illustrate that the one-dimensional profile likelihood approach results in too low coverage probabilities for the BMDL. A possible explanation for this is the extra randomness due to estimation of β_0. Transformation (11.4) expresses the dosage d as a function of both β_0 and β_d. Since β_0 is unknown, in practice we have to substitute an estimator for the unknown true parameter. This additional source of variability might largely explain why a 90% confidence interval for β_d does not transform into an "at least 90%" one-sided confidence interval for the BMD.

An alternative approach is to first reparameterize the likelihood in terms of the dose d. For a specified excess risk, either member of the pair (β_d, d) contains the same information, provided a monotonic relationship exists between β_d and d (Aerts, Declerck and Molenberghs 1997). For the Bahadur and beta-binomial models, (11.4) shows that the relationship between β_d and d is indeed monotone. For the conditional model, however, this transformation is most often not monotone. Therefore, the reparametrization approach will not be applied to that model.

In the above example of the Bahadur and beta-binomial model, a reparametrization of the likelihood in terms of d is obtained by replacing β_1 by

$$\beta_1 = \ln\{(1 + qe^{-\beta_0})/(1 - q)\}/d.$$

This equation is obtained by solving $r(d; \boldsymbol{\beta}) = q$ for β_1 in the linear-logit model. The likelihood function is now maximized directly with respect to (β_0, d, β_a). This leads to definition (11.8), which is, however, not equivalent to the previous profile method (11.7) (Claeskens, Aerts, Molenberghs and Ryan 2001):

$$\widetilde{BMDL}_{pl1} = \min\left\{d : 2[\ell\{\hat{d}, \hat{\beta}_0(\hat{d}), \hat{\beta}_a(\hat{d})\} - \ell\{d, \hat{\beta}_0(d), \hat{\beta}_a(d)\}]\right.$$

$$\left. \leq \chi_1^2(1 - 2\alpha)\right\}. \tag{11.8}$$

An advantage of this reparametrization is that the profile function is constructed directly in terms of dosage d. The presence of β_0 in the reparametrization (11.4) is automatically taken care of while obtaining the solution to the likelihood equations. Results of a simulation study shown in Claeskens, Aerts, Molenberghs and Ryan (2001) clearly show that the reparametrization increases the coverage percentages significantly.

This method is preferable to the simpler one-dimensional profile approach of Section 11.3.2.

11.3.5 A Two-Dimensional Profile Likelihood Determination of a Safe Dose

When reparametrization becomes too complicated, a second solution is a higher dimensional profile approach, where all parameters in the function $\pi(\cdot)$ are taken into account (Claeskens, Aerts, Molenberghs and Ryan 2001). For the model

$$\text{logit}[\pi(d)] = \beta_0 + \beta_1 d,$$

this means both parameters (β_0, β_d). This leads to the following definition,

$$BMDL_{p\ell2} \;=\; \min\left\{ d(\beta_0, \beta_1, \hat{\beta}_a) : r(d; \beta_0, \beta_1, \hat{\beta}_a) = q \text{ over all } (\beta_0, \beta_1) \right.$$

$$\text{such that } 2[\ell(\hat{\beta}_0, \hat{\beta}_1, \hat{\beta}_a) - \ell(\beta_0, \beta_1, \hat{\beta}_a(\beta_0, \beta_1))]$$

$$\left. \leq \chi_2^2(1 - 2\alpha) \right\}. \tag{11.9}$$

We now use one more degree of freedom; the degrees of freedom correspond to the length of the subset of the β-vector which effectively occurs in the dose-response model.

Comparing the results of the two-dimensional profile approach with those of the reparameterized one-dimensional approach, in the above mentioned article, it is found that the BMDL obtained via the two-dimensional method tends to be larger. Since there is not a 100% coverage it may happen that the BMDL is larger than the true BMD; in this situation, the distance between the BMDL and the true BMD is larger on average for the two-dimensional profile method than compared to the reparametrization method. A good method for BMDL determination would have a small mean/median distance between the BMDL and the true BMD.

11.4 A Profile Score Approach

The idea is to reconsider existing likelihood approaches for developmental toxicity risk assessment from a non-likelihood point of view; an important class of examples are the generalized estimating equations (Chapter 5). According to Williams and Ryan (1996) it is preferable to define the BMDL using the likelihood ratio statistic. This method is explained in Section 11.3.1. Of course, a full likelihood technique will perform best when the likelihood is correctly specified, but one might expect problems in case of misspecification; see Section 11.3. Therefore it is important to look at robust estimation methods such as quasi-likelihood, GEE and pseudo-likelihood. Another reason for extension is that the likelihood method is unavailable in quasi-likelihood settings, and hence also in GEE, since there is no analogue to the likelihood ratio statistic.

11.4.1 A Score or Robust Score-based Benchmark Dose

Because of the first order equivalence between likelihood ratio, Wald and score statistics, we can replace the likelihood ratio statistic in the previous definitions by either the Wald or the score statistic, without affecting the first order asymptotic properties. Several authors have indicated drawbacks of a Wald based approach (Krewski and Van Ryzin 1981, Crump 1984, Crump and Howe 1985). The non-invariance of Wald tests to parameter transformations (Phillips and Park 1988) makes a replacement by the score statistic more appealing. Score inverted confidence intervals are described in, for example, Hinkley, Reid and Snell (1991, p. 278).

Claeskens, Aerts, Molenberghs and Ryan (2001) define the following approach. Assume the parameter estimator $\hat{\boldsymbol{\beta}}$ is the solution of estimating equations

$$U(\boldsymbol{\beta}) = \sum_{i=1}^{N} \boldsymbol{\psi}(d_i, z_i, m_i; \boldsymbol{\beta}) = 0, \qquad (11.10)$$

where, as before, d_i is the dose, m_i the size and z_i the number of malformations of litter i and N is the total number of litters. For ML estimators $\boldsymbol{\psi}(d_i, z_i, m_i; \boldsymbol{\beta})$ is the derivative of the log-likelihood with respect to $\boldsymbol{\beta}$. The score statistic is defined as

$$\mathcal{S}(\boldsymbol{\beta}) = U(\boldsymbol{\beta})^T A(\boldsymbol{\beta})^{-1} U(\boldsymbol{\beta}),$$

where

$$A(\boldsymbol{\beta}) = -\sum_{i=1}^{N} \frac{\partial}{\partial \boldsymbol{\beta}} \boldsymbol{\psi}(d_i, z_i, m_i; \boldsymbol{\beta}).$$

A $100(1 - \alpha)\%$ score based lower limit for the BMD can be defined as

$$\min \left\{ d(\boldsymbol{\beta}) : r(d; \boldsymbol{\beta}) = q \text{ over all } \boldsymbol{\beta} \text{ such that } \mathcal{S}(\boldsymbol{\beta}) \leq \chi_p^2(1 - 2\alpha) \right\}, \quad (11.11)$$

with $\mathcal{S}(\boldsymbol{\beta})$ the score statistic at parameter value $\boldsymbol{\beta}$.

Within a likelihood framework the asymptotic χ_p^2 distribution is not valid anymore in case the probability density (likelihood) function is misspecified, e.g., when there is some overdispersion not correctly accounted for. This holds for the score statistic as well as for the log-likelihood ratio and Wald statistics. There exists an extensive literature on likelihood misspecification. For an overview and many related references, see White (1982, 1994).

When there is uncertainty about the correctness of the likelihood specification, it is better to use the so-called robust statistics, since these modified test statistics have an asymptotic chi-squared distribution, even when the assumed probability model is not correct. For full likelihood models, robust Wald and score tests can easily be modified by using the so-called sandwich variance estimator (Kent 1982 and Viraswami and Reid 1996). We are not aware of such a simple modification of the likelihood ratio test. An alternative is the use of bootstrap methods; see Aerts and Claeskens (1998, 1999). As indicated before, we focus on score tests. Another advantage of the robust score tests is that they are also defined in quasi-likelihood, GEE and pseudo-likelihood models (Liang and Zeger 1986, Rotnitzky and Jewell 1990, and Geys, Molenberghs and Ryan 1999).

The robustified BMDL is defined similarly as in (11.11) but with $\mathcal{S}(\boldsymbol{\beta})$ replaced by the robust score statistic

$$\mathcal{R}(\boldsymbol{\beta}) = \boldsymbol{U}(\boldsymbol{\beta})^T \boldsymbol{A}(\boldsymbol{\beta})^{-1} \left(\boldsymbol{A}(\boldsymbol{\beta})^{-1} \boldsymbol{B}(\boldsymbol{\beta}) \boldsymbol{A}(\boldsymbol{\beta})^{-1}\right)^{-1} \boldsymbol{A}(\boldsymbol{\beta})^{-1} \boldsymbol{U}(\boldsymbol{\beta})$$

with

$$\boldsymbol{B}(\boldsymbol{\beta}) = \sum_{i=1}^{N} \boldsymbol{\psi}(d_i, z_i, m_i; \boldsymbol{\beta}) \boldsymbol{\psi}(d_i, z_i, m_i; \boldsymbol{\beta})^T.$$

For details on the definition of $\mathcal{R}(\boldsymbol{\beta})$, we refer to the above mentioned papers. As explained before, all these methods (likelihood and score based) are expected to be too conservative. In the next section we turn to *profile likelihood* and *profile score* approaches.

11.4.2 A Profile Score or Profile Robust Score-based Benchmark Dose

Similar to definition (11.7), we now define a profile score based BMDL as

$$BMDL_{ps1} = \min \left\{ d(\beta_1, \hat{\boldsymbol{\beta}}_n(\beta_1)) : r(d; \beta_1, \hat{\boldsymbol{\beta}}_n(\beta_1)) = q \text{ over all } \beta_1 \right.$$
$$\left. \text{such that } \mathcal{S}(\beta_1, \hat{\boldsymbol{\beta}}_n(\beta_1)) \leq \chi_1^2(1 - 2\alpha) \right\} \qquad (11.12)$$

and, for the robust score approach,

$$BMDL_{pr1} = \min \left\{ d(\beta_1, \hat{\boldsymbol{\beta}}_n(\beta_1)) : r(d; \beta_1, \hat{\boldsymbol{\beta}}_n(\beta_1)) = q \text{ over all } \beta_1 \right.$$
$$\left. \text{such that } \mathcal{R}(\beta_1, \hat{\boldsymbol{\beta}}_n(\beta_1)) \leq \chi_1^2(1 - 2\alpha) \right\}. \qquad (11.13)$$

These profile BMDL definitions use only one degree of freedom for the chi-squared distribution; therefore, this profile approach is expected to be more efficient. In a similar fashion, definition (11.9) can be modified to score-based methods. Only the robustified profile score BMDL definition (11.12) is theoretically valid in case of misspecification.

In Claeskens, Aerts, Molenberghs and Ryan (2001) a comparison is made of the likelihood, score and robust score approaches. It turns out that all three yield very similar results. This confirms the applicability and usefulness of score and robust score definitions. Moreover, these latter two are computationally much easier to obtain since only estimates of the nuisance parameters need to be computed and not of the β_d parameter itself, which is necessary in the likelihood criterion (11.7).

An important question that arises is whether it still makes sense to define a BMDL when the estimated dose effect parameter β_d is not significantly different from zero, i.e., when the confidence interval for β_d contains zero. One argument against defining a BMDL could be that when there is no statistically significant effect of the dose on the outcome, there is no interest in the value

Table 11.7 *EG Study. BMD and BMDL determination.*

Model	Outcome	Benchmark Dose	Score	Robust	Likelihood
BB	External	0.00734	0.00447	0.00504	0.00385
Pr(2)	Visceral	0.03674	0.02244	0.01730	0.02242
	Skeletal	0.00056	0.00043	0.00045	0.00043
	Collapsed	0.00051	0.00041	0.00041	0.00040
GEE2	External	0.00824	0.00450	0.00486	
Pr(2)	Visceral	0.03797	0.01581	0.01683	
	Skeletal	0.00058	0.00044	0.00046	
	Collapsed	0.00053	0.00040	0.00042	
GEE2	External	0.00824	0.00347	0.00495	
Pr(1,d)	Visceral	0.03797	—	0.01228	
	Skeletal	0.00058	0.00041	0.00046	
	Collapsed	0.00053	0.00038	0.00042	

of a safe dose and one could just as well define the BMDL to be zero. On the other hand, it can still be of interest to have an idea on the value of the BMDL even if the dose effect happens to be statistically not significant.

Another case is when the estimated dose effect parameter $\hat{\beta}_d$ is negative. This implies a decreasing dose-response curve on the logit scale. A problem with this situation is that possibly $\pi(d; \boldsymbol{\beta}) \leq \pi(0; \boldsymbol{\beta})$ for $d \geq 0$, which implies that $r(d; \boldsymbol{\beta}) \leq 0$. In other words, it could be "healthy" to be exposed to a certain level of the particular exposure. The above methods do not apply to this case.

11.4.3 Toxicity Study on Ethylene Glycol

As before, we will consider external, skeletal, and visceral malformation, as well as a collapsed outcome. We will take a univariate approach and estimate BMD and determine BMDL's for each of these malformation types in turn using the two-dimensional profile approach Pr(2) and the one-dimensional reparametrization method Pr(1, d).

For GEE2, only the robust profile BMDL's should be considered. Note that for GEE2, the last column is empty since the likelihood ratio based BMDL does not exist.

Table 11.7 shows BMD and BMDL values, which, for visceral malformation are much higher than for the other types of malformations. For visceral and skeletal malformation, and for a collapsed outcome variable (any of those 3 malformation types), profile likelihood and score BMDL values nearly coincide for the BB model. For skeletal and collapsed outcomes, the results by GEE2 (both profile methods) do not differ much from the corresponding BB-values. For skeletal malformation and a collapsed outcome, the largest BMDL values are obtained by the GEE2 profile robust score method; for external malformation by the BB profile robust score method and for visceral by the EB profile likelihood approach. For skeletal malformation and the collapsed outcome, the BMDL values obtained by the robust score approach are, up to rounding, the same for both profile GEE2 methods. For visceral malformation, the one-dimensional profile method Pr(1,d) yields a somewhat smaller value, in comparison with the Pr(2) method. The reverse is true for external malformation, although both values are rather comparable.

11.4.4 Concluding Remarks

Throughout this section, we considered a linear dose-response function on the logit scale, which, for the marginal models under study, resulted in an explicit formula for the BMD; see equation (11.4). In general, one might have to deal with more flexible functional relationships $\pi(d; \boldsymbol{\beta})$, such as non-linear dose-response models, and/or with other link functions. The BMD is defined in the same way as before, or equivalently, as the solution d to the equation

$$\pi(d; \boldsymbol{\beta}) = q + (1 - q)\pi(0; \boldsymbol{\beta}) \tag{11.14}$$

which, in general, will require numerical methods. In case expression (11.14) has more than one solution, one could define the BMD to be the smallest positive solution. BMDL determination in these models can proceed through a profile method, similar to the one in Section 11.3.5; one then only needs to adapt the degrees of freedom of the chi-squared distributed random variable according to the number of components of $\boldsymbol{\beta}$ actually appearing in (11.14).

To address the coverage percentage issue, one possibility to increase the coverage percentages, in particular for the two-dimensional profile method and for the robust score approach, is to use bootstrap critical points instead of the χ^2 critical values. This could be advantageous if the reason for small coverage is the use of the critical value from the asymptotic distribution. A naive application of bootstrap methods for the construction of confidence intervals would perform a bootstrap test at each value of the grid, which would make the method computationally very unattractive. Alternatively, other methods for obtaining confidence intervals could be considered; see Davison and Hinkley (1997) for an overview. None of those techniques have been studied in the context of quantitative risk assessment, and their theoretical and practical properties in this context are yet unknown.

CHAPTER 12

Exact Dose-Response Inference

Chris Corcoran

Utah State University, Logan, UT

Louise M. Ryan

Harvard School of Public Health, Boston, MA

Methods for analyzing correlated binary data have been well established in recent decades. Pendergast *et al.* (1992) offer a review of methods for correlated binary data, with a focus on cluster-correlated observations. Some of these methods, including marginal, random effects, and Markovian or conditional models, have been introduced in Chapter 4 and studied in subsequent chapters. However, the justification of inferences that rely on such methods usually rests upon the approximate normality of the statistics of interest. Such a distributional assumption may be untenable when samples are small or sparse. If a normal approximation is not accurate, the result might be tests that do not preserve the *a priori* testing level established by the investigator. Likewise, actual coverage probabilities for confidence intervals may be much lower or higher than the nominal confidence level. Moreover, where likelihood or quasi-likelihood methods are applied, inference can be further complicated when parameter estimates lie at or near the boundary of the parameter space. The following two examples illustrate these perils of approximate unconditional inference for cluster-correlated binary data. In this chapter, we will use the examples introduced in Sections 2.7 and 2.8.

12.C.5 Issues

What are the options when faced with the practical problems posed by these data sets? In the first example, no inference is possible due to the separability inherent in the data. In the second example, the sparseness of the data (due to the apparently very low baseline probability of malformation) may render a normal approximation suspect.

While much progress has been made in deriving large-sample methods for correlated binary data, there are comparatively few nonparametric options available to those faced with analyzing data such as those found in Tables 2.6 and 2.7. One *ad hoc* method for small or sparse samples includes using summary response measures. For example, one might treat the observed proportion of responses within each cluster as the outcome, and then use a generalized Wilcoxon test, such as the Kruskal-Wallis or Jonckheere-Terpstra test, to assess whether the median cluster-level response probability differs by dose level. While generally more conservative, such methods based upon summary response measures do not necessarily guarantee preservation of the significance level established *a priori* by the researcher. This preservation is a primary aim of exact tests. Another approach might be to treat cluster membership as a fixed effect, and then use exact conditional logistic regression to condition out the cluster effects. However, including a nominal covariate whose levels increase with increasing sample size may lead to poor testing properties. Moreover, exact logistic regression is computationally intensive, and may be infeasible when a great number of fixed effects, such as clusters, are included in the model.

In the next section, we introduce an exponential family model for clustered binary data that allows an exact permutation test comparing either two populations of clustered binomial observations, or K ordered clustered binomial populations. The resulting hypothesis test is "exact" in the sense that the test statistic of interest has a known distribution that is free of any unknown parameters. In Section 12.2 we compare this permutation test to some large-sample procedures in the context of a common teratology study design. Finally, in Section 12.3 we revisit the data of Table 2.7, and discuss some general issues regarding the exact test.

12.1 Exact Nonparametric Dose-Response Inference

In the case of independent, uncorrelated observations, an exact trend test is fairly straightforward: we have K ordered binomial populations, each with n_i subjects who have been exposed to a level x_i, a given compound.

Suppose that we have N clusters, where the ith cluster has n_i subjects with an associated ordinal covariate x_i, for $i = 1, \ldots, N$. Let Y_{ij} represent the binary response of the jth observation in the ith cluster, and $Z_i = \sum_{j=1}^{n_i} Y_{ij}$ represent the total number of responses within the ith cluster.

The Molenberghs and Ryan (MR) model (1999), introduced in Section 4.2, specifies the density of $\boldsymbol{Y}_i = (Y_{i1}, Y_{i2}, \ldots, Y_{in_i})^T$ as

$$Pr(\boldsymbol{Y}_i = \boldsymbol{y}_i) = \exp \left\{ \theta_i z_i - \delta_i z_i (n_i - z_i) + A_i(\theta_i, \delta_i) \right\},$$

where δ_i represents the dispersion parameter and $A_i(\theta_i, \delta_i)$ is the normalizing constant, summing over all possible realizations of $\boldsymbol{Y}_i$. The parameter δ_i reflects intracluster correlation. It is easy to show that the model reduces to a product of independent binary probabilities when $\delta_i = 0$. Using the logit link, a linear predictor of the form $\theta_i = \alpha + \beta x_i$, and assuming that $\delta_i = \delta$ for each

i, the density is expressed as

$$Pr(\boldsymbol{Y}_i = \boldsymbol{y}_i | x_i) = \\ \exp\{(\alpha + \beta x_i)z_i - \delta z_i(n_i - z_i) + A_i(\alpha, \beta, \delta)\}. \quad (12.1)$$

Because of the exchangeability of the binary responses, this density depends on $\boldsymbol{Y}_i$ only through Z_i. Any permutation of the binary responses within $\boldsymbol{Y}_i$ yields the same probability. The density of Z_i can therefore be expressed as

$$Pr(Z_i = z_i | x_i) = \binom{n_i}{z_i} \exp\{(\alpha + \beta x_i)z_i - \delta z_i(n_i - z_i) + A_i(\alpha, \beta, \delta)\},$$

$z_i = 0, \ldots, n_i$. Assuming cluster independence, the joint distribution of $\boldsymbol{Z} = (Z_1, \ldots, Z_N)^T$ given $\boldsymbol{x} = (x_1, \ldots, x_N)^T$ is obtained as

$$Pr(\boldsymbol{Z} = \boldsymbol{z} | \boldsymbol{x}) = \prod_{i=1}^{N} \binom{n_i}{z_i}$$
$$\times \exp\{(\alpha + \beta x_i)z_i - \delta z_i(n_i - z_i) + A_i(\alpha, \beta, \delta)\}$$
$$= \left[\prod_{i=1}^{N} \binom{n_i}{z_i}\right]$$
$$\times \exp\left\{\alpha s_1 + \beta t - \delta s_2 + \sum_{i=1}^{N} A_i(\alpha, \beta, \delta)\right\}, \quad (12.2)$$

where $s_1 = \sum_i z_i$, $t = \sum_i x_i z_i$, and $s_2 = \sum_i z_i(n_i - z_i)$. Because this density is of the exponential family, s_1, t, and s_2 are sufficient for α, β, and δ.

We are interested in the null hypothesis H_0: $\beta = 0$. We can eliminate the nuisance parameters α and δ and obtain the exact distribution of $\boldsymbol{Z}$ under H_0 by conditioning on $\boldsymbol{s} = (s_1, s_2)$. Define the conditional reference set $\Gamma(s_1, s_2)$ such that

$$\Gamma(s_1, s_2) = \left\{\boldsymbol{z}^* : \sum_{k=1}^{N} z_k^* = s_1, \sum_{k=1}^{N} z_k^*(n_k - z_k^*)\right\},$$

where $\boldsymbol{z}^*$ is any generic table of the form $\boldsymbol{z}^* = (z_1^*, z_2^*, \ldots, z_N^*)^T$ and the cluster sizes n_k are held fixed. Then the density of $\boldsymbol{Z}$ given $\boldsymbol{s}$ is

$$Pr(\boldsymbol{Z} = \boldsymbol{z} \mid \boldsymbol{x}, \boldsymbol{s}) =$$

$$\frac{\left[\prod_{i=1}^{N} \binom{n_i}{z_i}\right] \exp\left\{\alpha s_1 + \beta t - \delta s_2 + \sum_{i=1}^{N} A_i(\alpha, \beta, \delta)\right\}}{\sum_{\boldsymbol{z}^* \in \Gamma(s_1, s_2)} \left[\prod_{k=1}^{N} \binom{n_k}{z_k^*}\right] \exp\left\{\alpha s_1 + \beta t - \delta s_2 + \sum_{k=1}^{N} A_k(\alpha, \beta, \delta)\right\}}$$

$$= \frac{\left[\prod_{i=1}^{N} \binom{n_i}{z_i}\right] \exp\{\beta t\}}{\sum_{\boldsymbol{z}^* \in \Gamma(s_1, s_2)} \left[\prod_{k=1}^{N} \binom{n_k}{z_k^*}\right] \exp\{\beta t\}}.$$

Table 12.1 *Congenital Ophthalmic Defects. Permutation distribution for data of Table 2.6.*

t	$Pr(T = t)$
0	0.09524
1	0.28571
2	0.28571
3	0.28571
4	0.04762

Under H_0 this reduces to

$$Pr(\mathbf{Z} = \mathbf{z}|\mathbf{s}, H_0) = \frac{\prod_{i=1}^{N} \binom{n_i}{z_i}}{\sum_{\mathbf{z}^* \in \Gamma(s_1, s_2)} \prod_{k=1}^{N} \binom{n_k}{z_k^*}}, \tag{12.3}$$

which is free of all unknown parameters.

An exact test is formed by ordering the tables within $\Gamma(s_1, s_2)$ according to some discrepancy measure. As we are interested in inference about β, we wish to find the null distribution of the trend statistic

$$T(\mathbf{Z}) = \sum_{i=1}^{N} x_i Z_i, \tag{12.4}$$

the sufficient statistic for β, conditional on the observed value of $\mathbf{s}$.

From (12.3), the probability under H_0 of observing a table

$$\mathbf{z}_{obs} = (z_1, \ldots, z_N)'$$

with associated sufficient statistic $\mathbf{s}$ and a realization of T denoted by $t_{obs} = t(\mathbf{z}_{obs})$ is

$$Pr(T > t_{obs}|H_0, \mathbf{s}) = \sum_{\substack{\mathbf{z}^* \in \Gamma(s_1, s_2): \\ T(\mathbf{z}^*) \geq t_{obs}}} \left\{ \frac{\prod_{i=1}^{N} \binom{n_i}{z_i}}{\sum_{\mathbf{z}^* \in \Gamma(s_1, s_2)} \prod_{k=1}^{N} \binom{n_k}{z_k^*}} \right\}. \tag{12.5}$$

A one-sided α-level test then rejects H_0 when $t_{obs} > t_\alpha$, where t_α is defined as the smallest value such that

$$Pr(T > t_\alpha|H_0, \mathbf{s}) \leq \alpha.$$

A two-sided p value can be obtained by doubling (12.5).

We now apply this test to the data of Table 2.6. Note that for this particular example, $N = 9$, $(n_1, n_2, \ldots, n_9) = (2, 2, 2, 2, 2, 2, 2, 1, 1)$, and $\mathbf{z} = (0, 0, 0, 0, 0, 1, 1, 1, 1)$. Assuming that $x_i = 0$ for children less than 3 years of age and $x_i = 1$ otherwise, we have $s_1 = 4$, $s_2 = 2$, and an observed trend test statistic of $t = 4$. Conditional on the number and sizes of the clusters, there are exactly 12 tables for which $s_1 = 4$ and $s_2 = 2$. For example, the tables $\boldsymbol{z}^* = (0,0,0,0,1,0,1,1,1)$ and $\boldsymbol{z}^{**} = (1,0,0,0,2,1,0,0,0)$ belong to this reference set. The tables in the conditional reference set give rise to the permutation distribution of t shown in Table 12.1. Based upon this distribution, the exact probability of having observed a test statistic with a value of at least 4 is 0.04762. This is the p value of a one-sided hypothesis test that β_1, the coefficient of the age indicator x_i, is equal to zero, versus the alternative that $\beta_1 > 0$. As previously noted, because none of the four children younger than 3 years of age experienced a graft rejection, large-sample testing procedures, particularly those requiring an estimate of the age effect, may be suspect if not impossible. The exact procedure introduced here, however, provides some evidence that older children are prone to a higher probability of graft rejection.

Computing the exact p value of (12.5) is made difficult by the necessity of obtaining its denominator. Corcoran *et al.* (2001) suggest one method of computing this tail distribution using a graphical approach. This algorithm is currently implemented in ToxTools (2001) and StatXact (2001).

12.2 Simulation Study

We present here the results of a small simulation study to evaluate the performance of the exact test. We compare the exact test to four other procedures: likelihood-ratio tests using the beta-binomial and the logistic-normal-binomial models, the score test using the conditional exponential family model of (12.1), and a closed-form GEE trend test statistic derived by Lefkopoulou and Ryan (1992).

Our experiment involves 4 dose groups of unequally-sized clusters, with relative dose levels set at 0, 1, 2, and 3. This scenario mimics that of the common teratology study of the National Toxicology Program (NTP), wherein 25 female rodents (dams) are randomized to one of four doses of an investigative compound, then impregnated and exposed to the compound during a critical period of gestation. The dams are sacrificed before gestation is complete and the foetuses observed for the presence or absence of some response, such as malformation. We randomly generated the cluster sizes themselves from the empirical distribution of Table 12.2, which is reproduced from Carr and Portier (1993). This distribution is constructed from historical control data of 21 NTP experiments involving mice.

We first generated data from the conditional model of (12.1), using a value for the dispersion parameter of $\delta_i = 0.1$ for each i. This is the value used by Molenberghs *et al.* (1998) in their study of the effect of model misspecification under a fully parametric setting.

Figure 12.1 compares the performance of the five tests. Results at each plotted point are based upon 500 samples. Figure 12.1 (a) plots the estimated level of the tests as a function of the baseline response rate under the null hypothesis. The horizontal axis in this case indexes the value of the constant under the

Table 12.2 *Empirical distribution of litter sizes from NTP historical control data (Carr and Portier 1993).*

Litter Size	Frequency	$\hat{p}$
1	1	.002
2	2	.003
3	4	.007
4	4	.007
5	9	.015
6	16	.027
7	15	.025
8	19	.032
9	32	.054
10	45	.076
11	93	.157
12	100	.169
13	104	.175
14	73	.123
15	39	.066
16	27	.046
17	8	.013
18	2	.003

logit link. Plots 12.1 (b) and 12.1 (c) show estimated power as a function of the covariate effect — plot 12.1 (b) was generated using a constant of -4.0 under the logit link, and plot 12.1 (c) a constant of -2.0 (yielding baseline response rates of approximately 1 and 10%, respectively). The horizontal axes under these plots represent the coefficient of dose under the logit link. Estimated group response rates generated by the simulation are included parenthetically along the horizontal axis to allow more straightforward interpretation of the model parameters.

Figure 12.1 (a) shows that both the GEE and MR-based procedures result in considerable violation of the Type I error over the entire range of baseline response probabilities. For 500 samples, assuming conservatively that each point along the horizontal axis is independent, an observed Type I error of about 0.073 or greater is considered significantly higher than 0.05. In light of the uniform invalidity of these tests, comparing them to the exact test in terms of relative power is unfair. Nevertheless, we see from plots 12.1 (b) and 12.1 (c) that the asymptotic tests have slightly higher power, although this is clearly at the cost of sacrificing the testing level. The exact test proves advantageous even under this design that employs such a large sample.

Figure 12.2 shows analogous results for an experiment with data generated from the beta-binomial model. Figure 12.2 compares the performance of the

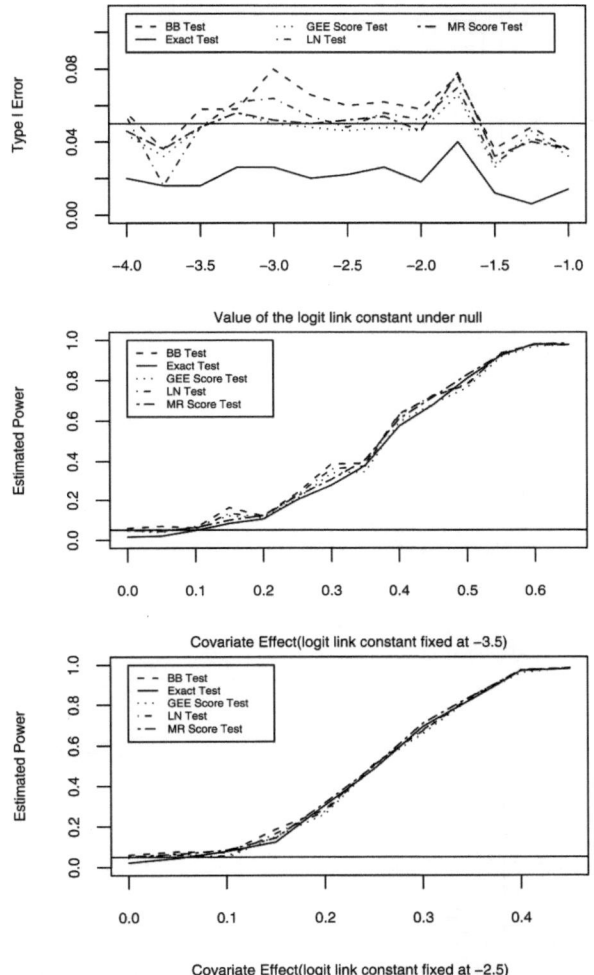

Figure 12.1 *Molenberghs and Ryan Random Data. (a) Estimated test size as a function of baseline response rate, and (b) and (c) estimated power as a function of group effect for baseline response rates of approximately 1.0% and 3.0%, respectively, when comparing four dose levels of 25 clusters each.*

five tests. Results at each plotted point are based upon 500 samples. Figure 12.2 (a) plots the estimated level of the tests as a function of the baseline response rate under the null hypothesis. The horizontal axis in this case indexes the value of the constant under the logit link. Plots 12.2 (b) and 12.2 (c) show estimated power as a function of the covariate effect, plot 12.2 (b) was generated using a constant of -4.0 under the logit link and plot 12.2 (c) a constant of -2.0 (yielding baseline response rates of approximately 1 and

202

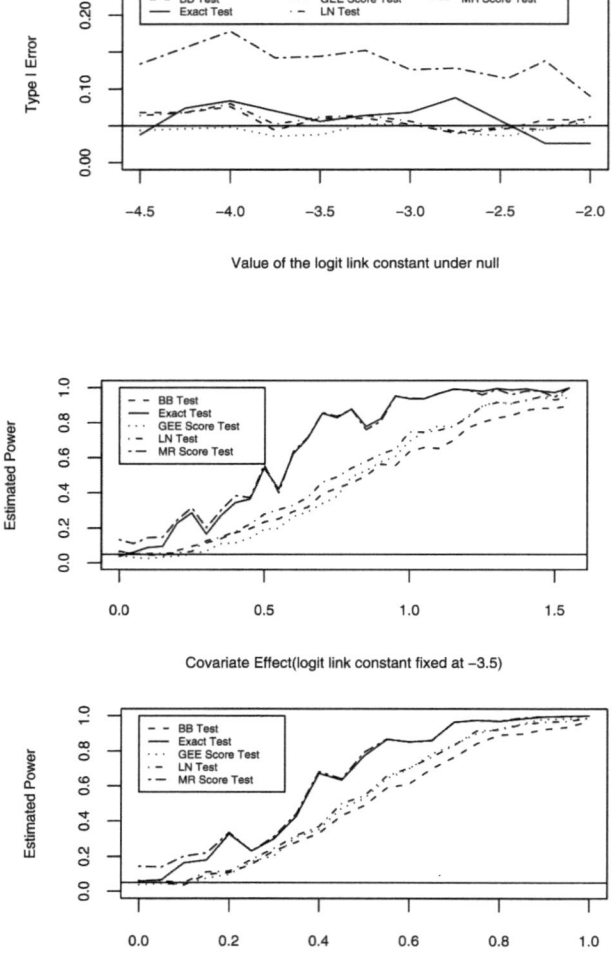

Figure 12.2 *Beta-binomial Random Data. (a) Estimated test size as a function of baseline response rate, and (b) and (c) estimated power as a function of group effect for baseline response rates of approximately 1.0% and 3.0%, respectively, when comparing four dose levels of 25 clusters each.*

10%, respectively). The horizontal axes under these plots represent the coefficient of dose under the logit link. Estimated group response rates generated by the simulation are included parenthetically along the horizontal axis to allow more straightforward interpretation of the model parameters.

Figure 12.2 (a) shows that both the GEE and MR-based procedures result in serious violation of the Type I error over the entire range of baseline response

Table 12.3 *Oral Teratogenic Data. Comparison of testing results of Table 2.7.*

Model	Test	p value
Logistic Normal	Likelihood ratio	0.131
	Wald	0.129
GEE	Score	0.145
	Wald	0.136
MR	Score	0.065
Nonparametric	Jonckheere-Terpstra	0.144
Nonparametric	Cochran-Armitage	0.103
Nonparametric	New Exact	0.059

probabilities. For 500 samples, assuming conservatively that each point along the horizontal axis is independent, an observed Type I error of about 0.073 or greater is considered significantly higher than 0.05. In light of the uniform invalidity of these tests, comparing them to the exact test in terms of relative power is unfair. Nevertheless, we see from plots 12.2 (b) and 12.2 (c) that the asymptotic tests have slightly higher power, although this is clearly at the cost of sacrificing the testing level. The exact test proves advantageous even under this design that employs such a large sample.

12.3 Concluding Remarks

It is useful to compare the five methods used in the previous section when applied to the data of Table 2.7. These findings are summarized in Table 12.3, along with the results of a method that uses the empirical cluster response rates as summary response measures. The exact test provides the strongest evidence of a dose-effect, while the likelihood and GEE-based tests seem more conservative. The Jonckheere-Terpstra test is a rank test of the null hypothesis that the median cluster-specific malformation rate is the same across dose groups, versus the ordered alternative that the median malformation rate is larger when the dose is greater. The results of this test yield weaker evidence of a dose effect than does the exact test. However, while a rank-based test using the sample proportions is relatively easy to use, it naively assumes that the 0/7 litter in the control group carries as much weight as does the 0/17 litter in the 8 mg/kg dose group.

Two of the asymptotic test statistics from Table 12.3, the GEE and MR score tests, directly approximate the distribution of T under H_0. Thus, particularly in light of the relatively anticonservative result of the exact test, it is instructive to compare these approximations to the exact conditional distribution of T given s. The conditional density of T is shown in Figure 12.3. The range of T is $(0, 2830)$, and its conditional mean and variance are $E(T|s, H_0) = 738.8$ and $\text{Var}(T|s, H_0) = 49,062$. We see further

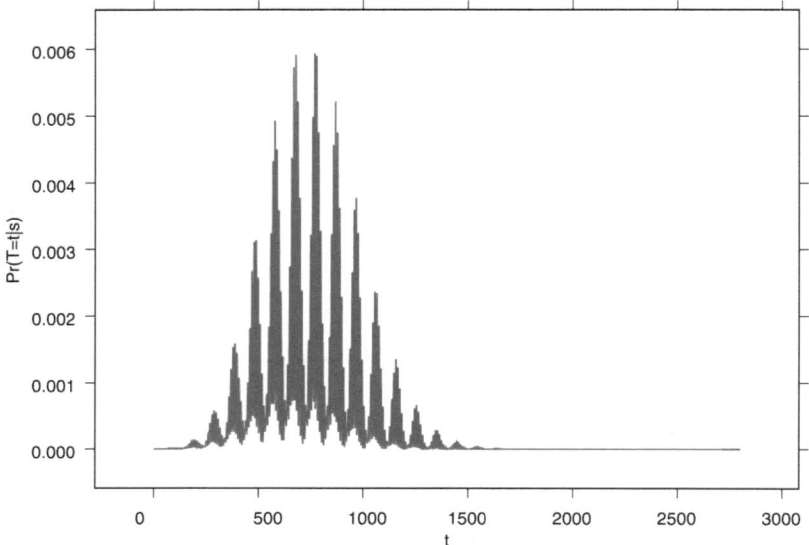

Figure 12.3 *Oral Teratogenic Data. Conditional density of T, given $\boldsymbol{s} = (s_1, s_2)'$, for data of Table 2.7.*

that this conditional density is highly discrete and decidedly nonnormal. On the other hand, the GEE and MR score test statistics are each asymptotically chi-square distributed with one degree-of-freedom, and are of the form $(T - E(T|H_0))^2/\text{Var}(T|H_0)$. However, the unconditional mean of this asymptotic distribution is approximated as 811.7 using the GEE method, and 743.1 under the MR score test. Furthermore, the GEE method approximates the variance of the asymptotic distribution as 67,234, while the MR score yields an estimate of 51,374. The reduced variance of the conditional distribution in this case indicates that the exact test is preferable as a basis for inference. There are several factors that can affect the conditional distribution, including imbalance in the data, unequal cluster sizes, the underlying variability between clusters in their average response probabilities, and the spacing of covariate levels. Future evaluation of the effects of all or some of these factors may explain further the anticonservatism of the conditional test for data such as these.

An additional feature of the exact test is that it can be extended naturally to handle exact logistic regression for clustered binary data. One need only condition on the additional sufficient statistics introduced by adding nuisance regression parameters to the model. However, extending the model, particularly with continuous-type covariates, can render an exact test computationally infeasible. Inclusion of continuous-type covariates may result in a degenerate permutation distribution. Even in situations where the distribution consists of meaningful support, additional covariates introduce additional layers of conditioning on sufficient statistics, which increases memory and storage requirements. Recent improvements in computational algorithms, such as those developed by Mehta, Patel and Senchaudhuri (2000) for exact logistic regression, may also be usefully applied to this problem.

Conditioning on the cluster sizes is possible only in cases where these marginal totals are ancillary for β. In practice, there may be some situations where this is not the case. In teratological research, for example, foetuses may be at higher risk of death before the end of the study due to the toxicity of the administered substance. The litter sizes hence provide some information regarding the dose effect. Another important consideration with regard to the covariate is its effect on the dispersion as well as the mean. Carr and Portier (1993) point out that, in practice, teratological studies that provide evidence in favor of a covariate effect tend also to show a positive relationship between dose and intracluster correlation. However, the assumption of homogeneous dispersion across clusters may also be relaxed. For example, one might wish to model the dispersion parameter δ as a function of covariates. As with the addition of nuisance regression parameters for the mean, this would require further conditioning on the sufficient statistics introduced by the augmented parameter vector, with the same possible computational consequences.

CHAPTER 13

Individual Level Covariates

Paige L. Williams

Harvard School of Public Health, Boston, MA

Helena Geys

transnationale Universiteit Limburg, Diepenbeek–Hasselt, Belgium

As discussed in Chapter 4, two main approaches to the analysis of clustered binary data are the cluster-specific (CS) approach (Section 4.3) and the population-averaged (PA) approach (grouping the conditional and marginal approaches of Sections 4.2 and 4.1). Cluster-specific models include cluster effects and thus are useful for assessing the effects of *individual-level* covariates. Individual-level covariates may take on different values, either by design or chance, for every unit in the cluster. These have also been referred to as *cluster-varying* covariates in the literature, since the values may vary within a cluster. Examples of CS models are mixed-effect logistic regression, with either parametric or nonparametric mixing distributions for the cluster effects, and conditional logistic regression. A number of cluster-specific approaches have been introduced in Section 4.3. In contrast, population-averaged models do not include cluster effects, and thus are most useful for assessing the effects of *cluster-level* covariates. Cluster-level covariates take on the same values for every unit in the cluster. The effects of individual-level covariates can also be estimated from population-averaged models, but their interpretations are based on the overall population, without adjusting for cluster effects. Quasi-likelihood models and models based on generalized estimating equations (GEEs, Chapter 5) fall under the heading of PA models. Excellent reviews of these modeling approaches for clustered binary data are provided by Prentice (1988), Fitzmaurice, Laird and Rotnitzky (1993), Diggle, Liang and Zeger (1994), and Pendergast *et al.* (1996).

Three examples will serve to illustrate the concepts of cluster-level versus individual-level covariates. First, consider a developmental toxicity study which evaluates the occurrence of fetal malformations in response to an environmental or chemical exposure. Clustered data result from the fact that binary outcomes (malformation versus no malformation) are evaluated on the offspring, while the exposure is administered to the pregnant female. In most developmental toxicity studies, as for most toxicity studies in general, the primary interest is in evaluating dose-response effects. Since the exposure level is a cluster-level covariate, many models encountered in the developmental toxicity literature are "population-averaged" (PA) models. In particular, GEEs have become a very popular choice for the analysis of developmental toxicity studies (Ryan 1992).

In contrast, consider a study conducted in 52 human subjects, 23 of whom were HIV-infected, in order to determine whether the lymphocyte proliferation assay (LPA) could be run on blood samples which had been shipped or stored rather than requiring fresh blood samples (Weinberg et al. 1998, Betensky and Williams 2001). The LPA measurements were performed on up to 36 combinations of conditions on each subject's blood sample, reflecting the three possible storage methods (fresh, shipped, or stored blood samples), three different anticoagulants, and four possible stimulants. In this study, the individual subject defines the cluster and the repeated LPA measurements on each subject form the clustered outcomes. The primary interest focused on the handling method, which is a cluster-varying (i.e., individual-level) covariate. Anticoagulant and stimulant are also individual-level covariates, since they pertain to the processing of an individual blood sample within a study subject. However, HIV infection status is a cluster-level covariate since it remains constant for each study subject. For this type of study, cluster specific models are likely to be more appropriate.

Last of all, suppose a multicenter clinical trial has been conducted in HIV-infected subjects to compare the effects of two combination antiretroviral regimens on HIV-1 RNA viral load. The viral load may be analyzed as a continuous outcome after log-transformation (ignoring, for the moment, the issue of censored data resulting from measurements below the limit of quantification of the viral load assay), or by dichotomizing as above or below the limit of quantification. In either case, such studies often measure the viral load at each clinic visit, resulting in repeated measurements for each subject. One of the primary interests of such a study might be to compare the trajectories of viral load over follow-up time between patients randomized to the two regimens. In this scenario, treatment regimen is a cluster-level covariate, while week on therapy is a cluster-varying covariate. However, because primary interest revolves around identifying treatment differences, a population-averaged model would be appropriate.

In contrast to models for dependent continuous outcomes, the two approaches for dependent binary data produce parameters with different interpretations and actually address different questions. From the above examples, it should be clear that the choice between one modeling approach or another depends primarily on the scientific question of utmost importance to the study. However, there may be instances in which both cluster-level and individual-level covariates are of interest within the same study. For example, in the heatshock studies described in Section 2.2, the embryos are explanted from the uterus of the maternal dam and exposed *in vitro* to various combinations of heat stress (increased temperature) and exposure duration; thus, the exposure covariate does not remain constant within a litter. Yet the genetic similarity of offspring from the same litter may still induce an intralitter correlation. Analysis of such data with a CS model may allow distinction between the genetic and environmental components of the intralitter effect, whereas studies with only cluster-level covariates can account for a "litter effect" but cannot disentangle this any further. In some such cases it may be reasonable to consider both approaches as equally valid. In these situations, issues of efficiency and robustness should be considered.

In this chapter, modeling approaches are described for addressing individual-level covariates in the context of clustered binary outcomes. Then, cluster-specific models for binary data are addressed, and are further broken down into conditional and marginal inferential approaches (Section 13.1). In Section 13.2, population-averaged models for binary data are reviewed, and are similarly subdivided into conditional and marginal model forms. The marginal models are further classified as likelihood-based versus those based on generalized estimating equations. Issues of efficiency are discussed in Section 13.3, for situations in which more than one modeling approach might produce estimates with valid interpretations. An example of analysis by these various modeling approaches for the particular case of the heatshock data (See Section 2.2) is provided in Section 13.4. In Section 13.5, cluster-specific (or random-effects) models are discussed when the outcomes of interest are continuous.

Let us first turn attention to binary outcomes.

13.1 Cluster-Specific Models

If primary interest of a study lies in within-cluster comparisons, then cluster-specific approaches are most appropriate. Cluster-specific approaches can be further subdivided into conditional and marginal models. The methods we consider below are both based on likelihood approaches.

13.1.1 Marginal Likelihood Approach

Consider a clustered data experiment in which the response of interest is a binary random variable Y_{ij} which is measured on the jth unit, $j = 1, ..., n_i$, of the ith cluster, $i = 1, ..., N$. Both the mixed effect and conditional logistic regression models are derived from the same general form for a cluster-specific model:

$$\text{logit}[P(Y_{ij} = 1 | \mathbf{X}_{ij}, \alpha_i, \boldsymbol{\beta}_{cs})] = \alpha_i + \mathbf{X}'_{ij} \boldsymbol{\beta}_{cs}, \qquad (13.1)$$

where α_i is a random intercept term for the ith cluster, $\mathbf{X}_{ij}$ is a vector of covariates of interest for the jth individual within the ith cluster, and $\boldsymbol{\beta}_{cs}$ is the corresponding vector of parameters. The vector $\mathbf{X}_{ij}$ may contain both a cluster-level component, $\mathbf{X}_{(f)i}$, and a cluster-varying component, $\mathbf{X}_{(v)ij}$, where the subscript (f) denotes "fixed" within a cluster, and (v) denotes

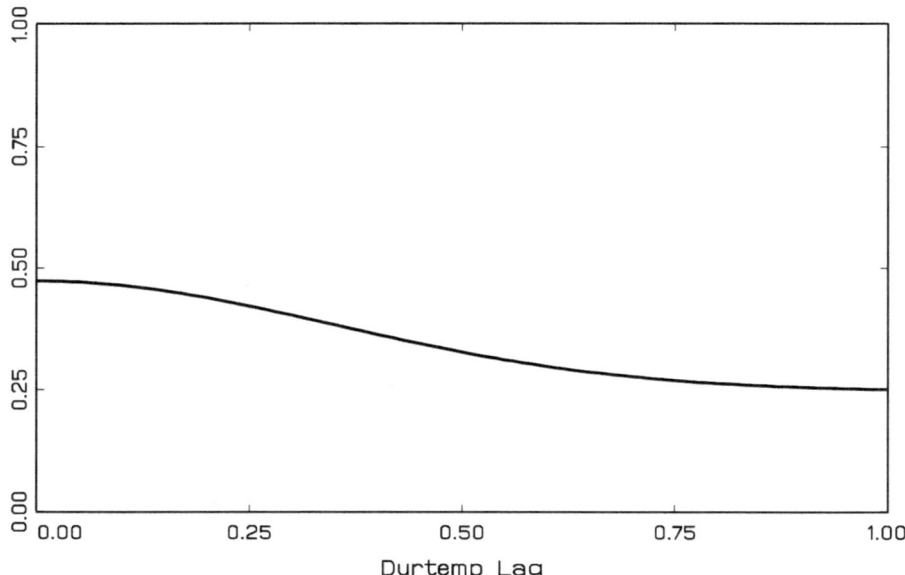

Figure 13.1 *Heatshock Study. Fitted correlation function.*

"varying". The corresponding components of $\boldsymbol{\beta}_{cs}$ are $\boldsymbol{\beta}_{cs}^{(f)}$ and $\boldsymbol{\beta}_{cs}^{(v)}$. With this notation, model (13.1) is written

$$\text{logit}[P(Y_{ij} = 1|\mathbf{X}_{ij}, \alpha_i, \boldsymbol{\beta}_{cs})] \;\;=\;\; \alpha_i + \mathbf{X}'_{(f)i}\boldsymbol{\beta}_{cs}^{(f)} + \mathbf{X}'_{(v)ij}\boldsymbol{\beta}_{cs}^{(v)}. \quad (13.2)$$

The likelihood under the mixed effect logistic regression model is obtained by integrating (13.2) over the mixing distribution G for the α_i within each cluster, and then taking the product over the independent clusters:

$$L_F \;\;=\;\; \prod_i^m P(\mathbf{Y}_i = \mathbf{y}_i | \mathbf{X}_i, \boldsymbol{\beta}_{cs}, G) \qquad (13.3)$$

$$=\;\; \prod_i^m \left\{ \int \prod_j p_{ij}^{y_{ij}} \, (1 - p_{ij})^{(1-y_{ij})} dG(\alpha) \right\},$$

where $\text{logit}[p_{ij}] = \alpha_i + \mathbf{X}'_{(f)i}\boldsymbol{\beta}_{cs}^{(f)} + \mathbf{X}'_{(v)ij}\boldsymbol{\beta}_{cs}^{(v)}$. (See also Section 4.3.2.) With respect to the mixing distribution G, both parametric distributions such as $\alpha_i \sim N(0, \sigma^2)$ (Stiratelli, Laird and Ware 1984) and semi-parametric mixture models in which G is estimated nonparametrically in conjunction with $\boldsymbol{\beta}_{cs}$ (Lindsay and Lesperance 1995) have been suggested.

The parameter α_i in the cluster-specific model (13.2) represents the log-odds of response for any of the n_i units in cluster i given that $X_{(f)i} = 0$ and $X_{(v)ij} =$

0. The parameter $\beta_{cs}^{(f)}$ in (13.2) measures the change in the *conditional* logit of the probability of response with a unit change in the covariate $X_{(f)}$ for a given value of $X_{(v)}$, with corresponding odds ratio $\exp(\beta_{cs}^{(f)})$. Note that this within-cluster change is unobservable, since $X_{(f)}$ by definition stays constant within a cluster. For example, for the LPA study described in Section 13.1, this would be akin to comparing HIV-infected to noninfected within an individual subject; yet HIV status was assumed to remain constant for each subject, and so this within subject comparison is not possible. In contrast, the parameter $\beta_{cs}^{(v)}$ in (13.2) measures the change in the *conditional* logit of the probability of response with a unit change in the covariate $X_{(v)}$ for a given value of $X_{(f)}$, with corresponding odds ratio $\exp(\beta_{cs}^{(v)})$, and is based on observable data. For the LPA study, an example would be the comparison of the LPA results for a shipped sample versus a sample stored overnight within an individual subject.

13.1.2 Conditional Likelihood Approach

The conditional likelihood (CL) is based on factoring the likelihood in (13.3) as $L_F = L_C \times L_S$, where L_C is a conditional term that depends on the sufficient statistics, $S_i = \sum_j Y_{ij}$, $i = 1, \ldots, m$, and L_S is the product of the marginal distributions of the sufficient statistics, as follows:

$$
\begin{aligned}
L_F &= \prod_i^m P(\mathbf{Y}_i = \mathbf{y}_i | S_i, \mathbf{X}_i; \boldsymbol{\beta}_{cs}) P(S_i = s_i | \mathbf{X}_i, \boldsymbol{\beta}_{cs}, G) \\
&= \left[\prod_i^m P(\mathbf{Y}_i = \mathbf{y}_i | S_i, \mathbf{X}_i; \boldsymbol{\beta}_{cs}) \right] \left[\prod_i^m P(S_i = s_i | \mathbf{X}_i, \boldsymbol{\beta}_{cs}, G) \right] \\
&= L_C \times L_S.
\end{aligned}
$$

The conditional MLEs for $\boldsymbol{\beta}_{cs}$ are then obtained by maximizing L_C. The resulting conditional likelihood is:

$$
L_C = \prod_i^m \frac{\prod_j^{n_i} \left[\exp(\mathbf{X}_{ij}' \boldsymbol{\beta}_{cs}) \right]^{y_{ij}}}{\sum_w \prod_{j_w} \left[\exp(\mathbf{X}_{ij}' \boldsymbol{\beta}_{cs}) \right]^{y_{ij}}},
$$

where w indexes the number of possible permutations of the n_i repeated measurements per cluster and j_w indexes the repeated measurements under assignment w (Ten Have, Landis and Weaver 1995). The conditional approach does not allow for estimation of any cluster-level covariates; they are removed from the likelihood with the cluster effects. Thus, for example, the effect of HIV infection status for the LPA study could not be estimated on the basis of a conditional logistic regression model. Similarly, for a developmental toxicity study, the effects of any covariates specific to the pregnant female, such as maternal body weight, could not be estimated.

13.2 Population-averaged Models

If the primary interest of a study lies in assessing overall effects of exposure (or treatment) on repeated outcomes, then population-averaged models are most appropriate. PA approaches model the average response to changes in the covariates, and are thus best-suited for evaluating between-cluster effects. There are several examples of population-averaged (PA) models, including the beta-binomial, quadratic exponential, quasi-likelihood, and generalized estimating equation (GEE) approaches. In general, PA models can be subdivided as conditionally specified (Section 4.2) models or marginal models (Section 4.1).

13.2.1 Conditionally Specified Models

In a conditionally specified model, the parameters describe a function (e.g., probability, odds, or logit) of a set of outcomes, given values for the other outcomes. These models have been introduced and studied in Section 4.2, primarily in the context of cluster-level covariates. Here, we will indicate some peculiar aspects when individual-level covariates are observed.

The most familiar example of such a model is the log-linear model. For binary outcomes which are correlated within clusters, Molenberghs and Ryan (1999) proposed a likelihood-based model which relies on the multivariate exponential family. Using conditionally specified models can be somewhat controversial. On one hand, the conditional approach has been criticized because the interpretation of the covariate effect on one outcome is conditional on the responses of other outcomes for the same individual, outcomes of other individuals, and the cluster size (Diggle, Liang and Zeger 1994, pp. 147–148). On the other hand, it has also been shown that conditionally specified models, such as that proposed by Molenberghs and Ryan, are very flexible for exchangeable clustered binary data (Geys, Molenberghs and Ryan 1999). However, with *only* individual-level covariates, the limitations of conditionally specified marginal models become severe.

For the conditional model (Section 4.2), we will follow the slightly non-traditional notation used by Geys, Molenberghs and Ryan (1999). Assume that $y_{ij} = 1$ when the jth individual in cluster i exhibits the response of interest and -1 otherwise. Although the $0/1$ coding is more common for binary outcomes, this alternative coding provides a better parameterization under variable cluster sizes since it leads to invariance properties when the role of success and failure are reversed (Cox and Wermuth 1994). Molenberghs and Ryan (1999) proposed a joint distribution for exchangeable clustered binary data based on a multivariate exponential family model by setting the higher order interactions to zero. Extending their model to individual-level covariates is, at the formal level, straightforward. Similarly to Zhao and Prentice (1990), the joint density for the ith cluster is given by:

$$f_{\mathbf{Y}_i}(\mathbf{y}_i; \boldsymbol{\Theta}_i) = \exp\left\{\sum_{j=1}^{n_i} \theta_{ij} y_{ij} + \sum_{j<j'} \delta_{ijj'} y_{ij} y_{ij'} - A(\boldsymbol{\Theta}_i)\right\}, \quad (13.4)$$

where $A(\boldsymbol{\Theta}_i)$ is the normalizing constant, obtained by summing (13.4) over all 2^{n_i} possible outcomes. The parameters θ_{ij} refer to the main effects, whereas the parameters $\delta_{ijj'}$ refer to the association between two individuals within the same cluster. Grouping the parameters in the $(q \times 1)$ vector $\boldsymbol{\Theta}_i$ $(q = n_i + \binom{n_i}{2})$ and applying a linear link function, we assume a model $\boldsymbol{\Theta}_i = X_i\boldsymbol{\beta}$, with X_i a $(q \times p)$ design matrix and $\boldsymbol{\beta}$ a $(p \times 1)$ vector of unknown regression coefficients.

While model (13.4) benefits from the elegance and simplicity of exponential family theory, it is not entirely appropriate for analysis of clustered binary data with covariates specific to each observation. To illustrate this point, consider a cluster of size 2 yielding two outcomes (Y_1, Y_2). The marginal probability of response for the first individual then equals

$$P(Y_1 = 1) = \exp(\theta_1 - \theta_2 - \delta_{12} - A(\boldsymbol{\Theta})) + \exp(\theta_1 + \theta_2 + \delta_{12} - A(\boldsymbol{\Theta})),$$

which depends on the covariates of the second individual. For the particular case of the heatshock data, this would imply that the response for one embryo would depend on the exposure group to which a separate littermate was randomized. This seems to be an undesirable property for modeling the effects of individual-level covariates.

13.2.2 Marginal Models

Marginal models have been introduced at length in Section 4.1. Here, we will focus on the Bahadur model, in particular on the general form that needs to be used when individual-level covariates are used.

The Bahadur model has been used by several authors (Kupper and Haseman 1978, Altham 1978), especially in the context of toxicological experiments, and can thus be considered an important representative of the marginal family. Bahadur (1961) describes the joint distribution of clustered binary data for a single outcome in terms of marginal means $\boldsymbol{\mu}_i = (\mu_{i1}, \ldots, \mu_{in_i})^T$ and marginal correlations $\boldsymbol{\rho}_i = (\rho_{i12}, \rho_{i13}, \ldots, \rho_{i12\ldots n_i})^T$. The closed form probability mass function is given as:

$$f(\mathbf{y}_i, \boldsymbol{\mu}_i, \boldsymbol{\rho}_i) = \prod_{j=1}^{n} \mu_{ij}^{y_{ij}} (1 - \mu_{ij})^{(1-y_{ij})}$$

$$\times \left(1 + \sum_{j_1 < j_2} \rho_{ij_1 j_2} r_{ij_1} r_{ij_2} + \sum_{j_1 < j_2 < j_3} \rho_{ij_1 j_2 j_3} r_{ij_1} r_{ij_2} r_{ij_3} \right.$$

$$\left. + \cdots + \rho_{i1\ldots n} r_{i1} \cdots r_{in} \right).$$

In practice, three-way and higher order associations are often small in magnitude and difficult to interpret. If we set all three- and higher way correlations equal to zero, Bahadur's representation simplifies to:

$$f(\mathbf{y}_i|\boldsymbol{\pi}_i, \boldsymbol{\rho}_i) = \prod_{j=1}^{n_i} \mu_{ij}^{y_{ij}} (1 - \mu_{ij})^{1-y_{ij}} \left(1 + \sum_{j<k} \rho_{ijk} e_{ij} e_{ik} \right), \quad (13.5)$$

with $e_{ij} = \frac{y_{ij} - \mu_{ij}}{\sqrt{\mu_{ij}(1-\mu_{ij})}}$. Using appropriate link functions, the marginal mean parameters μ_{ij} ($j = 1, \ldots, n_i$), as well as the marginal correlations $\rho_{ijk}(j < k)$, can be modeled as a function of a $(n_i (n_i + 1)/2 \times p)$ covariate matrix $\mathbf{X}_i$ and a parsimonious $(p \times 1)$ vector of regression parameters $\boldsymbol{\beta}^*$. The logistic link function is a natural choice for μ_{ij}, while Fisher's z-transform is convenient to model ρ_{ijk}. This leads to the following generalized linear model:

$$\left(\begin{array}{c} \boldsymbol{\eta}_{i1} \\ \boldsymbol{\eta}_{i2} \end{array} \right) \left(\begin{array}{c} \ln\left(\frac{\mu_{ij}}{1-\mu_{ij}}\right)_{j=1}^{n_i} \\ \ln\left(\frac{1+\rho_{ijk}}{1-\rho_{ijk}}\right)_{j<k} \end{array} \right) = \mathbf{X}_i \boldsymbol{\beta}^* = \left(\begin{array}{c} X_i \boldsymbol{\beta} \\ Z_i \boldsymbol{\alpha} \end{array} \right).$$

The maximum likelihood estimator $\hat{\boldsymbol{\beta}}$ for $\boldsymbol{\beta}$ is defined as the solution to $\mathbf{U}(\hat{\boldsymbol{\beta}}) = 0$ with $\mathbf{U}(\boldsymbol{\beta})$ the score function. A Fisher scoring or Newton-Raphson algorithm can be used to obtain the maximum likelihood estimate $\hat{\boldsymbol{\beta}}$.

A drawback of this approach is the fact that the correlation parameters are highly constrained when the higher order correlations have been set to zero. Even when higher order parameters are included, the parameter space of marginal parameters and correlations has a very peculiar shape. We refer to Appendix A for more details. Despite the attendant restrictions on model parameters, they are satisfied for the heatshock studies. Therefore we can fit model (13.5) to these studies.

13.2.3 Generalized Estimating Equations

Even though a variety of flexible models exist, maximum likelihood can be unattractive due to excessive computational requirements, especially when high dimensional vectors of correlated data arise. As a consequence, there has been a demand for alternative methods. Liang and Zeger (1986) proposed use of generalized estimating equations (GEEs) which require only the correct specification of the univariate marginal distributions provided one is willing to adopt "working" assumptions about the association structure. Generalized estimating equations are studied in detail in Chapter 5.

The GEE approach yields consistent estimates of covariate effects even when the association structure is misspecified. However, severe misspecification may seriously affect the efficiency of the GEE estimators (Liang, Zeger and Qaqish 1992). A second-order extension of these estimating equations, which has been referred to as "GEE2", includes marginal pairwise associations as well and specifically models these associations rather than treating them as nuisance parameters (Heagerty and Zeger 1996, Liang, Zeger and Qaqish 1992, Molenberghs and Ritter 1996). The GEE2 approach is nearly fully efficient, although bias may occur in the estimation of the main effect parameters when the association structure is misspecified. The GEE2 approaches proposed by Heagerty and Zeger (1996) Liang, Zeger and Qaqish (1992), and Molenberghs and Ritter (1996) all use odds ratios as measures of association. However, the estimating equations can just as easily be specified in terms of correlation parameters, which has the advantage of enabling an easy comparison with the Bahadur model.

The PA model based on the GEE approach has become very popular due to its ease of implementation in many popular statistical software packages (e.g., SAS, Stata, and SPlus). The general GEE population-averaged model can be written as:

$$\text{logit}[P(Y_{ij} = 1|\mathbf{X}_{ij}, \boldsymbol{\beta}_{pa})] = \alpha_{pa} + \boldsymbol{\beta}_{pa}\mathbf{X}_{ij}. \tag{13.6}$$

When both cluster-level and individual-level covariates, denoted by $\mathbf{X}_{(f)i}$ and $\mathbf{X}_{(v)ij}$, respectively, are included in the model, the basic model in (13.6) can be expanded as:

$$\text{logit}[P(Y_{ij} = 1|\mathbf{X}_{ij}, \boldsymbol{\beta}_{pa})] = \alpha_{pa} + \mathbf{X}'_{(f)i}\boldsymbol{\beta}^{(f)}_{pa} + \mathbf{X}'_{(v)ij}\boldsymbol{\beta}^{(v)}_{pa}. \tag{13.7}$$

The probability $P(Y_{ij}|\mathbf{X}_{ij}, \boldsymbol{\beta}_{pa})$ represents the marginal distribution of the Y_{ij}, averaged over the clusters. In fitting this model, it is also necessary to make some assumption about the working correlation structure of $\mathbf{Y}_i$ to account for intracluster correlation. Common assumptions include independence, exchangeable (equi-correlated), auto-regressive, and unstructured correlation. Liang and Zeger (1986) derive consistent estimates for $\boldsymbol{\beta}_{pa}$ based on solving a set of estimating equations, and provide estimates for both model-based and robust standard errors.

The interpretation of parameters in cluster-specific and population-averaged models has been discussed by Neuhaus, Kalbfleisch and Hauck (1991), Neuhaus and Kalbfleisch (1998), Graubard and Korn (1994), Zeger, Liang and Albert (1988), and Ten Have, Landis and Weaver (1995), among others. For the PA model, the parameter α_{pa} represents the population log-odds (or logit) of response in the baseline group, i.e., those in the population with $X_{(f)} = 0$ and $X_{(v)} = 0$:

$$\alpha_{pa} = \log\left[\frac{P(Y = 1|X_{(f)} = 0, X_{(v)} = 0)}{1 - P(Y = 1|X_{(f)} = 0, X_{(v)} = 0)}\right].$$

The parameters $\beta^{(f)}_{pa}$ and $\beta^{(v)}_{pa}$ in model (13.7) reflect *unconditional* logits of the overall population prevalences. Specifically, the population-averaged effect in the log odds from a unit increase in $X_{(f)i}$ for fixed $X_{(v)}$ is defined as

$$\beta^{(f)}_{pa}(X_{(f)}, X_{(v)}) = \log\frac{P(Y = 1|X_{(f)} + 1, X_{(v)})/P(Y = 0|X_{(f)} + 1, X_{(v)})}{P(Y = 1|X_{(f)}, X_{(v)})/P(Y = 0|X_{(f)}, X_{(v)})}.$$

In the population-averaged model, this quantity is independent of both $X_{(f)}$ and $X_{(v)}$.

Since a mixed effect model specifies a marginal model for response, it is possible to define population-averaged effects from the mixed effect models. Betensky and Williams (2001) show that, for small values of α, the PA parameter for the baseline response can be approximated by:

$$\alpha_{pa} = \log\left(\frac{2 + E_G(\alpha)}{2 - E_G(\alpha)}\right),$$

where expectation is taken with respect to the mixing distribution, G, for the cluster effects. Based on the mixed effect model, the unconditional logits of population prevalences reflected by $\beta_{pa}^{(f)}$ and $\beta_{pa}^{(v)}$ depend on both $X_{(f)}$ and $X_{(v)}$; for example, at values of $\beta_{cs}^{(f)}$ close to zero, the approximate relationship below holds:

$$\beta_{pa}^{(f)}(X_{(v)}) \approx \beta_{cs}^{(f)}[1 - \rho_y(0|\beta_{cs}^{(v)}, X_{(v)})],$$

where

$$\rho_y(0|\beta_{cs}^{(v)}, X_{(v)}) = \text{Corr}(Y_{ij}, Y_{ij'}|\beta_{cs}^{(f)} = 0, \beta_{cs}^{(v)}, X_{(v)ij} = X_{(v)ij'} = X_{(v)}).$$

Thus, in the cluster-specific model, the population-averaged effect of a cluster-level covariate is approximately independent of that covariate, but not of the cluster-varying covariates. The analogous result for the population-averaged effect in the log odds from a unit increase in $X_{(v)}$ for fixed $X_{(f)}$ is

$$\beta_{pa}^{(v)}(X_{(f)}) \approx \beta_{cs}^{(v)}[1 - \rho_y(0|\beta_{cs}^{(f)}, X_{(f)})],$$

where

$$\rho_y(0|\beta_{cs}^{(f)}, X_{(f)}) = \text{Corr}(Y_{ij}, Y_{ij'}|\beta_{cs}^{(v)} = 0, \beta_{cs}^{(f)}, X_{(f)}).$$

Thus, the population-averaged effect of a cluster-varying covariate from a cluster-specific model is approximately independent of that covariate, but not of the cluster-level covariates. If both $\beta_{cs}^{(v)}$ and $\beta_{cs}^{(f)}$ are close to 0, then the approximations simplify to

$$\beta_{pa}^{(f)} \approx \beta_{cs}^{(f)}[1 - \rho_y(0)] \quad \text{and} \quad \beta_{pa}^{(v)} \approx \beta_{cs}^{(v)}[1 - \rho_y(0)] \tag{13.8}$$

where

$$\rho_y(0) = \text{Corr}(Y_{ij}, Y_{ij'}|\beta_{cs}^{(f)} = 0, \beta_{cs}^{(v)} = 0).$$

In this case, the population-averaged effects of any covariate are approximately independent of all covariates. This latter approximation (13.8) is implied by the results of Neuhaus, Kalbfleisch, and Hauck (1991).

13.3 Efficiency of Modeling Approaches

When both cluster-specific and population-averaged approaches give parameter estimates which have valid interpretations, issues of efficiency should be considered (Graubard and Korn 1994). For example, if interest centers on a cluster-varying covariate and the study design is balanced, both approaches may be valid. Previous research has shown that the relative efficiency of PA and CS approaches for clustered binary outcomes may depend on the intracluster correlation between covariate levels (e.g., Neuhaus and Lesperance

1996). For a cluster-level covariate, the intracluster correlation is $\rho_x = 1$. For a designed study, units or subjects within a cluster might be randomly assigned to different levels of a covariate, which would induce a negative intracluster correlation for that particular covariate. For example, a matched pairs study in which one unit is randomized to a new treatment and the other to a control has an intracluster correlation of -1 for the treatment covariate. For a cluster size of n, the maximal negative intracluster correlation of $\rho_x = -1/(n-1)$ is achieved by a balanced assignment of units within cluster to the possible covariate values.

For cluster-level covariates, Neuhaus (1993) and Neuhaus, Kalbfleisch and Hauck (1991) found that Wald tests involving cluster-level covariates based on cluster-specific models and population-averaged models were approximately equivalent. On the other hand, when there is positive intracluster correlation of responses, the cluster-varying covariate effects from population-averaged models are attenuated towards zero as compared to those from cluster-specific models. Further, tests of individual-level covariates based on cluster-specific models are more powerful than those based on population-averaged models, if the correlation structure is not modeled (i.e., a GEE independence model). However, tests of cluster-varying covariates from population-averaged models which specify a compound symmetry correlation structure (i.e., GEE exchangeable) are equally efficient to those from a cluster-specific model.

More specifically, for paired data, Neuhaus (1992, 1993) noted that the asymptotic relative efficiency (ARE) of $\hat{\beta}_{pa}$ under a GEE independence approach versus $\hat{\beta}_{cs}$ under a mixed effect model, for β_{cs} close to 0, is:

$$\mathrm{ARE}_{I,ME,n=2} = \frac{(1-\rho_y^2)}{(1-\rho_y^2\rho_x^2)},$$

where $\rho_x = \mathrm{Cov}(X_{ij}, X_{ij'})$ and $\rho_y = \mathrm{Cov}(Y_{ij}, Y_{ij'}|\beta_{cs} = 0)$. This approximation assumes equal cluster sizes ($n_i = n$), and that the mixing distribution G can be approximated by the distribution induced by a beta distribution on p (where $\mathrm{logit}(p) = \alpha_i + \mathbf{X}'_{ij}\boldsymbol{\beta}_{cs}$). Betensky and Williams (2001) generalized the paired data results to clusters of size n, and found that the ARE could be expressed as:

$$\mathrm{ARE}_{I,ME} = \frac{(1-\rho_y)\left[1+(n-1)\rho_y\right]}{[1+(n-1)\rho_x\rho_y][1+(n-2)\rho_y-(n-1)\rho_x\rho_y]}.$$

Based on this formula, the GEE independence model is consistently less efficient than the mixed effect model. For larger cluster sizes, in fact, the efficiency can be very low. In contrast, the minimum ARE for paired data is $1 - \rho_y^2$, or very close to 1 for small or moderate values of ρ_y.

It is also possible to calculate Pitman efficiencies for comparing the conditional logistic regression approach versus the GEE population-averaged approaches. The relationship between the parameters from the GEE and conditional logistic approaches can be expressed as $\beta_{pa} \approx (1-\rho_y)\beta_{cs}$ for β_{pa} and β_{cs} close to zero. Betensky and Williams (2001) showed that the Pitman ARE for the conditional logistic regression versus the GEE independence approach is:

$$\text{ARE}_{CL,I} = \frac{(1-\rho_x)}{(1-\rho_y)} \frac{(n-1)}{n} [1 + (n-1)\rho_y\rho_x]$$

and the ARE for comparing the conditional logistic regression approach versus the GEE exchangeable model is:

$$\text{ARE}_{CL,E} = \frac{(n-1)}{n} \frac{[1 + (n-1)\rho_y](1-\rho_x)}{[1 + (n-2)\rho_y - (n-1)\rho_y\rho_x]}.$$

The conditional logistic regression approach is always less efficient than the GEE exchangeable model. However, the conditional logistic regression approach can yield more efficient estimates than the GEE approach under independence. In fact, the conditional logistic estimation approach is only less efficient for fairly high values of ρ_x, which would not be anticipated in a designed study with individual-level covariates.

A distinction between the two popular cluster-specific models lies in their use of clusters with discordant outcomes; the conditional likelihood approach uses data only from clusters with discordant outcomes and covariates, whereas the mixed effect approach uses both concordant and discordant clusters. Thus, the conditional likelihood tends to be less efficient than the mixed effect approach. Neuhaus and Lesperance (1996) found that the approximate ARE of the conditional likelihood approach versus the full likelihood approach of the mixed effect model is:

$$\text{ARE}_{CL,ME} = \frac{(n-1)\{1 + (n-1)\rho_y\}(1-\rho_x)}{n\{1 + (n-2)\rho_y - (n-1)\rho_x\rho_y\}} \tag{13.9}$$

for β_{cs} close to 0. There are several implications of this ARE formula. First, for designed studies with maximal negative correlation $[\rho_x = -1/(n-1)]$, the ARE equals 1; in other words, conditional logistic regression is fully efficient. Second, the ARE is 0 for cluster-level covariates, and decreases as the covariate correlation ρ_x increases. The ARE increases as the cluster size increases, all other factors being held constant, and thus the conditional likelihood approach is least efficient relative to mixed effect estimation when data are paired $(n = 2)$. The latter result is fairly intuitive, since concordant clusters arise less frequently with increasing cluster size. For paired data, the above ARE approximation can be simplified to:

$$\text{ARE}_{CL,ME,n=2} = \frac{(1+\rho_y)(1-\rho_x)}{2(1-\rho_x\rho_y)}.$$

Figure 13.2 illustrates the efficiency of the conditional logistic regression model versus a full likelihood approach under a mixed effect model, for a value of $\rho_y = 0.14$, and cluster sizes of $n = 2$ (panel (a)) and $n = 23$ (panel (b)). From Figure 13.2, it is evident that the efficiency decreases almost linearly as a function of increasing covariate correlation when data are paired (panel (a)), but remains fairly high for larger cluster sizes with moderate covariate

(a)

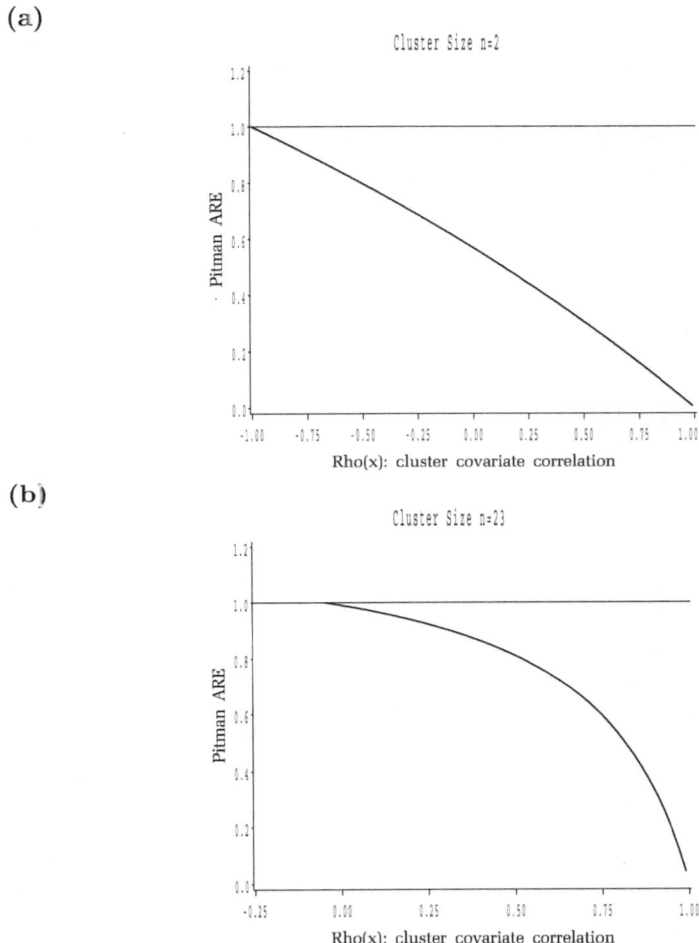

Figure 13.2 *Pitman ARE of conditional* versus *mixed effect logistic regression as a function of covariate correlation ρ_x, assuming $\rho_y = 0.14$, for cluster size (a) n=2 and (b) n=23.*

correlation (panel (b)). Thus, while efficiency of the conditional logistic model can be low for paired data with correlated intracluster covariates, it is a reasonably efficient estimation strategy for designed studies with larger cluster sizes.

An alternative interpretation of the relative efficiency of the conditional likelihood and mixed effect approaches is provided by Neuhaus and Kalbfleisch (1998). They note that the conditional likelihood approach measures within-cluster effects, and not between-cluster covariate effects. In contrast, mixed effect models typically assume that the within- and between-cluster effects of the covariates are the same, and thus may disagree with conditional likelihood estimates. When this assumption is met, such as in designed experiments in which the covariate pattern is the same for all clusters, the mixed effect and conditional likelihood approaches typically yield identical covariate effects. The increased efficiency of the mixed effect model compared with the conditional likelihood model arises solely through this assumption made by the mixed effect model, which may, in fact, be false. Thus, mixed effect models should be used only after checking this assumption or for designed experiments.

13.4 Analysis of Heatshock Data

The study design for the set of experiments conducted by Kimmel *et al.* (1994) is described in Section 2.2. Continuous outcomes have been analyzed in Section 13.5. A total of 375 embryos, arising from 71 initial dams, survived the heat exposure. These surviving embryos were further examined for malformations and delayed growth, and are included in our analysis. The distribution of cluster sizes ranged between 1 and 11, with a mean cluster size of 5. Note that, since embryos of questionable viability were not included, cluster sizes here are smaller than those observed in most other developmental toxicity studies and do not reflect the true original litter size.

Historically, comparison of response levels among exposures of different durations has relied on a conjecture called Haber's Law, which states that adverse response levels should depend only on cumulative exposure (dose × duration) (Haber 1924). For the heatshock studies, the vector of exposure covariates must incorporate both exposure level, d_{ij}, and duration, t_{ij}, for the jth embryo of the ith dam, and should be formulated in such a way that departures from Haber's premise can easily be assessed. The exposure metrics in these models are the cumulative heat exposure, $d_{ij} \times t_{ij}$, which will be denoted by dt_{ij}, and the effect of duration of exposure at temperatures above normal body temperature, $pd_{ij} = t_{ij} \times \delta_{dij}$, where $\delta_{dij} = 1$ if $d_{ij} > 37°C$ and 0 otherwise. While many outcomes were recorded in the heatshock study, we confine our interest to three of the most sensitive responses: midbrain, optic system and olfactory system.

The design of the heatshock studies allows us to quantify the association between different embryos from the same initial dam in terms of genetic as well as environmental components. We will consider three different modeling approaches for the pairwise associations. Model 1 assumes a constant value for the pairwise associations, ρ. Hence, the design matrix $\mathbf{X}_i$ for the ith cluster is a matrix with $n_i + \binom{n_i}{2}$ rows and 4 columns:

$$\mathbf{X}_i = \begin{pmatrix} 1 & pd_{i1} & dt_{i1} & 0 \\ \vdots & \vdots & \vdots & \vdots \\ 1 & pd_{in_i} & dt_{in_i} & 0 \\ 0 & 0 & 0 & 1 \\ \vdots & \vdots & \vdots & \vdots \\ 0 & 0 & 0 & 1 \end{pmatrix}, \text{ and } \boldsymbol{\beta} = \begin{pmatrix} \beta_0 \\ \beta_{pd} \\ \beta_{dt} \\ \alpha \end{pmatrix}, \qquad (13.10)$$

where $\alpha = \ln[(1 + \rho_{ijk})/(1 - \rho_{ijk})]$.

Model 2 allows separate parameters for pairwise association related to within-litter correlation and within-exposure group correlation; this allows an assessment of whether the pairwise association depends on the level of exposure. The pairwise correlations are modeled as:

$$\ln\left(\frac{1 + \rho_{ijk}}{1 - \rho_{ijk}}\right) = \begin{cases} \alpha + \gamma & \text{if } pd_{ij} = pd_{ik} \text{ and } dt_{ij} = dt_{ik}, \\ \alpha & \text{otherwise.} \end{cases}$$

Hence, the α-parameter reflects the "within-cluster" association reflecting a genetic component. The parameter γ reflects the extra association for embryos within the same exposure level. Note that such a model is not possible in conventional developmental toxicity studies, where exposure applies at the dam level, rather than at the individual fetus level.

In Model 3, the pairwise associations are modeled as a linear function of the "quadratic distances" between any two cumulative exposure values, i.e.,

$$\ln\left(\frac{1 + \rho_{ijk}}{1 - \rho_{ijk}}\right) = \alpha + \gamma(dt_{ij} - dt_{ik})^2. \qquad (13.11)$$

For the model matrix $\mathbf{X}_i$ and the vector of regression parameters $\boldsymbol{\beta}$, this implies:

$$\mathbf{X}_i = \begin{pmatrix} 1 & h_{i1} & dt_{i1} & 0 & 0 \\ \vdots & & & & \\ 1 & h_{in_i} & dt_{in_i} & 0 & 0 \\ 0 & 0 & 0 & 1 & (dt_{i1} - dt_{i2})^2 \\ 0 & 0 & 0 & 1 & (dt_{i1} - dt_{i3})^2 \\ \vdots & & & & \\ 0 & 0 & 0 & 1 & (dt_{i(n_i-1)} - dt_{in_i})^2 \end{pmatrix}, \qquad \boldsymbol{\beta} = \begin{pmatrix} \beta_0 \\ \beta_{pd} \\ \beta_{dt} \\ \alpha \\ \gamma \end{pmatrix}.$$

13.4.1 Population-averaged Models

If interest lies in the overall cumulative exposure and high temperature effects, then PA models are most appropriate. We will restrict attention to the Bahadur model and generalized estimating equations as representatives of PA models. Restrictions on the parameter space in the Bahadur representation present no problem for our data. As mentioned in the introduction, severe

Table 13.1 *Heatshock Study. Parameter estimates (standard errors) for the Bahadur model, applying different models for the association structure.*

Outcome	Par.	Model 1	Model 2	Model 3
Midbrain	β_0	-1.844 (0.211)	-1.846 (0.209)	-1.644 (0.226)
	β_{dt}	5.875 (1.713)	5.803 (1.699)	5.492 (1.456)
	β_{pd}	-3.826 (1.692)	-3.752 (1.689)	-4.033 (1.410)
	α	0.131 (0.078)	0.133 (0.076)	0.224 (0.084)
	γ		-0.049 (0.208)	-1.264 (0.576)
Optic syst.	β_0	-2.503 (0.228)	-2.413 (0.249)	-2.426 (0.250)
	β_{dt}	5.656 (1.577)	6.127 (1.695)	5.613 (1.583)
	β_{pd}	-3.690 (1.616)	-4.262 (1.720)	-3.931 (1.639)
	α	-0.059 (0.070)	-0.098 (0.072)	0.000 (0.094)
	γ		0.715 (0.280)	-0.541 (0.428)
Olfact. syst.	β_0	-1.469 (0.185)	-1.395 (0.202)	-1.430 (0.213)
	β_{dt}	6.688 (1.971)	7.349 (1.686)	7.118 (1.743)
	β_{pd}	-4.914 (2.003)	-5.612 (1.693)	-5.520 (1.799)
	α	0.258 (0.066)	0.221 (0.082)	0.294 (0.077)
	γ		0.488 (0.269)	-0.374 (0.511)

restrictions may arise for larger cluster sizes, but cluster sizes are relatively small for the heatshock studies.

Table 13.1 gives the parameter estimates (standard errors) using the Bahadur model for the three association models described above. For the midbrain response, an important cumulative exposure effect (dt) and an important additional effect of duration of exposure to temperatures above normal body temperature (pd) were observed. This indicates a departure from Haber's premise. The coefficients for β_{pd} were consistently negative, indicating that shorter acute exposures of the same temperature-duration combination cause more developmental damage than longer ones. As expected, the malformation probability tends to increase with increasing cumulative exposures.

The results of fitting Model 1 show no evidence of a significant intracluster correlation for midbrain responses. Direct exposure to the individual embryos seems to reduce the need to account for litter effects on midbrain malformations. Models 2 and 3 lead to similar conclusions. In addition, Model 2 does not support an "environmental" association component for midbrain responses. Comparing the likelihoods of Models 2 and 1 yields a deviance difference (D) of 0.052. In contrast, Model 3 yields a significant quadratic distance association parameter γ ($D=3.90$) in comparison with the constant association model (Model 1). This is evidence that the association between any two individuals decreases with the "distance" between their cumulative exposures. Goodness-of-fit statistics based on a Hosmer and Lemeshow approach (1989;

Table 13.2 *Heatshock Study. Goodness-of-fit deviance statistics (p values).*

	Model 1	Model 2	Model 3
Midbrain	6.784 (0.659)	6.905 (0.647)	6.361 (0.703)
Optic system	17.768 (0.038)	14.891 (0.094)	18.094 (0.034)
Olfactory system	25.181 (0.003)	22.652 (0.007)	25.451 (0.003)

see also Section 9.1) are provided in Table 13.2 for all fitted models. None of the deviance statistics indicates a lack-of-fit for the midbrain predicted probabilities.

For the optic system, important effects of cumulative exposure and an additional effect of duration of exposure to temperatures above normal body temperature were again observed. The clustering parameter α was never significant, indicating that there was no important intra-litter correlation working on the optic system. However, there was evidence for a dependence of the pairwise associations on the level of exposure, i.e., that animals within the same duration-temperature group behaved more similarly than animals from different groups (D=6.02). Parameter estimates for Models 1 and 2 were similar but dt and pd tend to be slightly more significant for Model 2, suggesting a gain in efficiency. There was no evidence of a quadratic distance effect (D=1.62), and Table 13.2 shows that Model 3 actually provided a poor fit to the data. Therefore, Model 2 is preferable for this data.

For the olfactory system, all models showed evidence of a significant intralitter correlation. This indicates that direct exposure of the embryos does not always reduce the need to account for litter effects. Model 2 also indicated a marginally significant extra contribution of association for individuals within the same duration-temperature group (D=3.68). The quadratic distance effect parameter γ in Model 3 was apparently unnecessary (D=0.56). However, Table 13.2 shows that all of these models fit poorly to the data. Including a quadratic effect for cumulative exposure, dt, improves the fit substantially. The best fit is then obtained for Model 2, yielding a goodness-of-fit deviance statistic of 12.31 (p value 0.193). For Models 1 and 3 the GOF statistics were 16.21 (p value 0.06) and 13.01 (p value 0.16), respectively. The intralitter correlation becomes even more significant, but the extra association component for individuals within the same duration-temperature group is reduced to a nonsignificant level.

Table 13.3 gives the parameter estimates together with model-based and empirically corrected (robust) standard errors of second order generalized estimating equations (GEE2), as described by Liang, Zeger and Qaqish (1992). In many cases, GEE2 models were difficult to fit. For instance, in the case of the more complicated Model 3, the GEE2 model could not be fit for any of the outcomes, and is therefore excluded from the table. For midbrain, the results of GEE2 are similar to those obtained using the Bahadur model. The

Table 13.3 *Heatshock Study. Parameter estimates (model-based standard error; empirically corrected standard error) for GEE2, applying different models for the association structure.*

Outcome	Par.	Model 1	Model 2
Midbrain	β_0	-1.813 (0.213;0.222)	-1.831 (0.210;0.219)
	β_{dt}	5.934 (1.679;1.941)	5.701 (1.636;1.956)
	β_{pd}	-3.928 (1.655;1.955)	-3.676 (1.615;1.971)
	α	0.095 (0.077;0.061)	0.108 (0.078;0.069)
	γ		-0.160 (0.181;0.170)
Optic system	β_0	-2.491 (0.241;0.224)	-2.415 (0.252;0.229)
	β_{dt}	5.635 (1.665;2.267)	5.644 (1.767;2.233)
	β_{pd}	-3.686 (1.703;2.217)	-3.710 (1.787;2.215)
	α	-0.049 (0.045;0.043)	-0.072 (0.056;0.056)
	γ		0.763 (0.252;0.269)
Olfactory system	β_0	-1.516 (0.223;0.275)	.
	β_{dt}	5.606 (1.502;1.527)	.
	β_{pd}	-3.701 (1.489;1.557)	.
	α	0.513 (0.079;0.136)	.
	γ		.

model-based standard errors correspond closely with those calculated by the likelihood method. Furthermore, model-based and empirically corrected (robust) standard errors are close to each other, indicating that complex association models need not be considered. In contrast, for the optic system outcome, there is a larger gap between model-based and empirically corrected standard errors, especially for Model 1. This could indicate that other more complex models should be considered.

To illustrate the importance of addressing complex association patterns, Table 13.4 presents the parameter estimates and estimated standard errors for each of the three binary outcomes (midbrain, optic system, and olfactory system) based on a logistic model, a standard first-order GEE procedure (GEE1), and the extended GEE1 approach described by Prentice (1988). All of these models are population-averaging approaches with an exchangeable correlation structure. Table 13.4 indicates a clear distinction between the logistic and correlated models. For the midbrain outcome, the logistic standard error was smaller than the model-based standard errors of the other two procedures, which were in turn smaller than any of the empirically corrected standard errors. More complex association designs which were previously obtained using the second-order generalized estimating equations (e.g., Table 13.3, Model 2) did not reduce the gap between model-based and empirically corrected standard errors. This may be due to the fact that the association parameters were

Table 13.4 *Heatshock Study. Parameter estimates (model-based standard error; empirically corrected standard error) for midbrain, optic system, and olfactory system outcome, comparing logistic regression with two different GEE1 procedures.*

		LOGISTIC	GEE1 (standard)	GEE1 (Prentice)
Outcome	Par.			
Midbr.	β_0	-1.799(0.199)	-1.808(0.203;0.226)	-1.822(0.214;0.222)
	β_{dt}	5.854(1.602)	5.910(1.618;1.953)	5.999(1.686;1.971)
	β_{pd}	-3.884(1.582)	-3.920(1.597;1.949)	-3.976(1.662;1.978)
	ρ	0	0.017	0.049
Optic	β_0	-2.475(0.251)	-2.466(0.236;0.222)	-2.468(0.239;0.222)
	β_{dt}	5.609(1.729)	5.678(1.649;2.304)	5.659(1.668;2.299)
	β_{pd}	-3.683(1.768)	-3.769(1.691;2.264)	-3.748(1.709;2.258)
	ρ	0	-0.034	-0.026
Olfact.	β_0	-1.435(0.183)	-1.544(0.216;0.216)	-1.573(0.231;0.214)
	β_{dt}	7.194(1.659)	6.649(1.767;2.152)	6.398(1.768;2.116)
	β_{pd}	-5.687(1.649)	-4.850(1.736;2.106)	-4.499(1.732;2.069)
	ρ	0	0.149	0.238

not significant for midbrain. Fitting more complex models would therefore not be very helpful.

In contrast, for the optic system outcome, model-based and empirically corrected standard errors tended to lie closer to each other for the GEE2 estimates (e.g., Table 13.3, Model 2) than for the exchangeably correlated PA procedures. The inclusion of the extra association parameter, γ, resulted in a significant improvement of the association model. Hence, it might be important to consider more complicated models for the association structure. Furthermore, the standard errors of the estimates obtained with the logistic model for this outcome were always smaller than those obtained with the empirically corrected version of the correlated models but larger than the model-based versions. Note that the estimated correlation parameter was always negative. For the olfactory system, the model-based standard errors for the logistic procedure were considerably smaller than the model-based standard errors for the correlated procedures. Furthermore, the discrepancy between model-based and empirically corrected standard errors for the correlated procedures is fairly large. Unfortunately, GEE2 was hard to fit for complex association models, such as Models 2 and 3. In conclusion, although GEE1 is much easier to fit than GEE2, it presents more difficulties when coping with complex association models.

Table 13.5 *Heatshock Study. Parameter estimates (standard errors; p values) for the mixed-effects logistic and conditional logistic models, applied on the midbrain, optic system and olfactory system outcomes.*

		mixed-effects logistic	conditional logistic
Outcome	Parameter		
Midbrain	β_0	-1.936 (0.225;0.000)	
	β_{dt}	6.375 (1.598;0.000)	6.839 (2.633;0.009)
	β_{pd}	-4.232 (1.522;0.005)	-4.639 (2.547;0.069)
Optic system	β_0	-2.475 (0.290;0.000)	
	β_{dt}	5.607 (1.364;0.000)	3.961 (3.009;0.188)
	β_{pd}	-3.680 (1.465;0.012)	-1.459 (3.036;0.631)
Olfact. system	β_0	-1.998 (0.319;0.000)	
	β_{dt}	8.061 (1.953;0.000)	6.303 (3.035;0.038)
	β_{pd}	-5.702 (1.933;0.003)	-3.399 (2.962;0.251)

13.4.2 Cluster-specific Approaches

If interest lies in within-cluster comparisons as opposed to overall effects, the CS approaches may be more appropriate since these effects are not confounded by cluster differences. The MIXOR software package (Hedeker and Gibbons 1993) was employed to fit mixed-effects logistic models to the binary outcomes midbrain, optic system and olfactory system. Alternatively, the SAS procedure NLMIXED could be used. The responses from the ith cluster are correlated by virtue of their sharing a common intercept.

Table 13.5 shows the parameter estimates (and p values) for the mixed-effects logistic model. Note that the parameter estimates for the corresponding PA models in Tables 13.1, 13.3, and 13.4 are consistently smaller than those of the CS model in Table 13.5; this "shrinkage effect" is consistent with the attenuation implied by equation (13.8). For all three outcomes, there is evidence of a significant effect of the cumulative exposure (dt) and a significant effect of duration of exposure at temperatures above normal body temperature (pd). Furthermore, the parameter estimate for pd is again negative, which is in agreement with earlier results.

Table 13.5 also shows the conditional logistic regression parameter estimates. All cluster level effects are conditioned out, so parameter estimates for the intercepts are not obtained. Clearly, there is a large discrepancy between the mixed-effects logistic models and conditional logistic parameter estimates, especially for the optic system and olfactory system responses. Neuhaus and Kalbfleisch (1998) note that a covariate has both a between-cluster component, which may be summarized in terms of $\overline{x}_i$, the cluster mean, and a within-cluster component $x_{ij} - \overline{x}_i$. The conditional logistic approach estimates the pure within-cluster covariate effect of $x_{ij} - \overline{x}_i$. However, the mixed-effects logistic approach estimates the effect of x_{ij}. Therefore, the results of conditional logistic and mixed-effects logistic are comparable only under the assumption of common between- and within-cluster covariate effects, in which case the mixed-effects logistic approach is more efficient.

For the heatshock studies the assumption of common between- and within-cluster covariate effects was satisfied for the midbrain response (a likelihood ratio test yields a value of 1.653 at 1 degree of freedom, p value= 0.198). That explains the similarity in mixed-effects logistic and conditional logistic parameter estimates for that response and the increased efficiency of mixed-effects logistic as opposed to the conditional logistic method. Whereas strong significant effects were identified for c and h by the mixed-effects logistic and PA approaches, the statistical significance was reduced for conditional logistic regression, as implied by equation (13.9). When evaluated at the mean cluster size of 5 and based on the estimated intracluster correlation from the standard GEE1 approach in Table 13.4 ($\rho_y = 0.017$), the ARE from formula (13.9) would be 81% for $\rho_x = 0$, and lower for higher values of ρ_x.

In contrast, for optic system and olfactory system, the assumption of equal between- and within-cluster covariate effects was not satisfied, explaining the large discrepancy between mixed-effects logistic and conditional logistic estimates. A comparison of standard errors or statistical significance is thus not appropriate here, unless we fit a mixed-effects logistic model with separate parameters for the between- and within-cluster covariate component. The within-cluster covariate effect estimates thus obtained for optic system were 3.929 (s.e. 2.647) and -1.255 (s.e. 2.779) for cumulative exposure dt and high temperature pd, respectively. Similarly, we found 5.804 (s.e. 3.044) and -2.839 (s.e. 2.968) for the within-cluster covariate effects of dt and pd on olfactory system. These estimates were similar to the conditional logistic estimates, but do not indicate a loss in efficiency.

13.5 Continuous Outcomes

As stated earlier, there are measurements on 13 morphological variables. Some are binary while others are measured on a continuous scale. In this section, we will focus on the continuous outcomes (Verbeke and Molenberghs 2000).

There are several continuous outcomes recorded in the heatshock study, such as size measures on crown rump, yolk sac, and head. We will focus on crown rump length (CRL). The linear mixed model, presented in Section 4.3.2, will be used to this effect (Verbeke and Molenberghs 2000).

It will be shown that the three components of variability customarily incorporated in a linear mixed-effects model of the form (4.38) can usefully be applied here as well, even in the absence of a repeated-measures structure. Although there will be no doubt that random effects are used to model interdam variability and also the role of the measurement error time is un-

Table 13.6 *Heatshock Study. Parameter estimates (standard errors) for initial and final model.*

Effect	Parameter	Initial	Final
Fixed effects:			
Intercept	β_1	3.622 (0.034)	3.627 (0.042)
Durtemp $(dt)_{ij}$	β_2	-1.558 (0.376)	-1.331 (0.353)
Posdur $(pd)_{ij}$	β_3	0.019 (0.006)	0.015 (0.006)
Random-effects parameters:			
$\mathrm{var}(b_{1i})$	d_{11}	0.010 (0.014)	0.046 (0.014)
$\mathrm{var}(b_{12})$	d_{12}	-0.038 (0.065)	
$\mathrm{cov}(b_{1i}, b_{2i})$	d_{22}	0.071 (0.032)	
Residual variance parameters:			
$\mathrm{var}(\varepsilon_{(1)ij})$	σ^2	0.097 (0.014)	0.097 (0.014)
$\mathrm{var}(\varepsilon_{(2)ij})$	τ^2	0.044 (0.017)	0.042 (0.017)
Spatial corr. parameter	ρ	4.268 (5.052)	4.143 (3.772)

ambiguous, it is less obvious what the role of the serial association would be. Generally, serial association results from the fact that within a cluster, residuals of individuals closer to each other are often more similar than residuals for individuals further apart. Although this distance concept is clear in longitudinal and spatial applications, it is less so in this context. However, covariates like duration and temperature, or relevant transformations thereof, can play a similar role. This distinction is very useful since random effects capture the correlation structure which is attributable to the dam and hence includes genetic components. The serial correlation, on the other hand, is entirely design driven. If one conjectures that the latter component is irrelevant, then translation into a statistical hypothesis and, consequently, testing for it are relatively straightforward. Note that such a model is not possible in conventional developmental toxicity studies, where exposure applies at the dam level, not at the individual fetus level.

The model we consider is based on Haber's Law (for a detailed discussion, see Section 13.4) and controlled deviations thereof, in the sense that the fixed-effects structure includes the interaction between duration of the experiment

Final Model   Posdur Removed

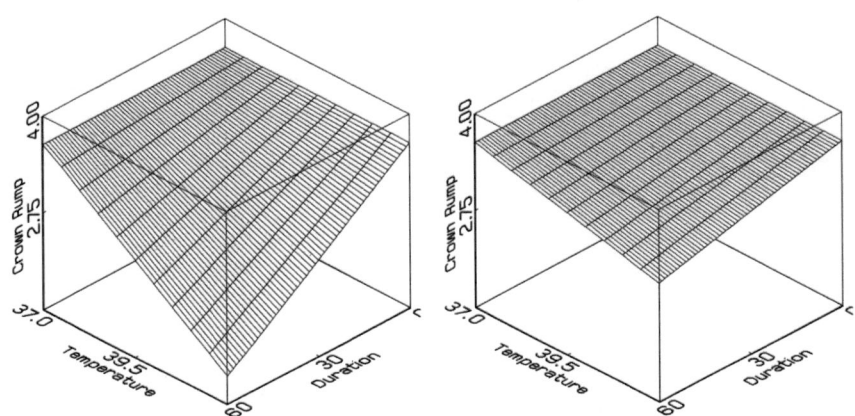

Figure 13.3 *Heatshock Study. Fixed-effects structure for (a) the final model and (b) the model with* posdur *removed.*

ard temperature elevation (*durtemp, dt*) and positive duration (*posdur, pd,* the duration in case there is a nonzero elevation of temperature, and zero otherwise). For computational convenience, the ranges of these covariates are transformed to the unit interval. The maximal values correspond to 225 min°C for *durtemp* and 60 minutes for *posdur*. The random-effects structure includes a random intercept and a random slope for *dt*. The residual covariance structure is decomposed into a Gaussian serial process in *dt* and measurement error. Formally,

$$Y_{ij} = (\beta_1 + b_{i1}) + (\beta_2 + b_{i2})(dt)_{ij} + \beta_3(pd)_{ij} + \varepsilon_{(1)ij} + \varepsilon_{(2)ij}, \quad (13.12)$$

where the $\varepsilon_{(1)ij}$ are uncorrelated and follow a normal distribution with zero mean and variance σ^2. The $\varepsilon_{(2)ij}$ have zero mean, variance τ^2, and serial correlation

$$h_{ijk} = \exp\left\{-\phi[(dt)_{ij} - (dt)_{ik}]^2\right\}.$$

The random-effects vector (b_{i1}, b_{i2}) is assumed to be a zero-mean normal variable with covariance matrix D. The initial model is reproduced in Table 13.6. Note that the variance of the random slopes is negative. This is allowed, in case one is prepared to focus on the marginal model only, ignoring the random-effects motivation of the model (Verbeke and Molenberghs 2000, Chapter 6).

First, the covariance model is simplified. The covariance between both random effects is not significant and can be removed ($G^2 = 3.35$ on 1 degree of freedom, $p = 0.067$). Next, the random *durtemp* effect is removed ($G^2 = 3.63$, 2 df, $p = 0.057$). The serial process cannot be removed ($G^2 = 6.19$, 2 df,

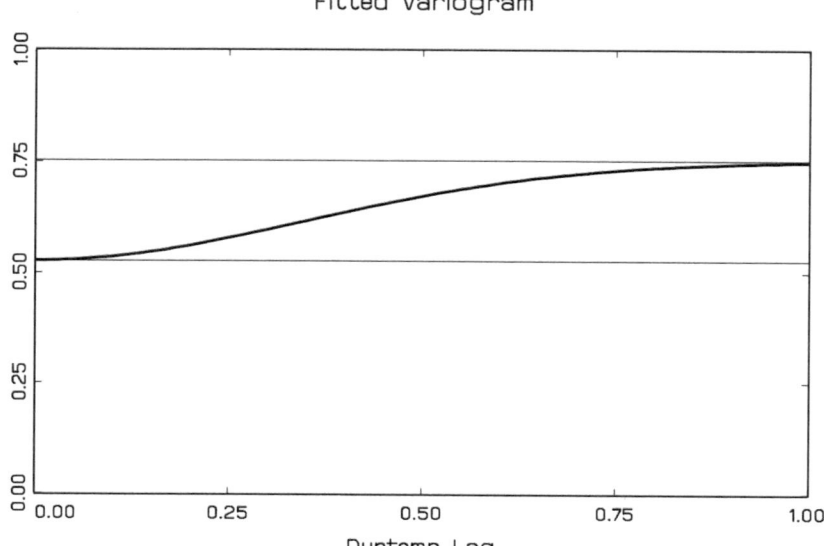

Figure 13.4 *Heatshock Study. Fitted variogram.*

$p = 0.045$). Finally, both fixed effects are highly significant and cannot be removed. The final model is given in Table 13.6.

The fixed-effects structure is presented in Figure 13.3. The left-hand panel shows the fixed-effects structure of the final model, as listed in Table 13.6. The coefficient of *durtemp* is negative, indicating a decreasing crown rump length with increasing exposure. This effect is reduced by a positive coefficient for *posdur*. Fitting this model with *posdur* removed shows qualitatively the same trend, but the effect of exposure is much less pronounced, underscoring that there is a significant deviation from Haber's Law.

The fitted variogram (Verbeke and Molenberghs 2000) is presented in Figure 13.4. Roughly half of the variability is attributed to measurement error, and the remaining half is divided equally over the random intercept and the serial process. The corresponding fitted correlation function is presented in Figure 13.1. The correlation is about 0.50 for two foetuses that are at the exact same level of exposure. It then decreases to 0.25 when the distance between exposures is maximal. This reexpresses that half of the correlation is due to the random effect, and the other half is attributed to the serial process in *durtemp*.

13.6 Concluding Remarks

In summary, the heatshock data example illustrated several issues encountered in the analysis of clustered binary outcomes. Marginal PA approaches using likelihood methods such as the Bahadur model or generalized estimating equations tended to yield similar parameter estimates, especially when the correlation structure was specifically modeled using a GEE2 approach. Failure to take into account the correlation structure yielded underestimates of the true standard errors. The parameter estimates for the PA models tended to be smaller than those of the corresponding CS models, supporting the "shrinkage effect" discussed by Neuhaus and Lesperance (1996). When between and within-cluster covariate effects were similar, as for the midbrain outcome, the conditional logistic regression approach yielded similar parameter estimates to those from a mixed effect logistic regression model, but was less efficient. However, the mixed effect model yielded similar tests of significance to the GEE1 approaches shown in Table 13.4, which assumed an exchangeable correlation structure. For this particular data, the Bahadur model and GEE2 approaches provided particular insight into the correlation structure, since they allowed the association to be modeled as a function of both exposure levels and clusters as defined by litters. However, in most cases it is advisable to fit several different classes of models and compare the results for consistency, rather than focus on only a single model. A complete analysis incorporating both CS and PA models can often provide insight into the scientific interpretation and impact of association structure when some covariates of interest are cluster-level while others are measured at the level of the individual.

CHAPTER 14

Combined Continuous and Discrete Outcomes

Meredith M. Regan

Beth Israel Deaconess Medical Center and Harvard Medical School, Boston, MA

Paul J. Catalano

Harvard School of Public Health, Boston, MA

For the analysis of data from Segment II developmental toxicity designs (Sections 1.2.1 and 2.2, and Chapter 13), focus has recently turned to better handling of the multivariate nature of the response. These studies seek to determine the overall adverse effects of dose on the offspring and it is most often not known *a priori* how effects will manifest. In developmental toxicity studies with a Segment II design, the uterus of each sacrificed dam is removed and examined for resorptions (very early deaths) and fetal deaths. The viable foetuses are measured for birth weight and length and examined carefully for the presence of different types of malformations. Among viable foetuses, the incidence of any malformation (binary) and reductions in fetal weight (continuous) are typically of primary concern, as both have been found to be sensitive indicators of a toxic effect (U.S. EPA 1991).

Often, dose-response relationships are characterized in each of the outcomes (death, weight, and malformation) separately, using appropriate methods to account for litter effects. Based on the dose-response patterns, the outcome that appears most sensitive to the exposure (called the critical effect) becomes the focus for risk assessment purposes (U.S. EPA 1991, 1995). This approach assumes that protecting against the most sensitive outcome protects against all other adverse outcomes; however there may be a more generalized pattern of effects. An approach that considers the multiple sources of risk and the relationship between them in quantifying an overall risk may be preferable. One approach considers the non-live and live outcomes as conditionally independent, and hence they can be modeled separately (Ryan 1992). Because the live outcomes are correlated (Chen and Gaylor 1992, Ryan, Catalano, Kimmel and Kimmel 1991), jointly modeling the live outcomes and using the bivariate outcome as a basis for risk assessment may be most appropriate.

This motivates the formulation of a joint distribution with mixed continuous and discrete outcomes. The joint model must allow different dose-response functions for each type of outcome and must account for the correlation between them, as well as the correlations that result from clustering. Methods for jointly modeling discrete and continuous outcomes, especially with clustering, are not well established (Regan and Catalano 1999a, 1999b, 2000, Geys, Regan, Catalano and Molenberghs 2001). Without an obvious multivariate distribution incorporating both types of outcomes, specifying a joint distribution of responses within a litter is not straightforward.

We discuss several models for jointly modeling discrete and continuous outcomes in the clustered data setting of developmental toxicity studies (Section 14.1). In Section 14.2, we then discuss the application of the models to quantitative risk assessment.

14.1 Models for Bivariate Data of a Mixed Nature

Early research, often motivated by psychology, focused on estimating the correlation between a discrete and continuous outcome and deriving the distribution of the estimator. Tate (1954) examined the case of a single binary outcome, assumed Bernoulli, and a continuous outcome with conditional distribution given the binary outcome assumed normal, and derived the asymptotic distribution of the sample point-biserial correlation coefficient. Olkin and Tate (1961) extended this work to a multivariate setting, assuming multinomial and conditional multivariate normal distributions.

Others have approached the problem by using latent variable, or tolerance distribution, ideas which presuppose the existence of an unobservable, continuous random variable underlying the discrete outcome. A binary event is assumed to represent an indicator that the latent variable exceeds some threshold value. It is assumed that the joint distribution of the observed and underlying continuous variables is bivariate normal. Tate (1955) focused on the estimation and asymptotic variances of the bivariate normal correlation and the point of dichotomy of the dichotomized normal variable; this was generalized to a multivariate normal with a single dichotomized variable (Hannan and Tate 1965) or a single discretized variable (Cox 1974).

Methods focusing on jointly analyzing discrete and continuous outcomes and estimating mean parameters have more recently been explored. The challenge stems from the lack of multivariate distributions for combining both types of outcomes, so that specification of a joint distribution of the responses is not straightforward. The multivariate exponential family model has a general form that easily incorporates both discrete and continuous outcomes and has been discussed in the unclustered setting (Prentice and Zhao 1991, Zhao, Prentice and Self 1992, Sammel *et al.* 1997). Otherwise, following the idea of Olkin and Tate (1961), a frequent approach in other, unclustered multivariate settings is to specify the joint distribution as the product of a marginal and a conditional distribution; this factorization leads to a model with two components that can be specified separately. Two versions of a conditional model are possible, depending on whether the conditioning is done on the continuous or the discrete outcomes (Cox and Wermuth 1992, Fitzmaurice and Laird 1997, Krzanowski 1988, Little and Schluchter 1985).

Conditional models have been proposed in the clustered data setting of developmental toxicity studies, where the joint distribution of the outcomes of the litter can be specified based on a clustered malformation model and a clustered fetal weight model, with one model conditional on the other outcomes. The choice of conditioning is mostly for statistical convenience rather than biologic rationale, as relatively little is understood about the biologic mechanisms of developmental toxicity.

The latent variable idea has been used to directly specify the joint distribution of the discrete and continuous outcomes of a litter. When the observed and latent continuous outcomes are assumed jointly normal, a mixed-outcome probit model arises which uses the correlation structure of the underlying multivariate normal distribution to characterize intrafoetus and intralitter correlation. An alternative is to specify a Plackett distribution, which allows flexibility for selection of marginal densities and uses the odds ratio as a measure of intrafoetus association.

14.1.1 Notation

Let N denote the total number of dams, and hence litters, in the study. For the ith dam ($i = 1, \ldots, N$), let n_i be the litter size, or the number of viable foetuses, of litter i. Each foetus is examined for the presence ($M_{ik} = 1$) or absence ($M_{ik} = 0$) of a certain malformation indicator and fetal weight (W_{ik}) is measured ($k = 1, \ldots, n_i$). Let M_i and W_i denote $n_i \times 1$ vectors of malformation and fetal weight outcomes, respectively, and let $(W_i^T, M_i^T)^T$ be the $2n_i \times 1$ vector of outcomes for the ith litter.

In models that appeal to the latent variable approach, the binary malformation outcome is assumed to arise from an unobservable continuous random variable, denoted M_{ik}^*; M_{ik} represents an indicator of whether this underlying variable exceeds some threshold, arbitrarily assumed to be 0. Let M_i^* denote the $n_i \times 1$ vector of latent malformation outcomes, and let $(W_i^T, M_i^{*T})^T$ be the $2n_i \times 1$ vector of observed and latent continuous outcomes for the ith litter.

Throughout this chapter, parameters corresponding to fetal weights will be subscripted with w and those corresponding to malformation with m.

Keeping with standard notation, I_i denotes an n_i-dimensional identity matrix, and J_i and 1_i denote an n_i-matrix and an n_i-vector of ones. $\phi_n(\cdot)$ denotes an n-dimensional multivariate normal density, and $\Phi(\cdot)$ and $\Phi_2(\cdot)$ are the standard univariate and bivariate normal distribution functions.

14.1.2 Conditional Models

In this section we consider two conditional models for mixed continuous-discrete outcomes: one model that conditions on the continuous outcomes and one that conditions on the discrete outcomes. Note that the term *conditional model* is used in line with its general meaning in the rest of the text (Section 4.2).

In a factorization that conditions on the continuous outcomes, Catalano and Ryan (1992) apply the latent variable concept to derive the joint distribution of continuous and binary outcomes and then use generalized estimating equation (GEE) ideas for estimation in the clustered data setting. The marginal distribution of the fetal weights is related to covariates using an identity link function, while for the conditional distribution of malformation given fetal weights, they use a probit link. The model has been extended to model an ordinal malformation variable (Catalano 1997).

Fitzmaurice and Laird (1995) reverse the factorization to condition on the discrete outcomes. Assuming independence among littermates, they write the joint distribution of the bivariate outcome as the product of a marginal Bernoulli distribution for the malformation response, and a conditional normal distribution for the fetal weight response given malformation; the correlation between outcomes is considered a nuisance parameter. They consider an extension of their model that allows for clustering, and use GEE methodology for estimation to avoid the computational complexity of maximum likelihood in the clustered setting.

Models that Condition on the Continuous Outcomes

Catalano and Ryan (1992) motivate their model by considering an unobservable continuous latent variable corresponding to malformation and then constructing a joint distribution between the latent variable and the observable fetal weight outcome. The $2n_i \times 1$ vector of observed and latent outcomes $(\boldsymbol{W}_i^T, \boldsymbol{M}_i^{*T})^T$ is assumed to follow a multivariate normal (MN) distribution with means $\mu_{w;ik}$ and $\mu_{m;ik}$ among the fetal weights and latent malformations, respectively. The assumed MN covariance structure allows for a constant correlation between observations on the same foetus (intrafoetus) and separate correlations between observations on littermates (intralitter). The correlation structure for littermates k and k' may be illustrated as:

$$
\begin{array}{ccccc}
W_{ik} & \longleftarrow & \varrho & \longrightarrow & M_{ik}^* \\
\uparrow & \searrow & & \nearrow & \uparrow \\
\rho_w & & \rho_{wm} & & \rho_m^* \\
\downarrow & \swarrow & & \searrow & \downarrow \\
W_{ik'} & \longleftarrow & \varrho & \longrightarrow & M_{ik'}^*
\end{array}
\qquad (14.1)
$$

which results in the following block equicorrelated covariance matrix:

$$\text{Cov}(\boldsymbol{W}_i^T, \boldsymbol{M}_i^{*T})^T =$$

$$\begin{bmatrix} \sigma_w^2[(1-\rho_w)\boldsymbol{I}_i + \rho_w\boldsymbol{J}_i] & \sigma_w\sigma_m[(\varrho-\rho_{wm})\boldsymbol{I}_i + \rho_{wm}\boldsymbol{J}_i] \\ \sigma_w\sigma_m[(\varrho-\rho_{wm})\boldsymbol{I}_i + \rho_{wm}\boldsymbol{J}_i] & \sigma_m^2[(1-\rho_m^*)\boldsymbol{I}_i + \rho_m^*\boldsymbol{J}_i] \end{bmatrix} .\tag{14.2}$$

Dose-response models are specified for each marginal mean, and may incorporate litter- and/or foetus-specific covariates, such as gender or litter size; a power parameter on dose can also be incorporated (Catalano et $al.$ 1993). For example, consider the following simple model for the means:

$$\mu_{w;ik} = \alpha_0 + \alpha_1 d_i + \alpha_2(n_i - \bar{n}), \tag{14.3}$$

$$\mu_{m;ik} = \beta_0^* + \beta_1^* d_i, \tag{14.4}$$

where d_i is the dose administered to the dam of the ith litter.

Following from MVN theory, the conditional distribution of the malformation latent variables given the fetal weight vector is also normal, where the ikth element of the conditional mean is given by

$$\mu_{m|w;ik} = \mu_{m;ik} + \left(\frac{\sigma_m}{\sigma_w}\right)\left(\frac{\varrho + (n_i-1)\rho_{wm}}{1 + (n_i-1)\rho_w}\right)\bar{e}_{w;i}$$

$$+ \left(\frac{\sigma_m}{\sigma_w}\right)\left(\frac{\varrho - \rho_{wm}}{1 - \rho_w}\right)(e_{w;ik} - \bar{e}_{w;i}) \tag{14.5}$$

which depends on the average litter weight residual, $\bar{e}_{w;i} = (\overline{W}_i - \mu_{w;ik})$, and the individual fetal weight residuals, $e_{w;ik} - \bar{e}_{w;i}$, where $e_{w;ik} = (W_{ik} - \mu_{w;ik})$. The conditional distribution for the observable malformation indicator given fetal weights can be described by a correlated probit model, with average litter weight and individual weight residuals as covariates. Not all parameters in this model are estimable, but the model can be reparametrized to a fully estimable form:

$$\pi_{m|w;ik} = \boldsymbol{E}(M_{ik}|\boldsymbol{W}_i)$$

$$= \Phi\{\beta_0 + \beta_1 d_i + \beta_2 \bar{e}_{w;i} + \beta_3(e_{w;ik} - \bar{e}_{w;i})\} . \tag{14.6}$$

The dependence on the weight residuals indicates that both the litter average and individual weights affect the probability of malformation, a result of allowing different intrafoetus and intralitter correlations.

Note that the $\boldsymbol{\beta}$ parameters in the conditional probit model are directly related to variance and correlation parameters in the underlying latent variable model; so tests of model parameters and examination of their magnitudes lends insight into the validity of the assumed correlation structure. This is seen by comparing the probit model in (14.6) to the expression for the conditional mean in (14.5). A drawback, however, is that the $\boldsymbol{\beta}$ parameters do not have a marginal interpretation in the probit model.

As observed with the DYME data, live litter size may decrease with increasing dose levels due to an increasing fetal death rate with dose. In addition, fetal weight and malformation are often affected by litter size. Thus, it may be appropriate for an analysis of live outcomes to adjust for litter size. A reasonable approach in the current context is to begin with a model similar to (14.4) and incorporate a covariate for the deviation from the overall average litter size, $(n_i - \bar{n})$. Because the coefficient of the average weight residual in the conditional latent malformation mean (14.5) is a function of litter size, the resulting model for the correlated probit model is

$$
\begin{aligned}
\pi_{m|w;ik} \quad = \quad & \Phi\left\{\beta_0 + \beta_1 d_i + \beta_2 \bar{e}_{w;i} + \beta_3(e_{w;ik} - \bar{e}_{w;i})\right. \\
& \left. + \beta_4(n_i - \bar{n}) + \beta_5(n_i - \bar{n})\bar{e}_{w;i}\right\} .
\end{aligned} \tag{14.7}
$$

That is, incorporating litter size leads to a probit model similar to (14.6), but with a covariate for litter size and a litter size-by-average fetal weight interaction term.

Parameter estimation proceeds in two steps, thus implementing the GEE methodology of Liang and Zeger (1986) (see also Chapter 5) to each component of the conditional model separately. First, a correlated regression of fetal weight on dose and other covariates is fit to obtain estimates of $\boldsymbol{\alpha}$; moment-based estimates of the working correlation parameter, ρ_w, and of σ_w^2 are obtained. Next, a correlated probit regression of malformation conditional on weight, with dose, average, and individual fetal weight residuals, and other covariates is fit to obtain estimates of $\boldsymbol{\beta}$; a moment-based working correlation parameter, ρ_m, is also estimated. The corresponding GEEs for the regression parameters $\boldsymbol{\theta} = (\boldsymbol{\alpha}, \boldsymbol{\beta})$ can be written

$$
\sum_{i=1}^{N} \boldsymbol{D}_i^T \boldsymbol{V}_{w;i}^{-1} \left(\boldsymbol{W}_i - \boldsymbol{\mu}_{w;i}(\boldsymbol{\alpha})\right) \quad = \quad \boldsymbol{0},
$$
$$
\sum_{i=1}^{N} \boldsymbol{E}_i^T \boldsymbol{V}_{m;i}^{-1} \left(\boldsymbol{M}_i - \boldsymbol{\pi}_{m|w;i}(\boldsymbol{\beta})\right) \quad = \quad \boldsymbol{0},
\tag{14.8}
$$

where $\boldsymbol{D}_i = \partial\boldsymbol{\mu}_{w;i}/\partial\boldsymbol{\alpha}$ with $\boldsymbol{\mu}_{w;i}$ denoting the marginal mean of $\boldsymbol{W}_i$, and $\boldsymbol{E}_i = \partial\boldsymbol{\pi}_{m|w;i}/\partial\boldsymbol{\beta}$ with $\boldsymbol{\pi}_{m|w;i}$ denoting the conditional mean vector with elements from (14.7). As working covariance matrices, they use:

$$
\mathrm{Cov}(\boldsymbol{W}_i) \quad = \quad \boldsymbol{V}_{w;i} \quad \approx \quad \sigma_w^2\left[(1 - \rho_w)\boldsymbol{I}_i + \rho_w \boldsymbol{J}_i\right],
$$

$$
\mathrm{Cov}(\boldsymbol{M}_i \mid \boldsymbol{W}_i) \quad = \quad \boldsymbol{V}_{m;i} \quad \approx \quad \boldsymbol{\Delta}_i^{1/2}\left[(1 - \rho_m)\boldsymbol{I}_i + \rho_m \boldsymbol{J}_i\right]\boldsymbol{\Delta}_i^{1/2}, \tag{14.9}
$$

where $\boldsymbol{\Delta} = \mathrm{diag}[\pi_{m|w;ik}(1 - \pi_{m|w;ik})]$ from (14.7). Note the working correlation ρ_m is for the binary malformation endpoint after conditioning on the weight outcomes, and not ρ_m^* from (14.1). Model-based and robust estimates of the covariance of the regression parameters are computed following the GEE methodology of Liang and Zeger (1986) and Zeger and Liang (1986).

Due to the non-linearity of the link function relating the conditional mean of the binary response to the covariates, the regression parameters in the probit model have no direct marginal interpretation. Furthermore, if the model for the mean has been correctly specified, but the model for the association between the binary and continuous outcomes is misspecified, the regression parameters in the probit model are not consistent. The lack of marginal interpretation and lack of robustness may be considered unattractive features of this approach. An important advantage is that hypothesis tests of model parameters lend insight into the validity of the assumed correlation structure, as will be discussed for the example.

DYME in Mice

As a running example throughout the chapter, we will use the DYME data, as described in Section 2.1.3.

The parameter estimates obtained from fitting mean models (14.3) and (14.7) and working covariances (14.9) are displayed in Tables 14.1 and 14.2, in the column labeled *C-Cont*. Fitted model parameters are consistent with the summary data in Table 2.1. For fetal weight, the dose coefficient is significantly negative ($\hat{\alpha}_1 = -0.943$), but there appears to be little effect of litter size on weight ($\hat{\alpha}_2 = -0.0021$). The intralitter working correlation for weight is substantial ($\hat{\rho}_w = 0.572$), and the variance, estimated as the scale parameter of the GEEs, is $\hat{\sigma}_w^2 = 0.011$.

The dose coefficient is significantly positive for malformation ($\hat{\beta}_1 = 8.577$). The negative coefficients of the individual weight residual ($\hat{\beta}_2 = -2.164$) and average weight residual ($\hat{\beta}_3 = -4.366$) indicate that low fetal weight has an impact on risk of malformation at both the litter and foetus level for this substance. That is, foetuses from litters with a low average birth weight are at increased risk of malformation and, furthermore, an individual animal's reduced birth weight relative to its average litter weight increases its own risk of malformation. The negative coefficient of litter size ($\hat{\beta}_4 = -0.060$) suggests that larger litters had a lower risk of malformation, even after adjusting for individual and average fetal weight, though this effect is not statistically significant. The intralitter working correlation among the conditional binary malformation variables is small ($\hat{\rho}_m = 0.047$).

Although the model does not allow estimation of the intrafoetus correlation, denoted ϱ in (14.1), one can test whether $\varrho = \rho_{wm}$, that is, whether intrafoetus and intralitter correlation between weight and malformation are the same, by using the coefficient of the weight residual to test $H_0 : \beta_2 = 0$ in the malformation model (14.7). For these data this test has a value of $Z = -1.8$ ($p = 0.08$), suggesting separate intrafoetus and intralitter correlation parameters between the fetal weight and malformation outcomes. That is, an individual foetus's weight offers additional information over the litter average weight about the probability of malformation. If the intrafoetus and intralitter weight-malformation correlations were both zero then it would be the case that the coefficients related to fetal weight residuals (i.e., $\beta_2, \beta_3, \beta_5$ in (14.7)) would also be zero.

Other Models

In a similar modeling approach that conditions on the continuous outcomes, Chen (1993) uses conditional normal regression chain models (see Cox and Wermuth 1992) to model malformation using litter size and fetal weight as covariates. A difference from Catalano and Ryan (1992) is that Chen directly models litter size as a function of dose, resulting in a 3-step regression model. After modeling litter size as a function of dose, fetal weight is modeled as a function of dose and conditional on litter size, and finally malformation is modeled as a function of dose and conditional on both fetal weight and litter size. Ahn and Chen (1997) also propose a tree-structured logistic regression model in which fetal weight is similarly incorporated.

Models that Condition on the Discrete Outcomes

Fitzmaurice and Laird (1995) develop a bivariate model under an assumption of independence to motivate their model in the clustered setting. Assume temporarily that littermates are independent. Marginally, M_{ik} is Bernoulli and the mean response rate, $\pi_{m;ik} = \Pr(M_{ik} = 1)$, is related to covariates via a logit link function. The conditional distribution of W_{ik} given M_{ik} is assumed normally distributed with mean,

$$\mu_{w|m;ik} = \boldsymbol{E}(W_{ik} \mid M_{ik}) = \mu_{w;ik} + \zeta \left(M_{ik} - \pi_{m;ik}\right)$$

and variance σ_w^2, where $\mu_{w;ik} = \mathrm{E}(W_{ik})$ and ζ is an association parameter from the regression of W_{ik} on M_{ik}.

Dose-response models are specified for each marginal mean, and may incorporate litter- and foetus-specific covariates. The models are denoted:

$$\eta_{ik} = \mathrm{logit}(\pi_{m;ik}) = \boldsymbol{X}_{ik}\boldsymbol{\beta} = \beta_0 + \beta_1 d_i + \beta_2(n_i - \bar{n}), \quad (14.10)$$

$$\mu_{w;ik} = \boldsymbol{X}_{ik}\boldsymbol{\alpha} = \alpha_0 + \alpha_1 d_i + \alpha_2(n_i - \bar{n}), \quad (14.11)$$

where $\boldsymbol{X}_{ik}$ is a covariate vector, which in general may differ for each mean model, though it does not in this example. For simplicity of notation, the covariate vector will be denoted without respect to its parameter.

The distribution of the bivariate outcome is specified as a product of the marginal and conditional distributions as

$$f(W_{ik}, M_{ik}) = f_m(M_{ik}) \times f_{w|m}(W_{ik} \mid M_{ik})$$

$$= \exp\{M_{ik}\,\eta_{ik} - \ln[1 + \exp(\eta_{ik})]\} \times$$

$$(2\pi\sigma_w^2)^{-\frac{1}{2}} \exp\{-\tfrac{1}{2}\sigma_w^2\,[W_{ik} - \mu_{w;ik} - \zeta\,e_{m;ik}]^2\},$$

where $e_{ik} = (M_{ik} - \pi_{m;ik})$. Note that, as a result of the parameterization of the model, $\boldsymbol{E}(W_{ik}) = \boldsymbol{X}_{ik}\boldsymbol{\alpha}$ so that both regression parameters, $\boldsymbol{\beta}$ and $\boldsymbol{\alpha}$, have marginal interpretations. This is an advantage of specifying the linear link on the conditional continuous model.

Table 14.1 *DYME in Mice. Model fitting results from different approaches. Estimates (standard errors; Z values).*

Parameter	C-Cont	C-Disk	Plackett-Dale	Probit Lik	Probit GEE
Fetal Weight:					
Intercept	1.024 (0.018;55.6)	1.026 (0.018;56.0)	1.012 (0.016;62.1)	1.023 (0.015;67.6)	1.023 (0.017;60.5)
Dose	−0.943 (0.069;−13.6)	−0.963 (0.069;−14.0)	−0.886 (0.057;−15.9)	−0.937 (0.057;−16.4)	−0.940 (0.061;−15.5)
$n_i - \bar{n}$	−0.0021 (0.004;−0.6)	−0.0019 (0.004;−0.5)	0.0005 (0.003;0.2)	−0.0006 (0.003;−0.2)	−0.0013 (0.003;−0.4)
$e_{m;ik}$	—	−0.036 (0.011;−3.1)	—	—	—
$\bar{e}_{m;i}$	—	−0.025 (0.043;−0.6)	—	—	—
Malformation:					
Intercept	−2.775 (0.234;−11.9)	−5.244 (0.491;−10.7)	−5.015 (0.498;−10.1)	−2.847 (0.212;−13.5)	−2.825 (0.205;−13.8)
Dose	8.577 (0.888;9.7)	15.869 (2.056;7.7)	15.094 (2.207;6.8)	8.551 (0.766;11.2)	8.479 (0.838;10.1)
$n_i - \bar{n}$	−0.060 (0.042;−1.4)	−0.177 (0.095;−1.9)	−0.229 (0.103;−2.2)	−0.089 (0.042;−2.1)	−0.108 (0.044;−2.5)
$\bar{e}_{w;i}$	−2.164 (1.231;−1.8)	—	—	—	—
$e_{w;ik} - \bar{e}_{w;i}$	−4.366 (1.358;−3.2)	—	—	—	—
$(n_i - \bar{n}) \times \bar{e}_{w;i}$	0.113 (0.477;0.2)	—	—	—	—

[a] The dose metric is g/kg/day

[b] C-Cont., C-Disc., Plackett-Dale, Probit GEE all report robust standard errors and Z scores

Table 14.2 DYME in Mice. Model fitting results from different approaches. Estimates (standard errors; Z values).

Parameter	C-Cont.	C-Disk	Plackett-Dale	Probit Lik	Probit GEE
Fetal Weight:					
Dose-dependent variance:					
0	0.011	0.107	0.014 (0.002;6.0)	0.015 (0.002;7.6)	0.013 (0.004;3.2)
0.0625	0.011	0.107	0.013 (0.002;7.6)	0.014 (0.002;9.2)	0.015 (0.004;3.6)
0.1250	0.011	0.107	0.012 (0.001;9.4)	0.013 (0.001;10.5)	0.013 (0.002;8.0)
0.2500	0.011	0.107	0.010 (0.001;9.0)	0.011 (0.001;8.8)	0.010 (0.002;4.8)
0.5000	0.011	0.107	0.007 (0.002;3.9)	0.007 (0.001;6.1)	0.008 (0.002;3.9)
Dose-dependent correlation:					
0	0.527	0.597	—	0.632 (0.056;11.3)	0.493
0.0625	0.527	0.597	—	0.657 (0.048;13.7)	0.493
0.1250	0.527	0.597	—	0.512 (0.056;9.1)	0.493
0.2500	0.527	0.597	—	0.671 (0.042;16.0)	0.493
0.5000	0.527	0.597	—	0.501 (0.092;5.4)	0.493
Malformation:					
Correlation	0.047	0.092	—	0.306 (0.079;3.9)	0.019
Fetal Weight / Latent Malformation Association:					
0	—	—	0.219 (0.128;1.7)	−0.269 (0.070;−3.8)	−0.184 (0.071;−2.6)
0.0625	—	—	0.219 (0.128;1.7)	−0.269 (0.070;−3.8)	−0.184 (0.071;−2.6)
0.1250	—	—	0.219 (0.128;1.7)	−0.269 (0.070;−3.8)	−0.340 (0.240;−1.4)
0.2500	—	—	0.219 (0.128;1.7)	−0.269 (0.070;−3.8)	−0.435 (0.110;−4.0)
0.5000	—	—	0.219 (0.128;1.7)	−0.269 (0.070;−3.8)	−0.517 (0.105;−5.0)

[a] The dose metric is g/kg/day

[b] C-Cont., C-Disc., Plackett-Dale, Probit GEE all report robust standard errors and Z scores

To simplify notation, let $\boldsymbol{H}_{ik} = \{\boldsymbol{X}_{ik}, e_{m;ik}\}$ and $\boldsymbol{\omega} = (\boldsymbol{\alpha}, \zeta)$, such that $\mu_{\boldsymbol{w}|m;ik} = \boldsymbol{H}_{ik}\boldsymbol{\omega}$, and let $S_{ik} = (W_{ik} - \mu_{w|w;ik})^2$. The score equations for joint estimation of the marginal mean and association parameters are derived and expressed as estimating equations:

$$\sum_{i=1}^{N} \begin{pmatrix} \boldsymbol{D}_{ik} & 0 & 0 \\ \boldsymbol{F}_{ik} & \boldsymbol{H}_{ik} & 0 \\ 0 & 0 & 1 \end{pmatrix}^{T} \begin{pmatrix} \Delta_{ik} & 0 & 0 \\ 0 & \sigma_w^2 & 0 \\ 0 & 0 & v \end{pmatrix}^{-1}$$

$$\times \begin{pmatrix} M_{ik} - \pi_{m;ik} \\ W_{ik} - \mu_{w|m;ik} \\ S_{ik} - \sigma_w^2 \end{pmatrix} = \boldsymbol{0}, \qquad (14.12)$$

with scalar parameters $\Delta_{ik} = \mathrm{Var}(M_{ik}) = \pi_{m;ik}(1 - \pi_{m;ik})$, $\quad v = \mathrm{Var}(S_{ik})$, $\sigma_u^2 = \boldsymbol{E}(S_{ik} \mid M_{ik})$, and $\boldsymbol{D}_{ik}$ and $\boldsymbol{F}_{ik}$ are vectors such that

$$\boldsymbol{D}_{ik} = \frac{\partial \pi_{m;ik}}{\partial \boldsymbol{\beta}} = \Delta_{ik} \boldsymbol{X}_{ik},$$

$$\boldsymbol{F}_{ik} = \frac{\partial \mu_{w|m;ik}}{\partial \boldsymbol{\beta}} = -\zeta \Delta_{ik} \boldsymbol{X}_{ik},$$

$$\boldsymbol{H}_{ik} = \frac{\partial \mu_{w|m;ik}}{\partial \boldsymbol{\omega}}.$$

In the clustered setting, for the $2n_i \times 1$ vector $(\boldsymbol{W}_i^T, \boldsymbol{M}_i^T)^T$ of responses for litter i, the model for the means, generalized from (14.10–14.11), is specified as

$$\mathrm{logit}(\pi_{m;ik}) = \beta_0 + \beta_1 d_i + \beta_2(n_i - \bar{n}), \qquad (14.13)$$

$$\mu_{w|m;ik} = E(W_{ik} \mid \boldsymbol{M}_i) = \alpha_0 + \alpha_1 d_i + \alpha_2(n_i - \bar{n}) \\ + \zeta_1 e_{m;ik} + \zeta_2 \bar{e}_{m;i}, \qquad (14.14)$$

where $\bar{e}_i = n_i^{-1} \sum_{k=1}^{n_i} (M_{ik} - \pi_{m;ik})$. The parameters (ζ_1, ζ_2) induce correlation between $\boldsymbol{W}_i$ and $\boldsymbol{M}_i$; the intralitter correlation is characterized by $n_i^{-1}\zeta_2$, while $\zeta_1 + n_i^{-1}\zeta_2$ characterizes the intrafoetus correlation.

Because maximum likelihood estimation becomes more complicated in the clustered setting, Fitzmaurice and Laird implement GEE methodology based on the score equations derived above. Working covariance matrices are specified as:

$$\mathrm{Cov}(\boldsymbol{M}_i) = \boldsymbol{V}_{m;i} \approx \Delta_i^{1/2}[(1 - \rho_m)\boldsymbol{I}_i + \rho_m \boldsymbol{J}_i]\Delta_i^{1/2},$$

$$\mathrm{Cov}(\boldsymbol{W}_i \mid \boldsymbol{M}_i) = \boldsymbol{V}_{w;i} \approx \sigma_w^2[(1 - \rho_w)\boldsymbol{I}_i + \rho_w \boldsymbol{J}_i], \qquad (14.15)$$

where now $\boldsymbol{\Delta}_i = \mathrm{diag}[\pi_{m;ik}(1 - \pi_{m;ik})]$. The GEEs implemented to estimate the regression parameters, $\theta = (\beta, \alpha, \zeta_1, \zeta_2)$, are:

$$\sum_{i=1}^{N} \begin{pmatrix} D_i & 0 \\ F_i & H_i \end{pmatrix}^T \begin{pmatrix} V_{m;i} & 0 \\ 0 & V_{w;i} \end{pmatrix}^{-1}$$

$$\times \begin{pmatrix} M_i - \pi_{m;i} \\ W_i - \mu_{w|m;i} \end{pmatrix} = 0, \qquad (14.16)$$

where $\pi_{m;i}$ and $\mu_{w|m;i}$ are the $n_i \times 1$ vectors of elements $\pi_{m;ik}$ and $\mu_{w|m;ik}$, and now D_i, F_i, H_i are matrices with $F_i = -(\zeta_1 + \zeta_2)\Delta_i X_i$, and H_i has rows $H_{ik} = \{X_{ik}, e_{m;ik}, \bar{e}_{m;i}\}$ corresponding to parameter vector $\omega = (\alpha, \zeta_1, \zeta_2)$. Moment-based estimates of $\sigma_w^2, \rho_w, \rho_m$ from (14.15) are obtained. The resulting GEE estimates of the mean parameters (β, α) will be consistent and asymptotically multivariate normal given only that the model for the marginal means is correctly specified. Model-based and robust estimates of the covariance of the regression parameters are computed following GEE methodology.

DYME in Mice

The parameter estimates obtained from fitting mean models (14.13)–(14.14) and working covariances (14.15) are displayed in Tables 14.1 and 14.2, in the column labeled *C-Disc*. The dose coefficient is significantly positive for malformation ($\hat{\beta}_1 = 15.869$), and the negative coefficient of litter size ($\hat{\beta}_2 = -0.177$) suggests that larger litters had a lower risk of malformation. The intralitter correlation is relatively large for a binary outcome ($\hat{\rho}_m = 0.092$).

For fetal weight, the dose coefficient is significantly negative ($\hat{\alpha}_1 = -0.963$), and there appears to be little effect of litter size on weight ($\hat{\alpha}_2 = -0.0019$). The negative coefficients of the malformation residual ($\hat{\zeta}_1 = -0.036$) and average malformation residual ($\hat{\zeta}_2 = -0.025$) indicate that fetal malformations are associated with lower fetal weight. As expected, the association between fetal weight and malformation is stronger ($\zeta_1 + n_i^{-1}\zeta_2$) for the same foetus than it is for different foetuses ($n_i^{-1}\zeta_2$) though the estimate for the average residual is not statistically significant ($Z = -0.6$). The intralitter correlation for weight (conditional on malformation) is substantial ($\hat{\rho}_w = 0.597$).

Other Models

Similar models that condition on the discrete outcomes can be derived by considering different marginal distributions of the binary outcomes in combination with a multivariate normal for the continuous responses conditional on the binary ones.

14.1.3 Unconditional Joint Models

In this section we consider latent variable models that directly specify the joint distribution of mixed continuous-discrete outcomes based on two approaches:

a Plackett-Dale approach and a probit approach. For an introduction to the standard versions of these models, see Section 4.1.

Geys *et al.* (2001) used a Plackett latent variable to specify the joint distribution of bivariate binary and continuous outcome. The Plackett distribution provides an alternative to the frequently used normal latent variable. A main advantage is the flexibility with which the marginal densities can be chosen (normal, logistic, complementary log-log, etc.). Furthermore, the odds ratio, being a natural measure of global association (Plackett 1965), is an attractive alternative to the correlation. Pseudo-likelihood (PL) methodology is the basis of estimation in the clustered setting.

Two mixed-outcome probit models have been proposed: a full-likelihood model that specifies the joint distribution of the binary and continuous outcomes in a litter (Regan and Catalano 1999a), and a bivariate model that relies on GEE methodology to describe the litter effects (Regan and Catalano 1999b). Both maintain marginal dose-response interpretations for the continuous and binary outcomes, and use the correlation of the underlying multi- or bivariate normal distribution to characterize the intrafoetus correlation.

Plackett-Dale Approach

The Plackett-Dale idea was first proposed for bivariate outcomes of independent subjects (Molenberghs, Geys and Buyse 2001) then extended to the clustered data setting using pseudo-likelihood ideas (Geys *et al.* 2001). Assume temporarily that littermates are independent. The density function $f_{W_{ik}}$ is assumed normal with mean $\mu_{w;ik}$ and variance $\sigma^2_{w;ik}$. The success probability, $\Pr(M_{ik} = 1)$, is denoted by $\pi_{m;ik}$.

The cumulative distributions of the fetal weight (W_{ik}) and the binary malformation (M_{ik}) are given by $F_{W_{ik}}$ and $F_{M_{ik}}$. Their dependence can be defined using a global cross-ratio at cutpoint (w, m) ($m = 0, 1$):

$$\psi_{ik} = \frac{F_{W_{ik},M_{ik}}(w,m)\left[1 - F_{W_{ik}}(w) - F_{M_{ik}}(m) + F_{W_{ik},M_{ik}}(w,m)\right]}{\left[F_{W_{ik}}(w) - F_{W_{ik},M_{ik}}(w,m)\right]\left[F_{M_{ik}}(m) - F_{W_{ik},M_{ik}}(w,m)\right]}.$$

This expression can be solved for the joint cumulative distribution $F_{W_{ik},M_{ik}}$ (Plackett 1965):

$$F_{W_{ik},M_{ik}}(w,m) = \begin{cases} \frac{1+[F_{W_{ik}}(w)+F_{M_{ik}}(m)](\psi_{ik}-1)-S(F_{W_{ik}}(w),F_{M_{ik}}(m),\psi_{ik})}{2(\psi_{ik}-1)} \\ \quad \text{if } \psi_{ik} \neq 1, \\[2mm] F_{W_{ik}}(w)\, F_{M_{ik}}(m) \\ \quad \text{if } \psi_{ik} = 1, \end{cases} \tag{14.17}$$

where

$$S(F_{W_{ik}}, F_{M_{ik}}, \psi_{ik}) = \sqrt{[1 + (\psi_{ik} - 1)(F_{W_{ik}}(w) + F_{M_{ik}}(m))]^2 + 4\psi_{ik}(1 - \psi_{ik})F_{W_{ik}}(w)F_{M_{ik}}(m)}.$$

Based upon this distribution function, a bivariate Plackett density function $g_{ik}(w, m)$ for mixed continuous-binary outcomes is derived. Define $g_{ik}(w, m)$ by specifying $g_{ik}(w, 0)$ and $g_{ik}(w, 1)$ such that they sum to $f_{W_{ik}}(w)$.

If $g_{ik}(w,0) = \partial F_{W_{ik},M_{ik}}(w,0)/\partial w$, then this leads to specifying g_{ik} by:

$$
g_{ik}(w,0) \;=\; \begin{cases} \dfrac{f_{W_{ik}}(w)}{2}\left[1 - \dfrac{1+F_{W_{ik}}(w)(\psi_{ik}-1)-(1-\pi_{m;ik})(\psi_{ik}+1)}{S\big(F_{W_{ik}},1-\pi_{m;ik},\psi_{ik}\big)}\right] \\ \qquad \text{if } \psi_{ik} \neq 1, \\[2mm] f_{W_{ik}}(w)\,(1-\pi_{m;ik}) \\ \qquad \text{if } \psi_{ik} = 1, \end{cases}
\tag{14.18}
$$

$$
g_{ik}(w,1) \;=\; f_{W_{ik}}(w) - g_{ik}(w,0).
\tag{14.19}
$$

Note that $g_{ik}(w,0)$ naturally factors as a product of the marginal density $f_{W_{ik}}(w)$ and the conditional density $f_{M_{ik}|W_{ik}}(0|w)$ (and similarly for $g_{ik}(w,1)$). Interesting special cases are obtained when $\psi_{ik} = 1$ (independence), $\psi_{ik} = 0$ (perfect negative association), or when $\psi_{ik} = \infty$ (perfect positive association).

Dose-response models that incorporate litter- and foetus-specific covariates can be considered for each of the parameters by using appropriate link functions. For the DYME data we fit,

$$
\begin{aligned}
\operatorname{logit}(\pi_{m;ik}) &= \beta_0 + \beta_1 d_i + \beta_2(n_i - \bar{n}), \\
\mu_{w;ik} &= \alpha_0 + \alpha_1 d_i + \alpha_2(n_i - \bar{n}), \\
\ln(\sigma^2_{w;ik}) &= \boldsymbol{X}_{ik}^T \boldsymbol{\varsigma}, \\
\ln(\psi_{ik}) &= \boldsymbol{X}_{ik}^T \boldsymbol{\xi},
\end{aligned}
\tag{14.20}
$$

which, in contrast to conditional models, allows fetal weight variances to differ across doses and directly models dependence between fetal weight and malformation.

In the case of clustering, rather than considering the full likelihood contribution for litter i, computational complexity is avoided by replacing the full likelihood by a pseudo-likelihood function that is easier to evaluate. The contribution of the ith litter to the log pseudo-likelihood function is defined as:

$$
p\ell_i = \sum_{k=1}^{n_i} \ln g_{ik}(w_{ik}, m_{ik}),
\tag{14.21}
$$

where $g_{ik}(\cdot)$ is defined in (14.18)–(14.19). With this approach, the correlation between weight and malformation outcomes for an individual foetus is modeled explicitly, but for outcomes from different littermates independence is taken as a working assumption. A sandwich variance estimator, formulated below (14.23), is then used to adjust for potential bias in the variance estimates. If the amount of clustering is of interest as well, then the pseudo-likelihood (14.21) can be extended to this case as well (Geys et al. 2001). For a detailed discussion of pseudo-likelihood, see Chapters 6 and 7.

Estimates of the regression parameters, $\boldsymbol{\theta} = (\boldsymbol{\alpha}^T, \boldsymbol{\varsigma}^T, \boldsymbol{\beta}^T, \boldsymbol{\xi}^T)^T$, are obtained by solving the estimating equations corresponding to the pseudo-likelihood

(14.21):

$$\sum_{i=1}^{N} \boldsymbol{U}_i(\boldsymbol{\theta}) = \sum_{i=1}^{N} \sum_{k=1}^{n_i} \left(\frac{\partial \ln g_{ik}(x, y)}{\partial \boldsymbol{\theta}} \right) = \boldsymbol{0}. \qquad (14.22)$$

Arnold and Strauss (1991) showed that the PL estimator $\hat{\boldsymbol{\theta}}$, obtained by maximizing (14.22), is consistent and asymptotically normal with covariance matrix estimated by:

$$\widehat{\mathrm{Cov}}(\hat{\boldsymbol{\theta}}) = \left(\sum_{i=1}^{N} \frac{\partial \boldsymbol{U}_i}{\partial \boldsymbol{\theta}} \right)^{-1} \left(\sum_{i=1}^{N} \boldsymbol{U}_i(\boldsymbol{\theta}) \boldsymbol{U}_i(\boldsymbol{\theta})^T \right)$$

$$\left(\sum_{i=1}^{N} \frac{\partial \boldsymbol{U}_i}{\partial \boldsymbol{\theta}} \right)^{-1} \Bigg|_{\boldsymbol{\theta} = \hat{\boldsymbol{\theta}}}. \qquad (14.23)$$

A key difference between conditional and unconditional models is that there is an estimated measure of association. The generality of (14.20) is an important advantage of this approach; as for developmental toxicity data, the assumptions of constant variance and constant association are often not tenable. Another advantage is the close connection of pseudo-likelihood with likelihood, which enabled Geys, Molenberghs and Ryan (1999) to construct pseudo-likelihood ratio test statistics that have easy-to-compute expressions and intuitively appealing limiting distributions. The pseudo-likelihood ratio tests are summarized in Chapter 7.

LYME in Mice

The parameter estimates obtained from fitting dose-response models (14.20) are displayed in Tables 14.1 and 14.2, in the column labeled *Plackett-Dale*. The selection of a parsimonious model for the log odds ratio and fetal weight variance relied on the pseudo-likelihood procedure for the Plackett-Dale approach; Z-statistics are displayed in Tables 14.1 and 14.2 to facilitate comparison with other approaches. Similarly, the fetal weight and malformation models fit were dictated by the other approaches.

For fetal weight, the dose coefficient is significantly negative ($\hat{\alpha}_1 = -0.886$), and there appears to be little effect of litter size on weight ($\hat{\alpha}_2 = 0.0005$). The variances were initially fit separately by dose, but even though the variances are not monotonically decreasing when calculated without regard to clustering, a PLR test ($G_a^{*2}(H_0)$, Chapter 7) showed this could be reduced to a more parsimonious model where variances (with ln link function) were modeled as a linear function of dose (PLR statistic=1.157, 3 degrees of freedom). The linearly decreasing parameterization of the variance accommodates the decreased variance at the highest two doses as compared with the control and lower two doses. Because of the working independence assumption, there is no estimated intralitter correlation for fetal weight.

The dose coefficient is significantly positive for malformation ($\hat{\beta}_1 = 15.094$), and the significantly negative coefficient of litter size ($\widehat{\beta}_2 = -0.229$) suggests that larger litters had a lower risk of malformation. Again, because of the working independence assumption, there is no estimated intralitter correlation for malformation.

The log odds ratio characterizing the association between fetal weight and malformation was initially fit separately by dose as well, except that the lowest two doses were combined to avoid difficulty in fitting when few malformations are observed. However, a PLR test shows that comparing 4 separate to a common parameter was not a significant improvement in fit (PLR=0.708, $3df$). The estimated odds ratio is $\widehat{\psi}_{ik} = 0.219$ and is less than 1 because of the negative association between weight and malformation; the value close to zero indicates the strength of the association.

Full-likelihood Mixed-Outcome Probit Model

Regan and Catalano (1999a) introduced a mixed-outcome probit model that extends a correlated probit model for binary outcomes (Ochi and Prentice 1984) to incorporate continuous outcomes. The correlated binary model assumes that the latent malformation variables $\boldsymbol{M}_i^*$ for litter i share a MVN distribution, denoted by $\phi_{n_i}(\boldsymbol{M}_i^*; \gamma_{m;i}, 1, \rho_{m;i}^*)$, with mean $\gamma_{m;i}\mathbf{1}_i$ and equicorrelated covariance $\boldsymbol{\Sigma}_{m;i} = [(1 - \rho_{m;i}^*)\boldsymbol{I}_i + \rho_{m;i}^*\boldsymbol{J}_i]$. The n_i binary malformation outcomes are defined according to whether the n_i latent malformation variables exceed a common threshold, arbitrarily assumed to be zero. The joint distribution of the binary malformation variables is written,

$$\Pr(M_{i+}) = \begin{pmatrix} n_i \\ M_{i+} \end{pmatrix} \overbrace{\int_{-\infty}^{-\gamma_{m;i}} \cdots \int_{-\infty}^{-\gamma_{m;i}}}^{n_i - M_{i+}} \overbrace{\int_{-\gamma_{m;i}}^{\infty} \cdots \int_{-\gamma_{m;i}}^{\infty}}^{M_{i+}} \phi_{n_i}(\boldsymbol{Z}_i^*; 0, 1, \rho_{m;i}^*) \, d\boldsymbol{Z}_i^*,$$

where M_{i+} is the number of malformed foetuses in litter i and $Z_{ik}^* = (M_{ik}^* - \gamma_{m;i})$. The marginal probability of malformation is $\pi_{m;ik} = \Pr(Z_{ik}^* > -\gamma_{m;i}) = \Phi(\gamma_{m;i})$. Dose-response models are incorporated by expressing $\gamma_{m;i}$ and $\rho_{m;i}$ as functions of dose and other litter-specific covariates, for which maximum likelihood (ML) estimates of the regression parameters are obtained.

To extend the correlated probit model to both binary and continuous outcomes, the $2n_i \times 1$ vector of observed and latent continuous outcomes, written as $(\boldsymbol{W}_i^T, \boldsymbol{M}_i^{*T})^T$, is assumed to follow a MVN distribution,

$$\phi_{2n_i}(\boldsymbol{W}_i, \boldsymbol{M}_i^*; \mu_{w;i}, \gamma_{m;i}, \sigma_{w;i}^2, 1, \rho_{w;i}, \rho_{m;i}^*, \rho_{wm;i})$$

$$= (2\pi)^{-n_i} \mid \boldsymbol{\Sigma}_{2n_i} \mid^{-\frac{1}{2}}$$

$$\times \exp\left\{ -\frac{1}{2} \begin{pmatrix} \boldsymbol{W}_i - \mu_{w;i}\mathbf{1}_i \\ \boldsymbol{M}_i^* - \gamma_{m;i}\mathbf{1}_i \end{pmatrix}^T \boldsymbol{\Sigma}_{2n_i}^{-1} \begin{pmatrix} \boldsymbol{W}_i - \mu_{w;i}\mathbf{1}_i \\ \boldsymbol{M}_i^* - \gamma_{m;i}\mathbf{1}_i \end{pmatrix} \right\}$$

with means $\mu_{w;i}$ among the fetal weights and $\gamma_{m;i}$ among the latent malformations. The correlation structure is similar to (14.1). Among littermates both outcomes are assumed equicorrelated, with separate intralitter correlations $(\rho_{w;i}, \rho_{m;i}^*)$. A common correlation $(\rho_{wm;i})$ is assumed between fetal weight and latent malformation within a foetus and among different foetuses in the same litter (i.e., $\varrho_i = \rho_{wm;i}$) in (14.1). This complete exchangeability assumption is necessary to maintain a tractable likelihood. The correlation structure gives rise to the following covariance matrix:

$$\begin{aligned} \boldsymbol{\Sigma}_{2n_i} &= \mathrm{Cov}(\boldsymbol{W}_i^T, \boldsymbol{M}_i^{*T})^T \\ &= \begin{bmatrix} \sigma_{w;i}^2 \left[(1 - \rho_{w;i})\boldsymbol{I}_i + \rho_{w;i}\boldsymbol{J}_i\right] & \rho_{wm;i}\,\sigma_{w;i}\boldsymbol{J}_i \\ \rho_{wm;i}\,\sigma_{w;i}\boldsymbol{J}_i & \left[(1 - \rho_{m;i}^*)\boldsymbol{I}_i + \rho_{m;i}^*\boldsymbol{J}_i\right] \end{bmatrix}. \end{aligned}$$

As in the binary-only model, the n_i binary malformation outcomes are defined according to whether the n_i latent malformation variables exceed a common threshold. The resulting joint distribution of the vector of weights and binary malformations can be written

$$f_{2n_i}(\boldsymbol{W}_i, \boldsymbol{M}_i) \propto \overbrace{\int_{-\infty}^{-\gamma_{m;i}} \cdots \int_{-\infty}^{-\gamma_{m;i}}}^{n_i - M_{i+}} \overbrace{\int_{-\gamma_{m;i}}^{\infty} \cdots \int_{-\gamma_{m;i}}^{\infty}}^{M_{i+}}$$

$$\times \phi_{2n_i}(\boldsymbol{W}_i, \boldsymbol{Z}_i^*; \mu_{w;i}, 0, \sigma_{w;i}^2, 1, \rho_{w;i}, \rho_{m;i}^*, \rho_{wm;i})\, d\boldsymbol{Z}_i^* \qquad (14.24)$$

where again Z_{ik}^* is the standardized variate $(M_{ik}^* - \gamma_{m;i})$.

Dose-response models that incorporate litter-specific covariates are specified for each parameter. For the DYME mice, the following models were considered:

$$\begin{aligned} \gamma_{m;i} &= \beta_0 + \beta_1 d_i + \beta_2(n_i - \bar{n}), \\ \mu_{w;i} &= \alpha_0 + \alpha_1 d_i + \alpha_2(n_i - \bar{n}), \\ \ln(\sigma_{w;i}^2) &= \varsigma_0 + \varsigma_1 d_i, \\ FZ(\rho_{w;i}) &= \tau_{w;0}\mathcal{I}(d = 0) + \tau_{w;1}\mathcal{I}(d = 0.0625) \\ &\quad + \tau_{w;2}\mathcal{I}(d = 0.125) + \tau_{w;3}\mathcal{I}(d = 0.25) \qquad (14.25) \\ &\quad + \tau_{w;4}\mathcal{I}(d = 0.5), \\ FZ(\rho_{m;i}^*) &= \tau_m, \\ FZ(\rho_{wm;i}) &= \tau_{wm}, \end{aligned}$$

where FZ denotes Fisher's Z-transformation, $\ln[(1+\rho)/(1-\rho)]$, which is used as a link function for the correlations. The model not only specifies mean models, but also allows dose effects on weight variability and the correlations, which is often appropriate for developmental toxicity data.

From the log-likelihood based on (14.24), the first and second derivatives necessary to define the score function and information matrix are obtained and the Newton-Raphson algorithm implements ML estimation of the regression parameters, $\boldsymbol{\theta} = (\boldsymbol{\alpha}^T, \boldsymbol{\beta}^T, \boldsymbol{\varsigma}^T, \boldsymbol{\tau}_w^T, \boldsymbol{\tau}_m^T, \boldsymbol{\tau}_{wm}^T)^T$. Differentiation of the likelihood is not trivial and the reader is referred to the original article for details; the

complete exchangeability assumption between weight and latent malformation within a foetus and among different foetuses within the same litter is
necessary to maintain a tractable likelihood. However, the derivatives can be
expressed in terms of equicorrelated probit integrals of lower dimension so
their evaluation is not at all prohibitive. Multidimensional normal integrals
are evaluated using the approximation of Mendell and Elston (1974).

The model has several advantages and additional features compared with
conditional model approaches. As compared with Catalano and Ryan, all parameters maintain marginal dose–response interpretations, including malformation. In addition, the correlation between weight and malformation, $\rho_{wm;i}$,
may vary with dose, as well as the correlations accounting for litter effects
and the fetal weight variances. The likelihood model allows for assessment of
the model's goodness-of-fit; likelihood ratio tests are easily implemented to
assess the most parsimonious models for the six marginal parameters.

An assumption of the model is that of complete exchangeability between
weight and latent malformation outcomes, i.e., that intrafoetus and intralitter
correlation are the same. It may be the case that outcomes within a foetus are
more correlated than those between foetuses, but extending the model to allow for a separate intrafoetus correlation is not trivial. The tractability of the
multidimensional integral expressions is due to the exchangeability of foetuses
within a litter so that only litter–specific data are required for computations,
and not foetus–specific outcomes. In this same manner, only litter–specific covariates can be included in the dose–response models. The model presented in
the next section was designed to overcome the full exchangeability assumption
and to allow foetus–specific covariates.

DYME in Mice

The parameter estimates obtained from fitting dose-response models (14.25)
are displayed in Tables 14.1 and 14.2, in the column labeled *Probit Lik*. The fetal weight and malformation models fit were dictated by the other approaches.
Likelihood-ratio tests (LRT) selected a parsimonious model for the correlations and fetal weight variance; Z-statistics are displayed in Tables 14.1 and
14.2 to facilitate comparison with other approaches.

For fetal weight, the dose coefficient is significantly negative ($\widehat{\alpha}_1 = -0.937$),
and there appears to be little effect of litter size on weight ($\widehat{\alpha}_2 = -0.0006$).
The fitting of the fetal weight variance and correlation is interrelated and was
determined to be best with a separate correlation parameter for each dose, and
a simpler model for the variances. The variances, though not monotonically
decreasing when calculated without regard to clustering, fit well (with ln link
function) with a linear model of dose (LRT statistic=7.05, 1 degree of freedom,
versus common parameter), with no improvement of 3 additional parameters
to fit correlations separately by dose (LRT=1.92, 3 degrees of freedom). The
intralitter correlation estimates ($\widehat{\rho}_{w;i}$) are large and range from 0.501 to 0.671.

The dose coefficient is significantly positive for malformation ($\widehat{\beta}_1 = 0.8551$),
and the negative coefficient of litter size ($\widehat{\beta}_2 = -0.089$) suggests that larger

litters had a lower risk of malformation. The model for the latent malformation correlation fits a constant correlation with respect to dose and is estimated to be $\widehat{\rho}^*_{m;i} = 0.306$; fitting a second-order parameter of a latent variable can be difficult, especially when there are few events at the lowest doses as for the DYME data.

Whereas the correlations within each outcome are positive, the correlation between outcomes is negative ($\widehat{\rho}_{wm;i} = -0.269$), because fetal weight is decreasing while the probability of malformation is increasing. There was no improvement in model fit by estimating one correlation per dose (LRT=1.23, 4 degrees of freedom) versus a common correlation. Recall that this parameter characterizes correlation between (latent) malformation and weight for both an individual foetus and among littermates.

Semiparametric Mixed-Outcome Probit Model

To overcome the complete exchangeability assumption of the full-likelihood approach and to allow foetus-specific covariates in the dose-response models, Regan and Catalano (1999b) proposed an alternative mixed-outcome probit model. Similar to other models discussed, they developed a bivariate model under an assumption of independence to motivate their model in the clustered setting.

Assume temporarily that littermates are independent. To derive the marginal distribution of the bivariate response (W_{ik}, M_{ik}), the observed fetal weight and latent malformation variables for foetus ik are assumed to share a bivariate normal distribution,

$$f(W_{ik}, M^*_{ik}) = \phi_2(W_{ik}, M^*_{ik}; \mu_{w;ik}, \gamma_{m;ik}, \sigma^2_{w;ik}, 1, \varrho_{ik}),$$
(14.26)

where ϱ_{ik} is the intrafoetus correlation denoted in (14.1). To arrive at a convenient form of the bivariate distribution of the mixed outcomes, this density is rewritten as a product of the marginal density for fetal weight and conditional density of latent malformation given weight; so, the joint distribution of the bivariate fetal weight and binary malformation outcome for foetus ik can be written

$$
\begin{aligned}
f(W_{ik}, M_{ik}) &= f_w(W_{ik}) \times f_{m|w}(M_{ik} \mid W_{ik}) \\
&= \phi(W_{ik}; \mu_{w;ik}, \sigma^2_{w;ik}) \\
&\quad \times \pi^{M_{ik}}_{m|w;ik} [1 - \pi_{m|w;ik}]^{1-M_{ik}},
\end{aligned}
$$
(14.27)

where $\pi_{m|w;ik} = \Phi(\gamma_{m|w_{ik}})$ is the expectation of the conditional binary malformation outcome $E(M_{ik}|W_{ik})$, and from bivariate normal theory,

$$\gamma_{mw} = \frac{\gamma_{m;ik} + \varrho_{ik} \left(\frac{W_{ik} - \mu_{w;ik}}{\sigma_{w;ik}} \right)}{(1 - \varrho^2_{ik})^{1/2}}.$$

The marginal expectation is $\pi_{m;ik} = \Phi(\gamma_{m;ik})$.

Dose-response models are specified for all four parameters of the bivariate normal density, using appropriate link function. For the DYME data, the following dose-response models were specified:

$$
\begin{aligned}
\gamma_{m;ik} &= \beta_0 + \beta_1 d_i + \beta_2(n_i - \bar{n}) \\
\mu_{w;ik} &= \alpha_0 + \alpha_1 d_i + \alpha_2(n_i - \bar{n}) \\
\ln(\sigma^2_{w;ik}) &= \varsigma_0 \mathcal{I}(d = 0) + \varsigma_1 \mathcal{I}(d = 0.0625) + \varsigma_2 \mathcal{I}(d = 0.125) \\
&\quad + \varsigma_3 \mathcal{I}(d = 0.25) + \varsigma_4 \mathcal{I}(d = 0.5) \\
FZ(\varrho_{ik}) &= \tau_1 \mathcal{I}(d = 0, 0.0625) + \tau_2 \mathcal{I}(d = 0.125) \\
&\quad + \tau_3 \mathcal{I}(d = 0.25) + \tau_4 \mathcal{I}(d = 0.5)
\end{aligned}
\tag{14.28}
$$

where the ikth foetus has covariates that may be both foetus- and litter-specific and vectors of fixed regression parameters $\boldsymbol{\theta} = (\boldsymbol{\beta}^T, \boldsymbol{\tau}^T, \boldsymbol{\alpha}^T, \boldsymbol{\varsigma}^T)^T$. From the log-ikelihood based on the bivariate distribution (14.27), score functions for the regression parameters $\boldsymbol{\theta}$ can be written

$$
\sum_{i=1}^{N} \sum_{k=1}^{n_i} \left(\frac{\partial \ell_{ik}}{\partial \boldsymbol{\beta}^T} \quad \frac{\partial \ell_{ik}}{\partial \boldsymbol{\tau}^T} \quad \frac{\partial \ell_{ik}}{\partial \boldsymbol{\alpha}^T} \quad \frac{\partial \ell_{ik}}{\partial \boldsymbol{\varsigma}^T} \right)^T
$$

$$
= \sum_{i=1}^{N} \sum_{k=1}^{n_i}
\begin{pmatrix}
\Delta_{b_{ik}} \boldsymbol{X}_{ik} & \boldsymbol{0} & \boldsymbol{0} \\
\Delta_{t_{ik}} \boldsymbol{X}_{ik} & \boldsymbol{0} & \boldsymbol{0} \\
\Delta_{a_{ik}} \boldsymbol{X}_{ik} & \boldsymbol{X}_{ik} & \boldsymbol{0} \\
\Delta_{s_{ik}} \boldsymbol{X}_{ik} & \boldsymbol{0} & \sigma^2_{w;ik} \boldsymbol{X}_{ik}
\end{pmatrix}
\begin{pmatrix}
\{\pi_{m|w;ik} & 0 & 0 \\
[1 - \pi_{m|w;ik}]\}^{-1} & & \\
0 & \sigma^{-2}_{w;ik} & 0 \\
0 & 0 & \frac{1}{2} \sigma^{-4}_{w;ik}
\end{pmatrix}
$$

$$
\times
\begin{pmatrix}
M_{ik} - \pi_{m|w;ik} \\
W_{ik} - \mu_{w;ik} \\
S_{ik} - \sigma^2_{w;ik}
\end{pmatrix},
\tag{14.29}
$$

where $S_{ik} = (W_{ik} - \mu_{w;ik})^2$, and

$$
\Delta_{b_{ik}} = \frac{\partial \pi_{m|w;ik}}{\partial \gamma_{m;ik}}, \Delta_{t_{ik}} = \frac{\partial \pi_{m|w;ik}}{\partial \varrho_{ik}}, \Delta_{a_{ik}} = \frac{\partial \pi_{m|w;ik}}{\partial \mu_{w;ik}}, \Delta_{s_{ik}} = \frac{\partial \pi_{m|w;ik}}{\partial \sigma^2_{w;ik}},
$$

with $\boldsymbol{X}_{ik}$ generally denoting the appropriate covariate vector.

In the case of clustering, we avoid fully specifying the joint distribution of the n_i bivariate outcomes in litter i by using the score equations of the bivariate distribution derived under independence to motivate a set of GEEs for the clustered setting. We assume the marginal distribution of the bivariate outcome is defined by (14.27) and use the form of the score functions (14.29) to construct a set of GEEs for the regression parameters by replacing the diagonal covariance matrix by a working covariance matrix that incorporates correlation between littermates.

The regression parameters $\boldsymbol{\theta} = (\boldsymbol{\beta}^T, \boldsymbol{\tau}^T, \boldsymbol{\alpha}^T, \boldsymbol{\varsigma}^T)^T$ are estimated in the clustered setting from the following set of GEEs:

$$\sum_{i=1}^{N} \begin{pmatrix} \Delta_{b_i} \boldsymbol{X}_i & \Delta_{t_i} \boldsymbol{X}_i & \Delta_{a_i} \boldsymbol{X}_i & \Delta_{s_i} \boldsymbol{X}_i \\ 0 & 0 & \boldsymbol{X}_i & 0 \\ 0 & 0 & 0 & \Sigma_{w_i} \boldsymbol{X}_i \end{pmatrix}^T \begin{pmatrix} \boldsymbol{V}_{m_i} & \boldsymbol{V}_{wm_i} & 0 \\ \boldsymbol{V}_{wm_i} & \boldsymbol{V}_{w_i} & 0 \\ 0 & 0 & \boldsymbol{V}_{s_i} \end{pmatrix}^{-1}$$

$$\times \begin{pmatrix} \boldsymbol{M}_i - \boldsymbol{\pi}_{m|w;i} \\ \boldsymbol{W}_i - \boldsymbol{\mu}_{w;i} \\ \boldsymbol{S}_i - \boldsymbol{\sigma}^2_{w;i} \end{pmatrix} = 0 \qquad (14.30)$$

where the elements of these matrices are now matrix- or vector-versions of the elements from above (14.29): $\Delta_{b_i}, \Delta_{t_i}, \Delta_{a_i}, \Delta_{s_i}, \Sigma_{w_i}$ are $n_i \times n_i$ diagonal matrices; $\boldsymbol{\pi}_{m|w;i}, \boldsymbol{\mu}_{w;i}, \boldsymbol{\sigma}^2_{w;i}$ and $\boldsymbol{S}_i$ are $n_i \times 1$ vectors. $\boldsymbol{V}_{m_i}(\rho_m), \boldsymbol{V}_{w_i}(\rho_w)$, and $\boldsymbol{V}_{s_i}(\rho_w)$ are equicorrelated working covariance matrices based on the respective elements of (14.29). $\boldsymbol{V}_{wm_i}(\rho_{wm})$ is introduced to account for the correlation between the weight and malformation outcomes among littermates; the intrafoetus correlation is still characterized by ϱ_{ik}. The methodology of Liang and Zeger (1986) and Zeger and Liang (1986) is implemented to estimate $\boldsymbol{\theta}$: moment-based estimates of working correlation parameters, $(\rho_m, \rho_w, \rho_{wm})$ are obtained. The resulting GEE estimates of $\boldsymbol{\theta}$ will be consistent and asymptotically normal if the motivation of the bivariate distribution (14.27) has specified the correct moments for the GEEs (14.30). Model-based and robust estimates of the covariance of the parameter estimates are obtained following GEE methodology.

This model has been extended to handle multiple ordinal outcomes with a continuous outcome (Regan and Catalano 2000). This is useful because the malformation variable typically represents a binary indicator of any malformation, and malformations are sometimes ordinally measured such as (absent, signs of variation, full malformation). In the motivating example, the most frequent malformations occurred in the eyes, and were either anophthalmia (missing eye) or microphthalmia (small eye). Malformation was modeled as an ordinal outcome (normal, microphthalmia, anophthalmia) and each eye was considered separately. They compared this modeling strategy with the bivariate model in which malformation was a binary variable representing any malformation in either eye.

DYME in Mice

The parameter estimates obtained from fitting dose-response models (14.28) are displayed in Tables 14.1 and 14.2, in the column labeled *Probit GEE*.

For fetal weight, the dose coefficient is significantly negative ($\widehat{\alpha}_1 = -0.940$), and there appears to be little effect of litter size on weight ($\widehat{\alpha}_2 = -0.0013$).

The variances were fit separately by dose even though this may not be the most parsimonious choice. The model was chosen because a robust Wald test of whether all five parameters are equal was not significant ($Z^2 = 5.43$, 4 degrees of freedom) but a test comparing the variances of the three lower doses to the two highest doses was significant ($Z^2 = 4.4$, 1 degree of freedom). The variance estimates range from $\hat{\sigma}^2_{w;ik} = 0.015$ at 125 mg/kg/d to $\hat{\sigma}^2_{w;ik} = 0.008$ at 500 mg/kg/d. The working correlation parameter is $\hat{\rho}_w = 0.493$.

The dose coefficient is significantly positive for malformation ($\hat{\beta}_1 = 0.8479$), and the negative coefficient of litter size ($\hat{\beta}_2 = -0.108$) suggests that larger litters had a lower risk of malformation. The working correlation parameter for malformation, conditional on weight, is $\hat{\rho}_m = 0.019$.

The intrafoetus correlation between latent malformation and observed fetal weight was fit separately by dose, except the control and lowest dose were grouped because of difficulty fitting when few malformations were observed at lowest dose levels. A robust Wald test comparing separate to a common parameter was significant ($Z^2 = 7.9$, 3 degrees of freedom). The correlation estimates range from $\hat{\varrho}_{ik} = -0.184$ at 0 and 62.5 mg/kg/d to $\hat{\varrho}_{ik} = -0.517$ at 500 mg/kg/d.

14.2 Application to Quantitative Risk Assessment

Quantitative risk assessment was introduced in Section 3.4 and Chapter 10. We will use the same conventional notations. Recall that quantitative risk assessment involves the determination of a dose based on the experimental data in animals from which a safe level of exposure for humans can be estimated. Dose-response modeling is often the basis of this process. Based on a model, a dose corresponding to a specified level of increased response over background is estimated; in noncancer evaluations, this dose is often referred to as the benchmark dose (BMD_q) or the effective dose (ED_q). The subscript q corresponds to the level of increased response above background, also known as the benchmark response, and is typically specified as 1, 5, or 10%. To allow for estimation variability, a 95% lower confidence limit on this estimated dose (LED_q or LED_q) is the quantity suggested (Crump 1984, 1995) to be used as part of the quantitative risk assessment process for determining an acceptable low-risk exposure level for humans.

To determine the BMD, a quantity characterizing risk as a function of dose must be specified. For quantal outcomes, this is generally expressed as $P(d)$, the probability of response at dose d. To incorporate the background response into the BMD calculation, this probability is used in conjunction with a risk function, for example, additional risk $r(d) = P(d) - P(0)$, or extra risk $r(d) = \frac{P(d)-P(0)}{1-P(0)}$, where $P(0)$ is the background risk. Extra risk puts greater weight on outcomes with large background risks. The BMD_q (ED_q) is defined as the estimated dose satisfying $\hat{r}(d) = q$, where q is the benchmark response. With a univariate response, this expression can typically be solved explicitly for d, for example $d = P^{-1}(q + P(0))$; with a bivariate response, a closed-form solution cannot be determined and d must be obtained numerically.

One approach for determining a BMD from a continuous outcome is, after fitting the continuous dose-response model, to consider a dichotomized version of the outcome and use the framework for a quantal outcome. For example, we may consider $P(d)$ as the probability of low fetal weight, where a cutoff level is specified for determining a low weight extreme enough to be considered an adverse event. Because of the arbitrariness of the cutpoint, estimating a BMD from a continuous response has led to much discussion (Bosch *et al.* 1996, Crump 1984, 1995, Gaylor and Slikker 1990, Kavlock *et al.* 1995, Kodell and West 1993). A cutpoint of three standard deviations below the average fetal weight of control animals, corresponding to a low birth weight rate of 0.1% assuming normality, will be used here.

To incorporate the multiple outcomes evaluated in a study, a less integrative approach usually involves fitting dose-response models and determining BMDs (and LEDs) separately for each outcome. Typically, this would include embryolethality, malformation, and fetal weight; for illustration we will focus only on malformation and weight. The outcome that appears most sensitive to the exposure (i.e., the critical effect) becomes the focus for estimating risk in humans (U.S. EPA 1991, 1995). For example, a probit model could be fit to the malformation outcomes (using either likelihood or GEE estimation), and (omitting temporarily the litter and foetus indices)

$$P_m(d) = \Pr(\text{Malformation at dose d}) = \Phi\left(\gamma_m(d)\right) \qquad (14.31)$$

would be the basis of determining a BMD (and a LED). Fetal weight could be modeled with a random effects model or using GEEs, and assuming normality leads to a probit function,

$$P_w(d) = \Pr(\text{Low fetal weight at dose d}) = \Phi\left(\frac{W_c - \mu_w(d)}{\sigma_w(d)}\right) \qquad (14.32)$$

for determining a BMD and LED, where W_c denotes the cutpoint for determining low weight extreme enough to be considered an adverse event. The outcome resulting in the lower LED would be considered the critical effect. This method of selecting the smallest LED among several adverse events assumes that protecting against the most sensitive outcome protects against all other adverse outcomes. This approach to determining a BMD is unattractive when there are several adverse outcomes because there may be more generalized effects across the spectrum of adverse outcomes. It may therefore be preferable to determine the BMD from an expression that characterizes the combined risk to a foetus by incorporating the separate outcomes while accounting for the relationship between them.

To estimate the overall risk to an individual foetus for the bivariate outcome malformation and low fetal weight, $P(d)$ represents the probability that an individual foetus is malformed and/or of low birth weight. In other words, for the kth foetus in the ith litter:

$$P(d) = \Pr(W_{ik} < W_c \text{ and/or } M_{ik} = 1 \mid d).$$

Based on the methods presented in the previous section, expressions for the joint probability $P(d)$ can be specified. Fitzmaurice and Laird (1995) do not discuss the application of their model to risk assessment, and so it is not included in the discussion that follows.

For their model that conditions on the fetal weight outcomes, Catalano *et al.* 1993 proposed that risk assessment be based on the expression

$$
\begin{aligned}
P(d) &= 1 - [1 - P(M_{ik} \mid \boldsymbol{W}_i)] [1 - P(W_{ik})] \\
&= 1 - [1 - \pi_{m|w;ik}(d)] \left[1 - \Phi\left(\frac{W_c - \mu_{w;ik}(d)}{\sigma_w}\right)\right], \quad (14.33)
\end{aligned}
$$

based on the mean models (14.7) and (14.3). In this expression the model for weight is evaluated at the average litter size ($n_i = \bar{n}$); the conditional model for malformation is evaluated at the average litter size and zero fetal weight residuals, which has the interpretation as a prediction of an average animal from an average litter, after adjusting for the animal's and litter's fetal weight values. Thus, though the conditional modeling approach does not directly specify a measure of intrafoetus dependence, the conditioning argument provides an approximation of the joint risk.

An approach for the mixed-outcome probit models was proposed by Regan and Catalano (1999a). In these models, the observed weight and latent malformation variables for an individual foetus are assumed to share a bivariate normal distribution, as specified for the semiparametric model (14.26). The probability of an adverse outcome for an individual foetus as related to dose can then be expressed as:

$$
\begin{aligned}
P(d) &= 1 - \int_{-\infty}^{-\gamma_{m;ik}(d)} \int_{W_c}^{\infty} \\
&\quad \times \phi_2(W_{ik}, Z_{ik}^*; \mu_{w;ik}(d), 0, \sigma_{w;ik}^2(d), 1, \varrho_{ik}(d)) \, dW_{ik} \, dZ_{ik}^* \\
&= \Phi(\gamma_{m;ik}(d)) \\
&\quad + \Phi_2\left(-\gamma_{m;ik}(d), \left(\frac{W_c - \mu_{w;ik}(d)}{\sigma_{w;ik}(d)}\right); \varrho_{ik}(d)\right). \quad (14.34)
\end{aligned}
$$

Hence, $P(d)$ explicitly depends on the measure of association between fetal weight and malformation, which is the correlation of the underlying bivariate normal density. For the likelihood model, ρ_{wm} is used since it is assumed that $\varrho = \rho_{wm}$; for the semiparametric model ϱ itself is used.

For the Plackett-Dale approach, Geys *et al.* (2000) adopt the same approach. The observed weight and latent malformation variables for an individual foetus are assumed to share a bivariate Plackett distribution and the probability of an adverse outcome for an individual foetus is expressed as:

$$P(d) = \pi_{m;ik}(d) + F_{W_{ik}, M_{ik}}(W_c, 0) \tag{14.35}$$

which also depends explicitly on the odds ratio, ψ, as seen in (14.17).

Fetal death can be incorporated into risk assessment by using the notion of conditional independence between death and the live outcomes. Following Ryan (1992) and Catalano et al. (1993), one can specify:

$$P^*(d) = 1 - (1 - P_E(d))(1 - P_L(d)),$$

where $P_E(d)$ denotes the probability of embryolethality and $P_L(d)$ denotes the probability of an adverse live outcome at dose d, as specified by equations (14.33)–(14.35) above. We focus the rest of our discussion on the live outcomes; see Catalano et al. (1993) and Regan and Catalano (1999a) for details and examples.

Just as for a univariate outcome, three approaches to calculating a lower confidence limit on the BMD may be used; they are based on the asymptotic properties of the BMD, of the risk function, and of the likelihood ratio statistic. We implement the approach suggested by Kimmel and Gaylor (1988) and calculate the lower confidence limit based on the approximate normality of the estimated risk function, $\hat{r}(d)$, taking into account the variability associated with its estimation. This dose level is often referred to as the lower effective dose, LED_q. An upper confidence limit on the estimated risk function is computed, then the dose that corresponds to a $q\%$ increased response above background is determined from this upper limit curve. Thus the LED_q is the value of d that solves $q = \hat{r}(d) + 1.645\sqrt{\widehat{Var}(\hat{r}(d))}$, where the variance of the estimated risk function is estimated as

$$\widehat{Var}(\hat{r}(d)) = \left(\frac{\partial r(d)}{\partial \boldsymbol{\theta}}\right)^T \widehat{Var}(\boldsymbol{\theta}) \left(\frac{\partial r(d)}{\partial \boldsymbol{\theta}}\right)\Bigg|_{\theta=\hat{\theta}},$$

where, for models implementing GEEs and pseudo-likelihood estimation, the quantity $\widehat{Var}(\boldsymbol{\theta})$ is the robust, or sandwich, variance estimator.

14.2.1 DYME in Mice

To define the joint probability of an adverse live outcome, foetuses that weighed less than 0.673 g are considered to be of low fetal weight, which corresponds to a 0.9% rate in the control animals. The expressions (14.33)–(14.35) are evaluated at the GEE, PL, or ML estimates from the corresponding dose-response models, to estimate the probability that a live foetus is affected at each experimental dose. These probabilities are given in Table 14.3.

For illustration, we estimate the BMD corresponding to a 5% additional risk over the background, i.e., the dose satisfying $0.05 = P(d) - P(0)$. Here, results for 5% extra risk are similar since $P(0)$ is very small. From the bivariate probabilities (14.33)–(14.35), we obtain BMD_{05}s ranging from 122-128 mg/kg corresponding to the dose at which the risk of an adverse effect is 5% above

Table 14.3 *DYME in Mice. Quantitative risk assessment results from different approaches.*

Dose	Approach				
(mg/kg/d)	C-C	Pl-D	Prb-L	Prb-G	$\text{Min}(M, W)$
	$P(d)$				
0	0.0032	plik	0.0043	0.0032	–
62.5	0.0156	plik	0.0164	0.0199	–
125	0.0578	plik	0.0539	0.0553	–
250	0.3659	plik	0.3167	0.2990	–
500	0.9915	plik	0.9858	0.9748	–
	Benchmark Dose*				
BMD_{05}	cr	pl	125	122	146
LED_{05}	cr	pl	108	103	126

* Based on additional risk

the background risk of less than 0.5%. The corresponding LED_{05}s range from 103 to 109 mg/kg.

Finally, we will compare the joint modeling approaches with the usual approach for multiple outcomes in which the lower of the individual malformation and fetal weight LEDs is used as an overall LED (ignoring embryolethality for this illustration). We fit dose-response models and determined BMD_{05} and LED_{05} separately from each outcome; the minimum of the two LED_{05} can be compared with those obtained above from the outcomes jointly. Malformation and fetal weight were each modeled via: (1) GEEs using probit and identity link functions and exchangeable working correlation matrices, and (2) Plackett-Dale models with logit and identity link functions using pseudo-likelihood estimation. With d the dose administered to the dam (in g/kg/day), the BMDs and LEDs were estimated from

$$P_m(d) = \Phi(-2.86 + 8.50d),$$

$$P_w(d) = \Phi\left(\frac{0.673 - (1.02 - 0.94d)}{0.108}\right),$$

for GEE and

$$P_m(d) = \text{expit}(-5.67 + 17.23d),$$

$$P_w(d) = \Phi\left(\frac{0.673 - (1.01 - 0.88d)}{0.106}\right),$$

for Plackett-Dale using the same cutpoint for defining low fetal weight.

Let us describe the estimates. The BMD_{05} (LED_{05}) estimates are based on malformation is 146 (126) mg/kg/d for **GEE** and 162 (143) mg/kg/d for Plackett-Dale. Based on low fetal weight the BMD_{05} (LED_{05}) is 185 (168) mg/kg/d for **GEE** and 187 (178) mg/kg/d for Plackett-Dale.

Thus, using the definition above for low fetal weight, the minimum of the two LED_{05} is more than 20% higher than those obtained using the bivariate methods which incorporate the relationship between the two outcomes.

14.3 Discussion

In this chapter, a variety of approaches have been considered for the analysis of clustered data with a combination of continuous and discrete outcomes. They were applied to a case study (DYME in mice). It is useful to reflect on similarities and differences. Modeling approaches all need to consider the construction of multivariate or bivariate distributions for the combined continuous and discrete outcomes, thereby incorporating the clustering between littermates. Since such models are necessarily complicated, efficient computational tools are required.

The fetal weight mean model parameter estimates are almost identical regardless of modeling approach, as all approaches use an identity link function. The model for which the mean fetal weight conditions on the binary outcomes (column *C-Disc*) was parameterized to give marginal estimates of these parameters, so it can be compared directly with the other models. The corresponding marginal estimates of mean fetal weight by dose, and evaluated at $n_i = \bar{n}$, are given in Table 14.4.

The malformation regression parameters are not all comparable because of the different link functions, and the model that conditions on fetal weight (column labeled *C-Cont*) does not have a marginal interpretation. The estimates of the two unconditional probit models are almost identical. The *C-Disc* and Plackett-Dale models both use logistic link functions so the estimates are comparable and quite similar. Note that even though the *C-Cont* model does not have a marginal interpretation, the estimates are quite similar to the other unconditional probit models. The corresponding estimates of the probability of malformation by dose and evaluating at $n_i = \bar{n}$ and zero residuals for the *C-Cont* model are given in Table 14.4. Here, all four marginal estimates and the conditional estimates are very similar.

The main differences between models rest in the treatment of the second-order parameters, specifically fetal weight variance and the intrafoetus and intralitter correlations.

For the weight variance, the conditional models estimate a common moment-based scale parameter, based on GEE methodology. The unconditional models allow regression models to be specified. Often, with developmental toxicity data, the fetal weight variances are not constant across doses, as seen in the DYME example, and decrease with increasing dose where at the highest two doses the fetal weights were decreased in many foetuses across many litters.

Table 14.4 *DYME in Mice. Parameters estimated from different approaches.*

Dose (mg/g/day)	C-C	C-D	Pl-D	Pb-Lik	Pb-GEE
			μ_w^a		
0	1.02	1.03	pl	1.02	1.02
62.5	0.97	0.97	pl	0.96	0.96
125	0.91	0.91	pl	0.91	0.91
250	0.79	0.79	pl	0.79	0.79
500	0.55	0.54	pl	0.55	0.55
			$\Pr(M = 1)^a$		
0	0.003	0.005	pl	0.002	0.002
62.5	0.013	0.014	pl	0.010	0.011
125	0.044	0.037	pl	0.038	0.039
250	0.264	0.218	pl	0.239	0.240
500	0.935	0.936	pl	0.923	0.921
			Association $(W, M)^b$		
0	–	–	pl	-0.27	-0.18
62.5	–	–	pl	-0.27	-0.18
125	–	–	pl	-0.27	-0.34
250	–	–	pl	-0.27	-0.43
500	–	–	pl	-0.27	-0.52

[a] Evaluated at average covariates

[b] ψ for the P-D model, ρ_{wm} and ϱ for Pb-L and Pb-G models, respectively

The conditional models using GEE estimation and the Plackett-Dale model using PL estimation also treat clustering parameters as nuisance, as does one unconditional probit model using GEE estimation. GEE models estimate common moment-based parameters; the Plackett-Dale model uses independence as working assumption, and hence are assumed to be zero. The likelihood-based probit model allows modeling of these intralitter correlations; in the case of malformation they are, of course, on a latent scale.

The biggest difference is in the correlation between malformation and fetal weight. The conditional models do not directly specify a measure of association, though adding the residuals from the marginal model as covariates in the conditional model induces association. The associated parameters do lend insight into the strength of association on the one hand and whether an intrafoetus correlation is stronger than intralitter correlation on the other hand, even though they are not directly interpretable. The Plackett-Dale model uses the odds ratio as intrafoetus measure of association, which is readily interpretable. The probit models use the correlation of an underlying multivariate or bivariate normal distribution to characterize the intrafoetus and intralitter correlations. In the likelihood model one assumes complete exchangeability, i.e., intrafoetus and intralitter correlation are assumed constrained to be equal. Alternative approaches used on the example suggest this assumption is violated. Nevertheless, other parameter estimates seem fairly robust against this violation. The semiparametric probit model directly estimates the intrafoetus correlation. In the DYME example it was seen this then leads to different conclusions.

Several of the methods proposed have been applied to quantitative risk assessment. While there are subtle differences between all, one conclusion stands out. Ignoring the correlation between either littermates or outcomes leads to too conservative and hence unscientific safe doses.

CHAPTER 15

Multilevel Modeling of Complex Survey Data

Didier Renard, Geert Molenberghs

transnationale Universiteit Limburg, Diepenbeek–Hasselt, Belgium

As indicated in Chapter 1, many sets of data collected in human and biological sciences inherently have a hierarchical structure, where a hierarchy consists of units grouped at different levels. The denomination of 'clustered data' ordinarily refers to a two level hierarchy, where some basic units are grouped into clusters. Numerous examples have been given so far such as animals grouped into litters or subjects studied repeatedly over time.

A hierarchical structure can consist of more than two levels however, and examples also abound in practice. Schooling systems, for instance, present an obvious multilevel structure, with pupils grouped into classrooms, which are nested within schools which themselves may be clustered within education authorities. As a consequence, an important class of models, known under the generic name of multilevel models (Goldstein 1995), has been developed to represent such structures. These will be briefly reviewed in Section 15.1.

Often in sample surveys, for cost-related reasons or administrative considerations, multistage sampling schemes are adopted. In multistage sampling, the sample is selected in stages, with the sampling units at each stage being sub-sampled from the larger units drawn at the previous stage. Thus, it immediately becomes apparent that a sample obtained by multistage sampling is hierarchical in nature and, therefore, we may want to analyze such data using multilevel modeling techniques.

Based on the data from the Health Interview Survey (HIS) introduced in Section 2.3, we will illustrate the use of multilevel models with continuous (normally distributed) and discrete (binary) response variables in Section 15.2. In particular, the issue of weighting for unequal selection probabilities will be addressed as this is not, in principle, a simple extension of conventional weighing methods.

15.1 Multilevel Models

In this section, we briefly review multilevel models for normally-distributed and binary outcomes. While keeping an eye on the HIS, we consider a three-level population for notational convenience. To make matters more concrete, municipalities are termed level 3 units (or, in survey sampling terminology, the primary sampling units); households (secondary sampling units) correspond to level 2 units, while individuals (tertiary sampling units) form level 1.

For comprehensive accounts on multilevel modeling, the interested reader may consult Bryk and Raudenbush (1992), Longford (1993) and Goldstein (1995). Kreft and de Leeuw (1998) provide a more informal and introductory approach to the subject.

15.1.1 The Multilevel Linear Model

Suppose we have a sample consisting of K municipalities, with J_k households within the kth municipality ($k = 1, \dots, K$) and N_{jk} individuals within the jth household from the kth municipality ($j = 1, \dots, J_k$). Let y_{ijk} be the value of the response variable associated with the ith individual within the jth household from the kth municipality. We shall assume that the data y_{ijk} were generated according to the three-level model

$$y_{ijk} = x_{ijk}\boldsymbol{\beta} + z_{ijk}^{(3)}\boldsymbol{v}_k + z_{ijk}^{(2)}\boldsymbol{u}_{jk} + z_{ijk}^{(1)}\boldsymbol{e}_{ijk}, \tag{15.1}$$

where x_{ijk}, $z_{ijk}^{(3)}$, $z_{ijk}^{(2)}$, and $z_{ijk}^{(1)}$ are (possibly overlapping) fixed covariate row vectors, $\boldsymbol{\beta}$ is a fixed vector of parameters, and $\boldsymbol{v}_k$, $\boldsymbol{u}_{jk}$, and $\boldsymbol{e}_{ijk}$ are mutually independent normally distributed random variables. The $\boldsymbol{v}_k$'s and $\boldsymbol{u}_{jk}$'s are unobserved (sometimes called latent) variables that are essentially used to model variation in the data that is attributable to the clustering effect at the corresponding levels. These random variables can been thought of as representing various (cluster-specific) characteristics that are shared by all the elements of a cluster, thereby inducing some dependency between these units.

Note that there is an extremely close link between the multilevel modeling philosophy and random-effects models as introduced in Section 4.3 and studied in Chapter 13.

In its simplest form, the model will have $z_{ijk}^{(3)} \equiv 1$ and $z_{ijk}^{(2)} \equiv 1$, which reduces to a random intercept model. We will use such a model in the next section to investigate the clustering effect in the HIS. Also, it will commonly be the case that $z_{ijk}^{(1)} \equiv 1$, meaning that the model includes solely a simple residual error term, but the possibility of adding extra covariates permits the representation of complex variation at level 1, including subgroup variability or heteroscedasticity.

Parameter estimation in multilevel linear regression models can be carried out by maximizing the likelihood function. To this end, direct maximization, using Newton-Raphson or the EM algorithm (Little and Rubin 1987) for instance, can be performed. An equivalent procedure, called *iterative generalized least squares* (IGLS), was proposed by Goldstein (1986). His algorithm simply iterates between the estimation of the fixed and random parameters obtained by standard generalized least squares formulae, hence its name. IGLS is an attractive procedure as it tends to be quite efficient with large data sets typically encountered in the multilevel modeling framework. Note that the IGLS algorithm can be slightly modified (RIGLS) to perform similarly to residual (or restricted) maximum likelihood estimation, which yields unbiased estimates for variance components in random-effects models (Verbeke and Molenberghs 2000).

15.1.2 Multilevel Models for Discrete Response Data

We restrict attention to the case of a binary response, but the discussion below applies more generally to models for proportions or count data, for example. In fact, the theory can be developed for any non-linear multilevel model.

Consider the following multilevel logistic model:

$$\text{logit}(\pi_{ijk}) = x_{ijk}\beta + z_{ijk}^{(3)}v_k + z_{ijk}^{(2)}u_{jk}, \tag{15.2}$$

where $\pi_{ijk} = P[y_{ijk} = 1 | v_k, u_{jk}]$ and the same notation as in the previous section is followed. The model specification is completed by assuming that $y_{ijk} \sim \text{Bernouilli}(\pi_{ijk})$. In other words, the model can be written as

$$y_{ijk} = \pi_{ijk} + z_{ijk}^{(1)}e_{ijk},$$

where $z_{ijk}^{(1)} = \sqrt{\pi_{ijk}(1 - \pi_{ijk})}$, e_{ijk} has mean zero and variance 1, and π_{ijk} satisfies (15.2).

A few remarks are in place. First, the above model discriminates between level 1 and higher level variation. In particular, random disturbances at levels 2 and 3 appear on the logit scale, in contrast to the level 1 binomial variation which appears on the probability scale. Second, the conditional probability of observing a response is actually modeled, where conditioning takes place on the unobserved random variables v_k and u_{jk}.

Third, to proceed with inference, we can maximize the marginal likelihood function obtained after integrating out the random effects. Since the resulting expression is intractable, one needs to resort to numerical integration, such as Gaussian quadrature (Anderson and Aitkin 1985) or Markov Chain Monte Carlo techniques (Zeger and Karim 1991), but this can become computationally prohibitive. Approximate procedures have therefore been proposed to circumvent the problem. Breslow and Clayton (1993), for instance, exploit the penalized quasi-likelihood (PQL) estimator by applying Laplace's method for integral approximation. They also consider marginal quasi-likelihood (MQL), a name they give to a procedure previously proposed by Goldstein (1991). These two approaches entail iterative fitting of linear models based on first-order Taylor expansions of the mean function about the current estimated fixed part predictor (MQL) or the current predicted value (PQL).

As Rodríguez and Goldman (1995) demonstrate, these approximate procedures may be seriously biased when applied to binary response data. Their simulations reveal that both fixed effects and variance components may suffer from substantial, if not severe, attenuation bias in certain situations. Goldstein and Rasbash (1996) show that including a second-order term in the PQL expansion greatly reduces the bias described by Rodríguez and Goldman. Other authors have advised the introduction of bias-correction terms (Lin and Breslow 1996) or the use of iterative bootstrap (Kuk 1995) among other things.

15.1.3 Weighting in Multilevel Models

The issue of weighting in multilevel models has not been extensively investigated until quite recently (Pfeffermann et al. 1998). A reason might be that sampling schemes are commonly ignored in multilevel analyses of survey data since multilevel models enable the analyst to incorporate certain characteristics of the sampling design as covariates (e.g., stratification variables), although this argument breaks down when the relevant information is not made available or when it is not scientifically meaningful to be included in the model. When the sample selection probabilities are related to the response variable even after conditioning on covariates of interest, the conventional estimators of the model parameters may be biased, hence the need to study weighting procedures that attempt to correct for this bias.

It should be emphasized that weighting in multilevel models is not a trivial extension of conventional methods of weighting. One key feature of the multilevel approach is that sample inclusion probabilities can be defined at any stage of the hierarchy, conditionally on the membership to clusters from higher levels. Thus, municipality k is selected with inclusion probability π_k, household j is selected with probability $\pi_{j|k}$ within municipality k, and individual i is sampled with probability $\pi_{i|jk}$ within household j from municipality k. Unconditional selection probabilities can be derived from suitable products of conditional probabilities (e.g., $\pi_{jk} = \pi_k \pi_{j|k}$ denotes the probability that municipality k is sampled and that, within this municipality, household j is selected).

The approach Pfeffermann et al. advocate consists of substituting, in the IGLS sample estimators, each sum over units at a given level by a correspondingly weighted sum, using (inverse) conditional selection probabilities as defined above. When the sample inclusion probabilities (and hence the weights) are independent of the random effects, they show that a simple transformation of the variables specified in the random part of the model is sufficient. The appropriate transformation is:

- replace $z_{ijk}^{(1)}$ by $w_k^{-1/2} w_{j|k}^{-1/2} w_{i|jk}^{-1/2} z_{ijk}^{(1)} = w_{ijk}^{-1/2} z_{ijk}^{(1)}$,

- replace $z_{ijk}^{(2)}$ by $w_k^{-1/2} w_{j|k}^{-1/2} z_{ijk}^{(2)} = w_{ij}^{-1/2} z_{ijk}^{(2)}$,

- replace $z_{ijk}^{(3)}$ by $w_k^{-1/2} z_{ijk}^{(3)}$,

where the weights are defined by

$$w_k = \pi_k^{-1}, \qquad w_{j|k} = \pi_{j|k}^{-1}, \qquad w_{i|jk} = \pi_{i|jk}^{-1}.$$

Note that the weights should be rescaled in the above transformation in such a way that they have a mean of unity. The main advantage of this procedure is that it can easily be implemented with any standard software package which allows fitting of multilevel models.

If the weights are not independent of the random effects at a certain level (the sampling mechanism is then said to be informative), this leads to a more complicated procedure but the above authors conclude that, in this situation, the above step should produce acceptable results in many cases, though it can give biased estimates in some circumstances.

Finally, extra calculations are needed to obtain appropriate standard errors, but an easy solution is to use the robust or sandwich estimator instead, as advised in the MLwiN package (Section 15.1.4).

15.1.4 Specialized Software

This section is not devoted to general purpose software packages, such as SAS, SPlus, Stata, or SPSS, that have special built-in capabilities for multilevel modeling but, rather, to one specialized package named MLwiN. Details on other software packages specifically designed for multilevel modeling can be found, for example, on the Multilevel Models Project homepage at the address

http://multilevel.ioe.ac.uk/.

A review of many multilevel programs is also available at

http://www.stat.ucla.edu/deleeuw/software.pdf.

MLwiN and its older DOS incarnation MLn are the most extensive multilevel packages and were developed by researchers working on the Multilevel Models Project at the Institute of Education in London. It is of course beyond the scope of this chapter to provide a detailed description of all the functionalities available in MLwiN but we can emphasize a few interesting characteristics. Obviously, MLwiN allows fitting of linear models (using IGLS or RIGLS estimation) but it can also handle models for discrete responses (binomial data and counts) fitted using the MQL or PQL algorithm with first- or second-order approximation. Other interesting features include parametric and nonparameteric bootstrap estimation and Bayesian modeling using Markov chain Monte Carlo methods. Finally, the user can create his or her own macros for fitting special models, thus making it more flexible. A series of such macros are provided with the package to fit models for categorical outcomes, survival data, and time series, for example.

For analysis, data need to be prepared in rows corresponding to the cases observed. The MLwiN data structure is essentially that of a spreadsheet or worksheet with columns denoting variables and rows corresponding to observations. MLwiN can basically be operated in two modes. The user can either submit commands directly (or store them in a file that is run in batch), or the model be specified through a graphical interface. In the latter case, the user starts from the nucleus of the model and gradually builds it. This is a useful visual aid as one can immediately see whether the specified model is as one intended it to be.

Of the available multilevel packages, MLwiN is probably the most flexible and most complete one, but it might take some time before the user gets acquainted with all of its features. Additional information about the package can be currently obtained from the web site

$$\texttt{http://multilevel.ioe.ac.uk/features/}.$$

15.2 Application to the HIS

For the sake of illustration, we consider two response variables: body mass index (BMI), which will be log-transformed and analyzed as a normally distributed outcome, and a binary indicator for subjective or perceived health, which was originally rated by the interviewees on a 5-point scale and was dichotomized as good/very good versus other. As an attempt to find a parsimonious model for these data, the following covariates were examined: sex, age (eight categories), education (five categories), household income (5 categories), and smoking behavior. Note that the question about smoking behavior was addressed only to persons aged 15 or more, thus reducing the effective sample size from 10,221 to 8560.

In addition to the aforementioned covariates, information about the sample design can be taken into consideration:

- stratification variables: quarter and provinces;
- size variables: province, municipality, household;
- other variables: number of groups to be interviewed within a municipality, interviewee status (indicating whether he/she is the reference person or his/her partner).

15.2.1 Multilevel Linear Regression: An Example

Due to unit and item non-response, 7422 out of 8560 (87%) observations were available with complete information on the selected covariates and BMI. The model that we will fit can be written as follows:

$$y_{ijk} = x_{ijk}^T \boldsymbol{\beta} + v_k + u_{jk} + e_{ijk}, \tag{15.3}$$

with $v_k \sim N(0, \sigma_v^2)$, $u_{jk} \sim N(0, \sigma_u^2)$, and $e_{ijk} \sim N(0, \sigma_e^2)$. Thus, the total variation in (log) BMI can be decomposed into that between individuals within each household, that between households within municipalities, and that between municipalities. Among covariates listed above, only sex, age, education, and smoking behavior were found to have a significant effect and were included in the model. Second-order interaction terms of sex with age and education were also included. Among sampling-related variables, only province, household size, and interviewee status were retained.

The variance components (σ_v^2, σ_u^2, and σ_e^2) can be interpreted in terms of intra-unit correlation. Thus, the intra-municipality correlation is defined as

$$\rho_{MUN} = \frac{\sigma_v^2}{\sigma_v^2 + \sigma_u^2 + \sigma_e^2}, \qquad (15.4)$$

while the intra-household correlation is equal to

$$\rho_{HH} = \frac{\sigma_v^2 + \sigma_u^2}{\sigma_v^2 + \sigma_u^2 + \sigma_e^2}. \qquad (15.5)$$

The intra-unit correlation therefore reflects the proportion of the total variability in the outcome variable that is attributable to the clustering effect at a certain level and, as such, is a measure of within-group homogeneity.

Table 15.1 shows the results of fitting model (15.3) to the data, using weighted (as described in Section 15.1.3) and unweighted estimators. Robust standard errors are reported for parameter estimates. We see that there is generally good agreement (within standard error) between weighted and unweighted estimators, but that standard errors of the weighted estimators are subject to a sometimes severe loss of efficiency. Whether this is due to the use of the robust estimator for standard errors or to weighting itself is not entirely clear. As illustrated by Graubard and Korn (1994) for example, weighted estimates generally tend to be more variable than unweighted estimates, especially as sample weights become more variable. In the HIS, the unscaled weights w_k and $w_{i|jk}$ were characterized by a mean of 4.17 and 1.04 and standard deviation of 4.43 and 0.21, respectively, thus revealing substantial variability at the municipality level. Note that it is assumed there is no differential sampling at the household level.

The estimated variance components show that there is little clustering effect at the municipality level and a moderate effect at the household level, with an estimated value of 0.19 for ρ_{HH}. An estimate for the standard error of this parameter was obtained using the delta method for a ratio of two parameters (Herson 1975).

Finally, some diagnostic plots were examined (not shown). Thus, a plot of the level 1 residuals versus fixed part predictor did not reveal any special pattern, while normal probability plots of standardized residuals (at levels 1, 2, and 3) did not show any severe departures from the normality assumption, only pointing to a few extreme values.

15.2.2 Multilevel Logistic Regression: An Example

In this section, we consider an indicator for perceived health as our binary response variable. Due to unit and item non-response, 7254 out of 8560 (85%) observations were available for analysis. We will consider model (15.2), including random intercepts at the household and municipality levels. Variance components can no longer be interpreted in simple 'intra-unit correlation' terms, however, since such a measure now depends on the (fixed-effects) covariate matrix X. This follows from the mean-variance link, typical for gener-

Table 15.1 *Belgian Health Interview Survey. Multilevel linear regression model on* log(BMI). *Weighted and unweighted estimators are reported with empirical standard errors given in parentheses.*

	Unweighted	Weighted	Unweighted	Weighted
β Intercept	3.140(0.021)	3.120(0.034)		
Smoking (1=smoker):	-0.038(0.004)	-0.035(0.007)		
Age (categorical):	Males		Females	
15-24	-	-	-0.073(0.022)	0.076(0.033)
25-34	0.053(0.011)	0.041(0.018)	-0.038(0.022)	-0.014(0.034)
35-44	0.088(0.009)	0.087(0.023)	-0.035(0.023)	-0.017(0.038)
45-54	0.125(0.011)	0.115(0.019)	-0.041(0.022)	-0.027(0.033)
55-64	0.158(0.012)	0.152(0.018)	-0.067(0.027)	-0.040(0.037)
65-74	0.149(0.012)	0.124(0.016)	-0.068(0.022)	-0.033(0.037)
75+	0.079(0.014)	0.068(0.020)	-0.047(0.025)	-0.035(0.036)
Education:	Males		Females	
No diploma	-	-	-	-
Primary	-0.065(0.021)	-0.044 (0.033)	0.073(0.027)	0.055(0.035)
Lower secondary	-0.089(0.019)	-0.082(0.033)	0.095(0.021)	0.098(0.034)
Higher secondary	-0.106(0.019)	-0.091(0.033)	0.097(0.022)	0.088(0.035)
Higher	-0.151(0.020)	-0.135(0.034)	0.122(0.023)	0.108(0.034)
Provinces:				
Brussels	-	-		
Antwerpen	-0.001(0.007)	0.005(0.008)		
Vlaams Brabant	-0.005(0.008)	-0.003(0.009)		
Limburg	0.013(0.012)	0.012(0.021)		
Oost-Vlaanderen	0.007(0.008)	0.009(0.008)		
West-Vlaanderen	-0.017(0.007)	0.028(0.013)		
Brabant Wallon	0.022(0.014)	0.028(0.014)		
Hainaut	0.029(0.007)	0.036(0.010)		
Liege	0.022(0.009)	0.010(0.017)		
Luxembourg	0.025(0.010)	0.023(0.009)		
Namur	0.022(0.009)	0.025(0.014)		
German community	0.009(0.008)	0.008(0.010)		
Household size:	0.005(0.002)	0.006(0.002)		
Interviewee status:	0.034(0.007)	0.035(0.010)		
σ_v^2	0.000(0.000)	0.000(0.000)		
σ_u^2	0.004(0.000)	0.004(0.001)		
σ_e^2	0.019(0.001)	0.019(0.003)		
$\rho_{HH}^\dagger$	0.190(0.016)	0.199(0.025)		

† Standard errors were calculated using the delta method.

alized linear models, but absent in the normally distributed case. Approximate estimates can be produced nonetheless, albeit dependent on the set of covariates included in the model, as explained by Goldstein, Browne and Rasbash (2000)*. It should also be noted that a 'global' intra-unit correlation measure similar to (15.4) and (15.5) can be calculated, provided that the model can

* Unpublished manuscript that can be downloaded at
http://multilevel.ioe.ac.uk/team/currpap.html.

be formulated as a threshold model, i.e., if it can be assumed that the observed binary response arises from an underlying continuous unobserved (or latent) variable in such a way that a '1' is observed when a certain threshold is exceeded.

Another issue is related to weighting. The original paper by Pfeffermann *et al.* was framed in the context of multilevel linear models. In the MLwiN help, it is argued that a similar procedure applies to multilevel generalized linear models, the main difference being that level 1 is defined by the binomial variation and a method of incorporating the weights is to work with $w_{ijk}n_{ijk}$ instead of n_{ijk} as the denominator. The weighted explanatory variables at levels 2 and higher retain their meaning. We found this procedure to be numerically unstable in our case, however. The procedure was run on three different binary outcomes. With two of them (including the perceived health indicator), it did not converge, and in the last instance the final results were quite dramatically different from the unweighted analysis. It is not clear if the problem stems from the weighting scheme used in our example or from the procedure itself, but we cannot recommend its use in the multilevel logistic model until it is more thoroughly explored.

Table 15.2 shows the results of the unweighted analysis for the perceived health indicator. Among the original covariates, all were found important and included in the model. An interaction term between smoking behavior and age was also included. Among design-related covariates, only province indicator and size (household, municipality) variables were retained.

We allowed for an extra-dispersion parameter in the model, i.e., we assume that

$$y_{ijk} = \pi_{ijk} + z_{ijk}^{(1)} e_{ijk},$$

where e_{ijk} has mean zero and variance σ_e^2. As can be seen from Table 15.2, the estimated value of σ_e^2 strongly suggests that there is under-dispersion. Estimated variance components again reveal that there is almost no clustering effect at the municipality level compared to the household level. Note that a normal probability plot of standardized residuals at the household level showed a marked departure from normality, thereby making the appropriateness of the model to these data questionable.

15.3 Concluding Remarks

Multilevel data structures commonly occur in practice and sample survey data are no exception, especially given the widespread use of multistage sampling schemes. Multilevel regression models therefore constitute a natural tool that can be employed to account for the clustering aspect present in survey data. In addition to taking the hierarchical nature of the data explicitly into account, the great advantage of such models is that they allow the data analyst to easily incorporate covariates referring to certain characteristics of the sampling design, such as strata or size variables. If, conditionally on these characteristics, the sampling design is not informative, the analyst can simply proceed

Table 15.2 *Belgian Health Interview Survey. Multilevel logistic regression model on perceived health indicator. Unweighted estimators are reported with robust standard errors given in parentheses.*

Parameters	Est. (s.e.)	Parameters	Est. (s.e.)
β Intercept	1.427(0.290)		
Sex (1=male):	0.508(0.080)		
Age (categorical):	Non-smokers		Smokers
15-24	-		-0.540(0.303)
25-34	-0.531(0.221)		-0.081(0.245)
35-44	-1.298(0.204)		-0.368(0.166)
45-54	-1.701(0.204)		-0.226(0.180)
55-64	-2.250(0.203)		0.037(0.251)
65-74	-2.445(0.209)		-0.324(0.233)
75+	-2.814(0.217)		-0.112(0.424)
Education:		Household income:	
No diploma	-	< 20,000	-
Primary	0.240(0.206)	20,000-30,000	-0.180(0.177)
Lower secondary	0.560(0.191)	30,000-40,000	0.179(0.164)
Higher secondary	0.998(0.197)	40,000-60,000	0.510(0.170)
Higher	1.301(0.208)	> 60,000	0.982(0.213)
Provinces:		Household size:	0.110 (0.035)
Brussels	-	Municipality size:	-0.163(0.058)
Antwerpen	0.727(0.185)		
Vlaams Brabant	-0.080(0.186)	σ_v^2	0.054(0.036)
Limburg	0.118(0.224)	σ_u^2	3.050(0.157)
Oost-Vlaanderen	0.454(0.216)	σ_e^2	0.476(0.023)
West-Vlaanderen	0.664(0.228)		
Brabant Wallon	-0.118(0.141)		
Hainaut	-0.431(0.155)		
Liege	-0.417(0.204)		
Luxembourg	0.274(0.225)		
Namur	-0.155(0.151)		
German community	0.350(0.234)		

with the analysis. When this is not the case, conventional estimators may be biased and a weighting procedure can be used in an attempt to remove this bias.

APPENDIX A

Bahadur Parameter Space

Lieven Declerck

European Organization for Research and Treatment of Cancer, Brussels, Belgium

A.1 Restrictions on the Bahadur Model Parameters

In Section 4.1.1, it was suggested that the Bahadur model, unlike other models such as the beta-binomial model, has a heavily constrained parameter space. The beta-binomial model features all non-negative correlations, implying that there are only very mild constraints on the parameter space of this model. The restrictions on the Bahadur model parameters are much more complicated and stringent. Bahadur (1961) indicates that the sum of the probabilities of all possible outcomes is one, even when higher order correlations are set equal to zero. However, the requirement of having non-negative probabilities for all possible outcomes results in restrictions on the parameters. This holds even in the case of a Bahadur model with all higher order associations involved.

In this appendix, based on Declerck, Aerts and Molenberghs (1998), we study the nature of the restrictions of the parameter space of the Bahadur model. Throughout the appendix, the subscript referring to the cluster is omitted in order to simplify notation.

Bahadur (1961) discusses the restrictions on the second order correlation when all higher order associations are left out. He shows that the second order approximation is a probability distribution if and only if

$$-\frac{2}{n(n-1)} \min\left(\frac{\pi}{1-\pi}, \frac{1-\pi}{\pi}\right) \leq \rho_{(2)} \leq \frac{2\pi(1-\pi)}{(n-1)\pi(1-\pi) + 0.25 - \gamma_0}, \text{(A.1)}$$

where

$$\gamma_0 = \min_{z=0}^{n}\left\{[z - (n-1)\pi - 0.5]^2\right\}.$$

Bounds of the second order correlation $\rho_{(2)}$ are graphically represented in Figure A.1 for smaller litter sizes ($n = 2, 3, 4, 5$) and in Figure A.2 for larger litters ($n = 7, 10, 12, 15$). The lower bound for $\rho_{(2)}$ in a two-way Bahadur model attains its smallest value $-2/(n(n-1))$ at the malformation probability $\pi = 0.5$. This bound quickly approaches zero as the litter size n increases.

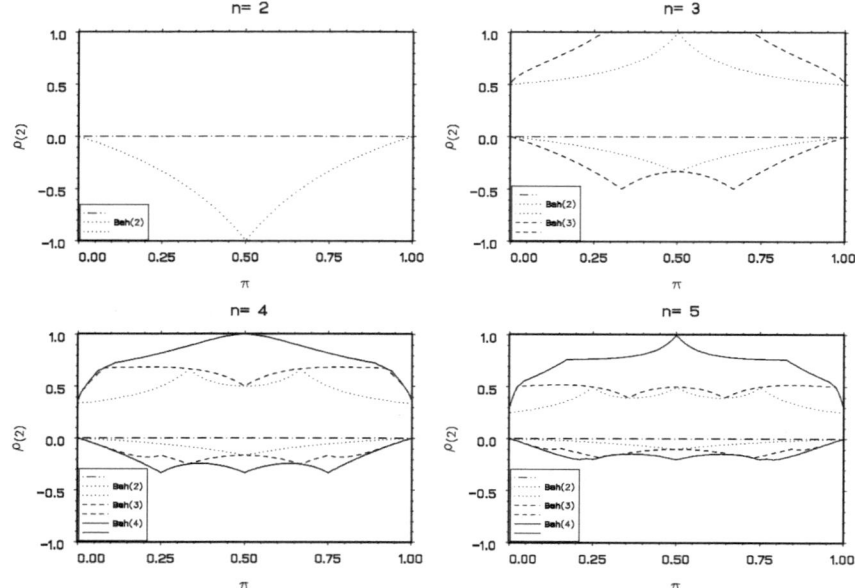

Figure A.1 *Boundaries for the second order correlation of the two-way, three-way and four-way Bahadur model for some smaller litter sizes.*

When $n = 2$, the upper bound for $\rho_{(2)}$ is one, independent of π. For larger values of n, the upper bound depends on π and varies between $1/(n-1)$ and $2/(n-1)$. As a consequence, the upper bound is in the range $(0.09; 0.18)$ for litters of size 12. As litter size increases, the restrictions on $\rho_{(2)}$ of Bah(2) become more severe.

Kupper and Haseman (1978) also consider the two-way Bahadur model and present numerical values for the constraints on $\rho_{(2)}$ for choices of π and n. Prentice (1988) studies the constraints in Bah(2) for any n. Furthermore, when the size of the clusters equals three, he argues that including the third order correlation removes the upper bound on $\rho_{(2)}$. However, it will be shown here that the requirements he verifies are necessary but not sufficient.

The parameter space of the general Bahadur model seems to be only partially known. The upper and lower bound of the second order correlation in Bah(3) and Bah(4) will be studied here. This leads to a clearer insight in the properties and usefulness of this model in general and in the behavior of the LR statistic in particular.

First, the focus is on the three-way Bahadur model. An analytical procedure that can handle any cluster size is developed. Explicit expressions for the bounds of $\rho_{(2)}$ are derived. These bounds are constructed such that for any value of $\rho_{(2)}$ between the lower and upper bound (both depending on the specified values of n and π), there exists at least one value of $\rho_{(3)}$ leading to a valid probability mass function.

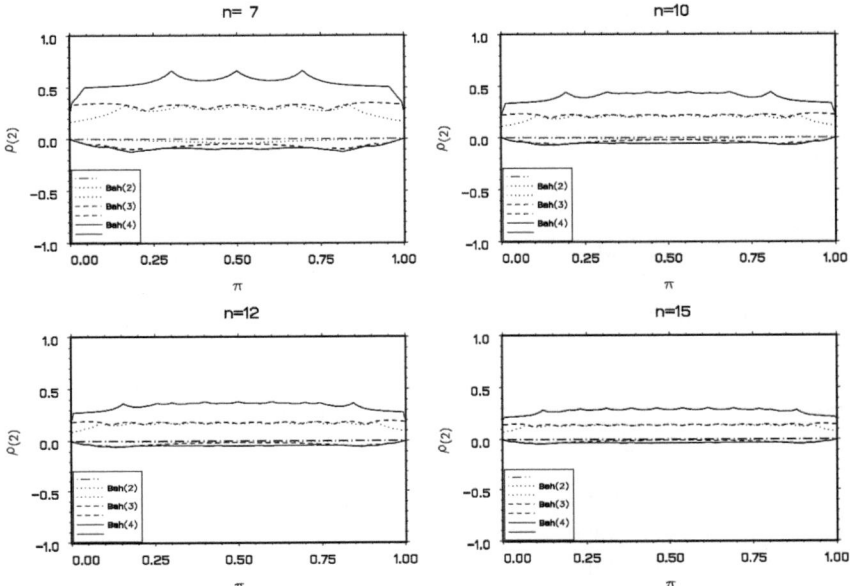

Figure A.2 *Boundaries for the second order correlation of the two-way, three-way and four-way Bahadur model for some larger litter sizes.*

The constraints on $\rho_{(2)}$ in Bah(3) for $n = 3$ are depicted in Figure A.1. Although this model is saturated in the sense that only clusters of size three are considered and two- and three-way correlations are included, still not all positive pairwise correlations are allowed. This is due to the condition that $\rho_{(3)} \in [-1, 1]$.

Constraints on $\rho_{(2)}$ for $n = 4$ are also shown in Figure A.1. The small values of the upper bound for extreme malformation probabilities are again due to the constraint $-1 \leq \rho_{(3)} \leq 1$.

For larger cluster sizes, boundaries for $\rho_{(2)}$ in a three-way Bahadur model are represented in Figures A.1 and A.2. For these clusters, the upper bound is very similar to the one of Bah(2), except for extreme values of π. Furthermore, it seems that the effect of adding a third order correlation to a two-way Bahadur model results in an upper bound for $\rho_{(2)}$ being almost independent on π. Compared to Bah(2), the range of negative $\rho_{(2)}$ is enlarged for small and large π.

Next, the focus is on the four-way Bahadur model. The analytical method described in Appendix A for Bah(3) can be extended to Bah(4). Developing first an expression for the constraints on $\rho_{(4)}$, the restrictions on $\rho_{(3)}$ are then derived, which finally result in the bounds for $\rho_{(2)}$. For any specified values of n and π, the lower and upper bound for the pairwise correlation is such that for any $\rho_{(2)}$ between these bounds, there exists at least one pair $(\rho_{(3)}, \rho_{(4)})$ leading to a valid probability mass function. Dealing with large clusters, Figure A.2 shows that compared to Bah(3), the range of allowable positive pairwise correlations of Bah(4) increases markedly, except for extreme values of π. The range of negative second order correlations remains very narrow.

In principle, constraints on $\rho_{(2)}$ for five- and higher-way Bahadur models can be calculated by generalizing the analytical procedure given for Bah(4).

Besides this analytical method, also a numerical procedure was developed to compute the bounds for the pairwise correlation in Bah(3) and Bah(4). This second procedure is used to check the calculations of the constraints on $\rho_{(2)}$. The numerical method is described here for Bah(3). First, the upper bound is calculated corresponding to some specified malformation probability. The starting value for the upper bound for $\rho_{(2)}$ is based on expression (A.1). Then, an increment is given to the starting value and by screening the interval $[-1, 1]$, a value of $\rho_{(3)}$ leading to a valid probability mass function is searched for. If such a value is found, an increment is given to the improved $\rho_{(2)}$ and the procedure is repeated. Otherwise, step halving is used and it is investigated whether a value of $\rho_{(3)}$ can be found resulting in non-negative probabilities for all outcomes. When improvements of the upper bound become smaller than some cut-off value, computations corresponding to the specified malformation probability are stopped. Next, an increment is given to π and the values of $\rho_{(2)}$ and $\rho_{(3)}$ corresponding to the previous π are used as starting values for the current malformation probability. The upper bound for $\rho_{(2)}$ is found for a grid of values of π. An analogous procedure is used to get lower bounds. It turns out that the results of the analytical and numerical procedures are essentially identical.

The findings are consistent with the results from both the asymptotic study and the analysis of the NTP data. Fitting a Bah(2) null model, the association parameter $\hat{\beta}_2$ captures part of the omitted dose effect. However, due to the (in general severe) restrictions on the second order association, this parameter is tied to a small range. This has some implications when dealing with strong dose effects in the underlying model. On the one hand, this results in values of $\hat{\beta}_2$ being smaller than for the beta-binomial model. On the other hand, the likelihood of the two-way Bahadur null model is smaller than the one of beta-binomial for which the constraints on the association parameter are very mild. In the case of the asymptotic study, the likelihood of the Bah(2) and beta-binomial alternative models is equal when there is no association in the underlying model. In the case of the NTP data, the difference between the likelihood of these two alternative models is minor relative to the null models. In conclusion, the likelihood of alternative and null models results in inflated values of the LR statistic when testing the null hypothesis of no dose effect.

In the previous discussion, artificial samples are generated without correlation in the underlying Bahadur model. Now, the focus is on the correlated case. With increasing dose effect, the association parameter of the two-way Bahadur null model will reach more quickly the boundary since there is already an association in the absence of dose effect. Here, an analogous explanation as for the case without association can be given for the behavior of the LR statistic.

Finally, the inclusion of a third order correlation into a two-way Bahadur model hardly changes the upper bound for $\rho_{(2)}$. This leads to values of the LR statistic being comparable to the ones under Bah(2). When adding a fourth order correlation, the constraints on $\rho_{(2)}$ are relaxed strikingly. As a consequence, the Bah(4) null model is much more likely than the Bah(2) and Bah(3) null models. Hence, the LR statistic results in values closer to the ones under beta-binomial. This finding clearly needs to be addressed carefully. In order to gain some additional insight, some binomial, beta-binomial and Bahadur distributions are displayed in Figure A.3. All distributions assume $\pi = 0.5$ and the cluster size is chosen to be $n = 20$. One striking observation is that the probability mass for the Bahadur model with only two-way association is bimodal for $\rho_{(2)}$ sufficiently large. It can be shown that when $\rho_{(2)}$ increases, the trough between the two modi reaches zero when the second order correlation reaches its upper bound, i.e., $\rho_{(2)} = 0.1$. When $\rho_{(3)}$ is added, the mass function is skewed as is obvious from definition (4.1). Considering the curve for $\rho_{(2)} = 0.05$, $\rho_{(3)} = 0$ and $\rho_{(4)} = 0.01$ is very insightful. Indeed, relative to the curve with only two-way association, the bimodal shape has disappeared, the curve is much closer to the binomial model, but the tails are thicker, which is in line with the concept of kurtosis. Thus, it seems that a plausible form of overdispersion is captured, not by merely adding $\rho_{(2)}$, but by adding $\rho_{(2)}$ and $\rho_{(4)}$. Observe, however, that the form of this distribution is still fairly different from the beta-binomial one. Since in the analysis of the NTP data, overdispersion seems to be more of an issue than skewness, $\rho_{(3)}$ adds little to the picture in this case. In general, since $\rho_{(3)}$ merely skews the distribution, rather than pulling up the trough, it is not surprising that $\rho_{(3)}$ only marginally relieves the bounds on $\rho_{(2)}$, whereas $\rho_{(4)}$ has a considerably stronger effect. This effect of $\rho_{(4)}$ is seen not only by the disappearance of the bimodal shape; in addition, this unimodal distribution is much closer to the binomial distribution.

A.2 Effect on the Likelihood Ratio Test

Fitting a two-way Bahadur model, an anomalous behavior of the LR test statistic for the null hypothesis of no dose effect is observed when analyzing data from artificial samples and from developmental toxicity studies. Dealing with artificial samples, the LR statistic inflates as the dose effect becomes stronger. Analyzing the NTP data, the values of this test statistic are in general strikingly larger than when fitting a beta-binomial model. Adding a third order correlation to the Bahadur model most often results in the same phenomena. However, considering Bah(4), the values of LR are more comparable to the ones under a beta-binomial model.

The behavior of the LR statistic when fitting a Bahadur model is explained by investigating the parameter space. Requiring a valid probability mass func-

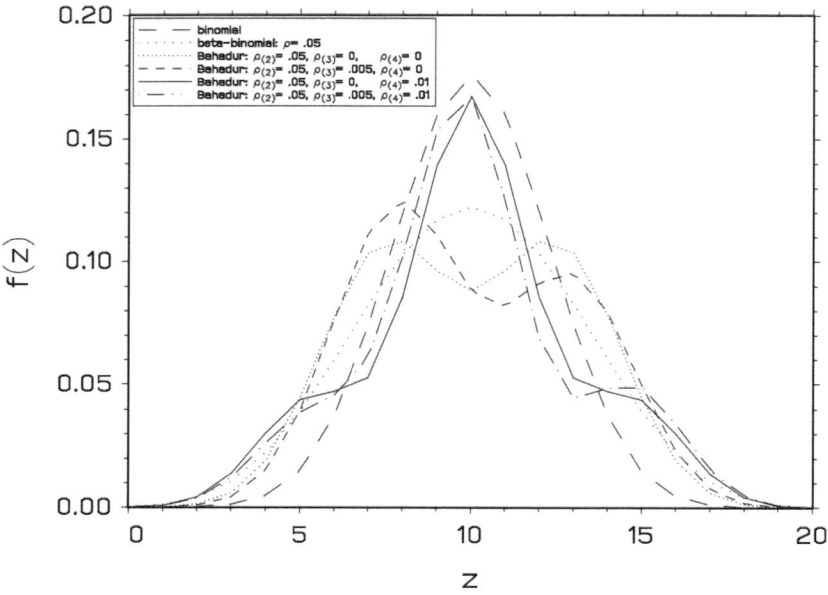

Figure A.3 *Distribution of binomial, beta-binomial and Bahadur models.*

tion, the parameters of the Bahadur model are subject to constraints. By means of analytical, numerical and graphical methods, it is shown that the inclusion of a third order association (playing the role of a skewness parameter) does not relax the upper bound of the second order correlation. In Bah(4), the range of positive second order associations is enlarged markedly. The combination of the second and fourth order correlations captures a more standard form of overdispersion (by means of a unimodal distribution) than a model with the second order association parameter only. This form shows a better resemblance with the overdispersion captured by a beta-binomial distribution, although there are still differences. Hence, in comparison with the LR statistic under Bah(2), the LR under Bah(4) is more comparable to the beta-binomial version.

The price to pay for including higher order associations is computational ease. While fitting Bah(2) is already more complex than fitting the beta-binomial model, the conditional model or GEE versions of the Bahadur model, even more difficulties are encountered with the Bah(3) and Bah(4) versions. This difficulty is not due to increased computation time, but to the complicated restrictions on the parameter space, which easily leads to divergence. It seems that even careful convergence monitoring is not able to fully relieve this problem.

A.3 Restrictions on $\rho_{(2)}$ in a Three-way Bahadur Model

An analytical method for the bounds of the second order correlation in a three-way Bahadur model is described. Let the coefficient of $\rho_{i(r)}$ in expression (4.1) be denoted by $g_r(\lambda, n, z)$. Hence, the three-way Bahadur model under exchangeability can be written as

$$f(\mathbf{y}) = \pi^z (1 - \pi)^{n-z} \left[1 + \sum_{r=2}^{3} \rho_{(r)} g_r(\lambda, n, z) \right].$$

Let the values for n and λ (or equivalently π) be arbitrary but fixed and drop them from notation. Hence, $g_r(\lambda, n, z)$ is abbreviated as $g_r(z)$. Restrictions on the Bahadur model parameters are resulting from the condition that the probability mass function has to be non-negative for all possible outcomes (Bahadur 1961), which for Bah(3) implies that

$$1 + \rho_{(2)} g_2(z) + \rho_{(3)} g_3(z) \geq 0, \tag{A.2}$$

for $z = 0, 1, \ldots, n$. Let $\mathbf{z}_P$, $\mathbf{z}_Z$ and $\mathbf{z}_N$ be the vectors containing the values of z for which $g_3(z)$ is positive, zero and negative respectively. Denote a general element of $\mathbf{z}_P$, $\mathbf{z}_Z$ and $\mathbf{z}_N$ by z_P, z_Z and z_N respectively. For each of the elements of $\mathbf{z}_P$, (A.2) can be expressed as

$$\rho_{(3)} \geq -\frac{1 + \rho_{(2)} g_2(z_P)}{g_3(z_P)}.$$

Analogously for $\mathbf{z}_N$, one obtains

$$\rho_{(3)} \leq -\frac{1 + \rho_{(2)} g_2(z_N)}{g_3(z_N)}.$$

Taking into account that $\rho_{(3)} \in [-1, 1]$, the constraints for $\rho_{(3)}$ are:

$$\max \left[\max_{z_P} \left(-\frac{1 + \rho_{(2)} g_2(z_P)}{g_3(z_P)} \right), -1 \right]$$

$$\leq \rho_{(3)} \leq$$

$$\min \left[\min_{z_N} \left(-\frac{1 + \rho_{(2)} g_2(z_N)}{g_3(z_N)} \right), 1 \right]. \tag{A.3}$$

In particular, in the case of clusters of size $n = 3$, expression (A.3) reduces to

$$\max(-\lambda^{-1} + (2\lambda^{-1} - \lambda)\rho_{(2)}, -\lambda^3 - 3\lambda\rho_{(2)}, -1) \leq \rho_{(3)} \leq$$
$$\min(\lambda^{-3} + 3\lambda^{-1}\rho_{(2)}, \lambda + (\lambda^{-1} - 2\lambda)\rho_{(2)}, 1). \tag{A.4}$$

For clusters of size $n = 4$, the expression analogous to (A.4) depends on π lying in the first, second, third, or fourth quarter of the $[0, 1]$-interval. The derivation is straightforward but lengthy and is omitted here.

Given $\rho_{(2)}$ in expression (A.3), there exists at least one value of $\rho_{(3)}$ leading to a valid probability mass function if and only if the lower bound is not larger than the upper bound. Equivalently, both terms on the left hand side have to be smaller than or equal to both terms on the right hand side. First, this implies for any pair (z_P, z_N) that

$$-\frac{1 + \rho_{(2)}g_2(z_P)}{g_3(z_P)} \leq -\frac{1 + \rho_{(2)}g_2(z_N)}{g_3(z_N)}. \tag{A.5}$$

Let

$$\Delta(z_P, z_N) = g_3(z_N) - g_3(z_P)$$

and

$$\tau(z_P, z_N) = g_2(z_N)g_3(z_P) - g_2(z_P)g_3(z_N).$$

Expression (A.5) can then be rewritten as

$$\rho_{(2)}\tau(z_P, z_N) \geq \Delta(z_P, z_N).$$

On the one hand, for any (z_P, z_N) for which $\tau > 0$, it implies that

$$\rho_{(2)} \geq \frac{\Delta(z_P, z_N)}{\tau(z_P, z_N)}.$$

On the other hand, for any (z_P, z_N) for which $\tau < 0$, it results in

$$\rho_{(2)} \leq \frac{\Delta(z_P, z_N)}{\tau(z_P, z_N)}.$$

Based on these inequalities, boundaries for $\rho_{(2)}$ are derived:

$$\max_{(z_P, z_N):\tau>0} \left(\frac{\Delta(z_P, z_N)}{\tau(z_P, z_N)}\right) \leq \rho_{(2)} \leq \min_{(z_P, z_N):\tau<0} \left(\frac{\Delta(z_P, z_N)}{\tau(z_P, z_N)}\right). \tag{A.6}$$

Expression (A.3) also implies for any element of z_P that

$$-\frac{1 + \rho_{(2)}g_2(z_P)}{g_3(z_P)} \leq 1.$$

On the one hand, for any z_P for which $g_2 > 0$, it leads to

$$\rho_{(2)} \geq -\frac{1 + g_3(z_P)}{g_2(z_P)}.$$

On the other hand, for any z_P for which $g_2 < 0$, it results in

$$\rho_{(2)} \leq -\frac{1 + g_3(z_P)}{g_2(z_P)}.$$

Based on these inequalities, boundaries for $\rho_{(2)}$ are derived:

$$\max_{z_P:g_2>0} \left(-\frac{1 + g_3(z_P)}{g_2(z_P)}\right) \leq \rho_{(2)} \leq \min_{z_P:g_2<0} \left(-\frac{1 + g_3(z_P)}{g_2(z_P)}\right). \tag{A.7}$$

Analogously, the condition that

$$-1 \leq -\frac{1 + \rho_{(2)}g_2(z_N)}{g_3(z_N)}$$

for any element of z_N implies that

$$\max_{z_N : g_2 > 0} \left(-\frac{1 - g_3(z_N)}{g_2(z_N)} \right) \leq \rho_{(2)} \leq \min_{z_N : g_2 < 0} \left(-\frac{1 - g_3(z_N)}{g_2(z_N)} \right). \qquad \text{(A.8)}$$

Also the effects of the elements of z_Z on the constraints on $\rho_{(2)}$ need to be studied. Expression (A.2) then reduces to

$$1 + \rho_{(2)} g_2(z_Z) \geq 0$$

for any z_Z. For z_Z for which $g_2 > 0$, it leads to

$$\rho_{(2)} \geq -\frac{1}{g_2(z_Z)},$$

while for z_Z for which $g_2 < 0$, it results in

$$\rho_{(2)} \leq -\frac{1}{g_2(z_Z)}.$$

Based on these inequalities, boundaries for $\rho_{(2)}$ are derived:

$$\max_{z_Z : g_2 > 0} \left(-\frac{1}{g_2(z_Z)} \right) \leq \rho_{(2)} \leq \min_{z_Z : g_2 < 0} \left(-\frac{1}{g_2(z_Z)} \right). \qquad \text{(A.9)}$$

From (A.6)–(A.9) and the constraint $-1 \leq \rho_{(2)} \leq 1$, it follows that the upper and lower bound for $\rho_{(2)}$ in a three-way Bahadur model can be written as

$$\max \left[\max_{(z_P, z_N) : \tau > 0} \left(\tfrac{\Delta}{\tau} \right), \max_{z_P : g_2 > 0} \left(-\tfrac{1 + g_3}{g_2} \right), \right.$$

$$\left. \max_{z_N : g_2 > 0} \left(-\tfrac{1 - g_3}{g_2} \right), \max_{z_Z : g_2 > 0} \left(-\tfrac{1}{g_2} \right), -1 \right]$$

$$\leq \rho_{(2)} \leq$$

$$\min \left[\min_{(z_P, z_N) : \tau < 0} \left(\tfrac{\Delta}{\tau} \right), \min_{z_P : g_2 < 0} \left(-\tfrac{1 + g_3}{g_2} \right), \right.$$

$$\left. \min_{z_N : g_2 < 0} \left(-\tfrac{1 - g_3}{g_2} \right), \min_{z_Z : g_2 < 0} \left(-\tfrac{1}{g_2} \right), 1 \right].$$

A.4 Restrictions on $\rho_{(2)}$ in a Four-way Bahadur Model

In contrast with Appendix A, this appendix deals with an analytical procedure for the derivation of the constraints on the second order correlation in a four-way Bahadur model. Again, represent the coefficient of $\rho_{i(r)}$ in formula (4.1) by $g_r(\lambda, n, z)$. One can then express the four-way Bahadur model under exchangeability as

$$f(\mathbf{y}) = \pi^z (1 - \pi)^{n-z} \left[1 + \sum_{r=2}^{4} \rho_{(r)} g_r(\lambda, n, z) \right].$$

Here, the values for n and λ are arbitrary but fixed and, hence, the coefficient $g_r(\lambda, n, z)$ is represented by $g_r(z)$. Constraints on the parameters of the Bahadur model are due to the condition that the density function needs to be non-negative for all outcomes (Bahadur 1961), which for the four-way Bahadur model results in

$$1 + \rho_{(2)}g_2(z) + \rho_{(3)}g_3(z) + \rho_{(4)}g_4(z) \geq 0, \tag{A.10}$$

for $z = 0, 1, \ldots, n$. The vectors containing the values of z for which $g_4(z)$ is positive, zero and negative are denoted by $\mathbf{z}_P$, $\mathbf{z}_Z$ and $\mathbf{z}_N$ respectively. Let a general element of $\mathbf{z}_P$, $\mathbf{z}_Z$ and $\mathbf{z}_N$ be represented by z_P, z_Z and z_N respectively. Then, expression (A.10) can be written as

$$\rho_{(4)} \geq -\frac{1 + \rho_{(2)}g_2(z_P) + \rho_{(3)}g_3(z_P)}{g_4(z_P)}$$

for each of the elements of $\mathbf{z}_P$. For $\mathbf{z}_N$, it follows from (A.10) that

$$\rho_{(4)} \leq -\frac{1 + \rho_{(2)}g_2(z_N) + \rho_{(3)}g_3(z_N)}{g_4(z_N)}.$$

Adding the constraint $\rho_{(4)} \in [-1, 1]$, the restrictions on the four-way correlation coefficient are:

$$\max\left[\max_{z_P}\left(-\frac{1 + \rho_{(2)}g_2(z_P) + \rho_{(3)}g_3(z_P)}{g_4(z_P)}\right), -1\right]$$

$$\leq \rho_{(4)} \leq \tag{A.11}$$

$$\min\left[\min_{z_N}\left(-\frac{1 + \rho_{(2)}g_2(z_N) + \rho_{(3)}g_3(z_N)}{g_4(z_N)}\right), 1\right].$$

For a particular value of $\rho_{(2)}$ and $\rho_{(3)}$ in expression (A.11), there exists at least one value of $\rho_{(4)}$ resulting in a valid density function if and only if the lower bound in (A.11) is not larger than the upper bound. Hence, each of the terms on the left hand side needs to be smaller than or equal to each of the terms on the right hand side. This implies, among others, that for any pair (z_P, z_N),

$$-\frac{1 + \rho_{(2)}g_2(z_P) + \rho_{(3)}g_3(z_P)}{g_4(z_P)} \leq -\frac{1 + \rho_{(2)}g_2(z_N) + \rho_{(3)}g_3(z_N)}{g_4(z_N)}. \tag{A.12}$$

Let

$$\Delta(z_P, z_N, \rho_{(2)}) =$$

$$g_4(z_N)\left[1 + \rho_{(2)}g_2(z_P)\right] - g_4(z_P)\left[1 + \rho_{(2)}g_2(z_N)\right] \tag{A.13}$$

and

$$\tau(z_P, z_N) = g_3(z_N)g_4(z_P) - g_3(z_P)g_4(z_N). \tag{A.14}$$

One can then re-express formula (A.12) as

$$\rho_{(3)}\tau(z_P, z_N) \geq \Delta(z_P, z_N, \rho_{(2)}).$$

For any (z_P, z_N) for which $\tau > 0$, it results in

$$\rho_{(3)} \geq \frac{\Delta(z_P, z_N, \rho_{(2)})}{\tau(z_P, z_N)}.$$

Also, for any (z_P, z_N) for which $\tau < 0$, it implies that

$$\rho_{(3)} \le \frac{\Delta(z_P, z_N, \rho_{(2)})}{\tau(z_P, z_N)}.$$

From these inequalities, constraints on the third order correlation coefficient are obtained:

$$\max_{(z_P, z_N) : \tau > 0} \left(\frac{\Delta(z_P, z_N, \rho_{(2)})}{\tau(z_P, z_N)} \right)$$

$$\le \quad \rho_{(3)}$$

$$\le \quad \min_{(z_P, z_N) : \tau < 0} \left(\frac{\Delta(z_P, z_N, \rho_{(2)})}{\tau(z_P, z_N)} \right). \qquad \text{(A.15)}$$

Formula (A.11) also results in

$$-\frac{1 + \rho_{(2)} g_2(z_P) + \rho_{(3)} g_3(z_P)}{g_4(z_P)} \le 1$$

for any element of z_P. On the one hand, for any z_P for which $g_3 > 0$, it implies that

$$\rho_{(3)} \ge -\frac{1 + g_4(z_P) + \rho_{(2)} g_2(z_P)}{g_3(z_P)}.$$

On the other hand, for any z_P for which $g_3 < 0$, it leads to

$$\rho_{(3)} \le -\frac{1 + g_4(z_P) + \rho_{(2)} g_2(z_P)}{g_3(z_P)}.$$

Based on these inequalities, restrictions on $\rho_{(3)}$ are found:

$$\max_{z_P : g_3 > 0} \left(-\frac{1 + g_4(z_P) + \rho_{(2)} g_2(z_P)}{g_3(z_P)} \right)$$

$$\le \quad \rho_{(3)}$$

$$\le \quad \min_{z_P : g_3 < 0} \left(-\frac{1 + g_4(z_P) + \rho_{(2)} g_2(z_P)}{g_3(z_P)} \right). \qquad \text{(A.16)}$$

In an analogous way, the condition that

$$-1 \le -\frac{1 + \rho_{(2)} g_2(z_N) + \rho_{(3)} g_3(z_N)}{g_4(z_N)}$$

for any element of z_N results in

$$\max_{z_N : g_3 > 0} \left(-\frac{1 - g_4(z_N) + \rho_{(2)} g_2(z_N)}{g_3(z_N)} \right)$$

$$\le \quad \rho_{(3)}$$

$$\le \quad \min_{z_N : g_3 < 0} \left(-\frac{1 - g_4(z_N) + \rho_{(2)} g_2(z_N)}{g_3(z_N)} \right). \qquad \text{(A.17)}$$

Also the effects of the elements of z_Z on the boundaries for the third order correlation have to be considered. Formula (A.10) then simplifies to

$$1 + \rho_{(2)} g_2(z_Z) + \rho_{(3)} g_3(z_Z) \geq 0$$

for any z_Z. For z_Z for which $g_3 > 0$, it implies that

$$\rho_{(3)} \geq -\frac{1 + \rho_{(2)} g_2(z_Z)}{g_3(z_Z)},$$

while for z_Z for which $g_3 < 0$, it results in

$$\rho_{(3)} \leq -\frac{1 + \rho_{(2)} g_2(z_Z)}{g_3(z_Z)}.$$

From these inequalities, restrictions on $\rho_{(3)}$ are derived:

$$\max_{z_Z : g_3 > 0} \left(-\frac{1 + \rho_{(2)} g_2(z_Z)}{g_3(z_Z)} \right) \leq \rho_{(3)} \leq \min_{z_Z : g_3 < 0} \left(-\frac{1 + \rho_{(2)} g_2(z_Z)}{g_3(z_Z)} \right). \quad (A.18)$$

Based on (A.15)–(A.18) and the restriction $-1 \leq \rho_{(3)} \leq 1$, the lower and upper bound for the third order correlation coefficient in a four-way Bahadur model can be expressed as

$$\max \left[\max_{(z_P, z_N) : \tau > 0} \left(\frac{\Delta}{\tau} \right), \ \max_{z_P : g_3 > 0} \left(-\frac{1 + g_4 + \rho_{(2)} g_2}{g_3} \right), \right.$$

$$\max_{z_N : g_3 > 0} \left(-\frac{1 - g_4 + \rho_{(2)} g_2}{g_3} \right), \ \max_{z_Z : g_3 > 0} \left(-\frac{1 + \rho_{(2)} g_2}{g_3} \right), -1 \right] \leq \rho_{(3)} \leq$$

$$\min \left[\min_{(z_P, z_N) : \tau < 0} \left(\frac{\Delta}{\tau} \right), \ \min_{z_P : g_3 < 0} \left(-\frac{1 + g_4 + \rho_{(2)} g_2}{g_3} \right), \right.$$

$$\min_{z_N : g_3 < 0} \left(-\frac{1 - g_4 + \rho_{(2)} g_2}{g_3} \right), \ \min_{z_Z : g_3 < 0} \left(-\frac{1 + \rho_{(2)} g_2}{g_3} \right), 1 \right].$$

Based on the previous formula, restrictions on the second order correlation coefficient can be derived. For a particular value of $\rho_{(2)}$, there exists at least one value of $\rho_{(3)}$ resulting in a valid distribution if and only if the lower bound for $\rho_{(3)}$ is smaller than or equal to the upper bound. Hence, each of the five terms in the lower bound of $\rho_{(3)}$ needs to be smaller than or equal to each of the five terms in the upper bound. More specifically, this implies, among others, that

$$\max_{(z_P, z_N) : \tau > 0} \left(\frac{\Delta(z_P, z_N, \rho_{(2)})}{\tau(z_P, z_N)} \right) \leq \min_{(z_P, z_N) : \tau < 0} \left(\frac{\Delta(z_P, z_N, \rho_{(2)})}{\tau(z_P, z_N)} \right). \quad (A.19)$$

Represent a pair (z_P, z_N) for which $\tau > 0$ by (z_{P1}, z_{N1}) and a pair (z_P, z_N) for which $\tau < 0$ by (z_{P2}, z_{N2}). Inequality (A.19) can be re-expressed as

$$\frac{\Delta(z_{P1}, z_{N1}, \rho_{(2)})}{\tau(z_{P1}, z_{N1})} \leq \frac{\Delta(z_{P2}, z_{N2}, \rho_{(2)})}{\tau(z_{P2}, z_{N2})} \quad (A.20)$$

for any combination of (z_{P1}, z_{N1}) and (z_{P2}, z_{N2}). Let

$$
\begin{aligned}
\nu(z_{P1}, z_{N1}, z_{P2}, z_{N2}) \;=\;\; & \tau(z_{P1}, z_{N1}) \left[g_2(z_{N2}) g_4(z_{P2}) - g_2(z_{P2}) g_4(z_{N2}) \right] \\
- \;\; & \tau(z_{P2}, z_{N2}) \left[g_2(z_{N1}) g_4(z_{P1}) - g_2(z_{P1}) g_4(z_{N1}) \right]
\end{aligned}
$$

and

$$
\begin{aligned}
\omega(z_{P1}, z_{N1}, z_{P2}, z_{N2}) \;=\;\; & \tau(z_{P1}, z_{N1}) \left[g_4(z_{N2}) - g_4(z_{P2}) \right] \\
- \;\; & \tau(z_{P2}, z_{N2}) \left[g_4(z_{N1}) - g_4(z_{P1}) \right].
\end{aligned}
$$

Using also expression (A.13) for $\Delta(z_P, z_N, \rho_{(2)})$, it follows from (A.20) that

$$
\rho_{(2)} \nu(z_{P1}, z_{N1}, z_{P2}, z_{N2}) \geq \omega(z_{P1}, z_{N1}, z_{P2}, z_{N2}).
$$

For any $(z_{P1}, z_{N1}, z_{P2}, z_{N2})$ for which $\nu > 0$, it implies that

$$
\rho_{(2)} \geq \frac{\omega(z_{P1}, z_{N1}, z_{P2}, z_{N2})}{\nu(z_{P1}, z_{N1}, z_{P2}, z_{N2})}.
$$

Furthermore, for any $(z_{P1}, z_{N1}, z_{P2}, z_{N2})$ for which $\nu < 0$, it follows that

$$
\rho_{(2)} \leq \frac{\omega(z_{P1}, z_{N1}, z_{P2}, z_{N2})}{\nu(z_{P1}, z_{N1}, z_{P2}, z_{N2})}.
$$

Based on these inequalities, constraints for $\rho_{(2)}$ are derived:

$$
\max_{(z_{P1}, z_{N1}, z_{P2}, z_{N2}) : \nu > 0} \left(\frac{\omega(z_{P1}, z_{N1}, z_{P2}, z_{N2})}{\nu(z_{P1}, z_{N1}, z_{P2}, z_{N2})} \right)
$$

$$
\leq \rho_{(2)} \leq
$$

$$
\min_{(z_{P1}, z_{N1}, z_{P2}, z_{N2}) : \nu < 0} \left(\frac{\omega(z_{P1}, z_{N1}, z_{P2}, z_{N2})}{\nu(z_{P1}, z_{N1}, z_{P2}, z_{N2})} \right).
$$

In an analogous way, the other constraints for the second order correlation coefficient in a four-way Bahadur model are obtained. The derivation is straightforward but tedious and, hence, that part is omitted here.

References

Aerts, M. and Claeskens, G. (1997). Local polynomial estimation in multi-parameter likelihood models. *Journal of the American Statistical Association*, **92**, 1536–1545.

Aerts, M. and Claeskens, G. (1999). Bootstrapping pseudolikelihood models for clustered binary data. *Annals of the Institute of Statistical Mathematics*, **51**, 515–530.

Aerts, M. and Claeskens, G. (2001). Bootstrap tests for misspecified models, with application to clustered binary data. *Computational Statistics and Data Analysis*, **36**, 383–401.

Aerts, M., Claeskens, G., and Hart, J.D. (1999). Testing the fit of a parametric function. *Journal of the American Statistical Association*, **94**, 869–879.

Aerts, M., Claeskens, G., and Hart, J.D. (2000). Testing lack of fit in multiple regression, *Biometrika*, **87**, 405–424.

Aerts, M., Claeskens, G., and Molenberghs, G. (1999). A note on the quadratic bootstrap and improved estimation in logistic regression. Technical Report, Limburgs Universitair Centrum, Diepenbeek.

Aerts, M., Claeskens, G., and Wand, M. (2002). Some theory for penalized spline additive models. *Journal of Statistical Planning and Inference*, **00**, 000–000.

Aerts, M., Declerck, L., and Molenberghs, G. (1997). Likelihood misspecification and safe dose determination for clustered binary data. *Environmetrics*, **8**, 613–627.

Agresti, A. (1990). *Categorical Data Analysis*. New York: John Wiley.

Ahn, H. and Chen, J.J. (1997). Tree-structured logistic model for over-dispersed binomial data with applications to modeling developmental effects. *Biometrics*, **53**, 435–455.

Akaike, H. (1974). A new look at statistical model identification. *I.E.E.E. Transactions on Automatic Control*, **19**, 716–723.

Altham, P.M.E. (1978). Two generalizations of the binomial distribution. *Applied Statistics*, **27**, 162-167.

Anderson, D.A. and Aitkin, M. (1985). Variance component models with binary response: interviewer variability. *Journal of the Royal Statistical Society, Series B*, **47**, 203–210.

Arnold, B.C., Castillo, E., and Sarabia, J.-M. (1992). *Conditionally Specified Distributions*. Lecture Notes in Statistics **73**. New York: Springer-Verlag.

Arnold, B.C. and Strauss, D. (1991). Pseudolikelihood estimation: some examples. *Sankhya B*, **53**, 233-243.

Ashford, J.R. and Sowden, R.R. (1970). Multivariate probit analysis. *Biometrics*, **26**, 535–546.

Autian, J. (1973). Toxicity and health threats of phthalate esters: Review of the literature. *Environmental Health Perspectives*, **4**, 3–26.

Bahadur, R.R. (1961). A representation of the joint distribution of responses of n dichotomous items. In: *Studies in item analysis and prediction*, H. Solomon (Ed.), Stanford Mathematical Studies in the Social Sciences VI. Stanford, California, Stanford University Press.

Besag, J. (1975). Statistical analysis of non-lattice data. *The Statistician*, **24**, 179–195.

Betensky, R.A. and Williams, P.L. (2001). A comparison of models for clustered binary outcomes: analysis of a designed immunology experiment. *Applied Statistics*, **50**, 43–61.

Bogdan, M. (1999). Data driven smooth tests for bivariate normality. *Journal of Multivariate Analysis*, **68**, 26–53.

Bosch, R.J., Wypij, D., and Ryan, L.M. (1996). A semiparametric approach to risk assessment for quantitative outcomes. *Risk Analysis*, **16**, 657–665.

Boos, D.D. (1992). On generalized score tests. *The American Statistician*, **46**, 327–333.

Bradstreet, T.E. and Liss, C.L. (1995). Favorite data sets from early (and late) phases of drug research–Part 4. *ASA Proceedings of the Section on Statistical Education*, 335–340.

Breslow, N.E. and Clayton, D.G. (1993). Approximate inference in generalized linear mixed models. *Journal of the American Statistical Association*, **88**, 9–25.

Breslow, N.E. and Day, N.E. (1987). *Statistical Methods in Cancer Research, Volume II*. Oxford: Oxford University Press.

Brown, C.C. (1982). On a goodness of fit test for the logistic model based on score statistics. *Communications in Statistics, Theory and Methods*, **11**, 1087–1105.

Brown, L.D. (1986). *Fundamentals of Statistical Exponential Families*. California: Institute of Mathematical Statistics.

Brown, N.A. and Fabro, S. (1981). Quantitation of rat embryonic development in vitro: a morphological scoring system. *Teratology*, **24**, 65–78.

Bryk, A.S. and Raudenbush, S.W. (1992). *Hierarchical Linear Models: Applications and Data Analysis Methods.* Newbury Park: Sage Publications.

Carey, V.C., Zeger, S.L., and Diggle, P.J. (1993). Modelling multivariate binary data with alternating logistic regressions. *Biometrika*, **80**, 517–526.

Carlin, B.P. and Louis, T.A. (1996). *Bayes and Empirical Bayes Methods for Data Analysis.* London: Chapman & Hall.

Carr, G.J. and Portier, C.J. (1993). An evaluation of some methods for fitting dose-response models to quantal-response developmental toxicology data. *Biometrics*, **49**, 779–791.

Carroll, R.J., Ruppert, D., and Welsh, A.H. (1998). Local estimating equations. *Journal of the American Statistical Association*, **93**, 214–227.

Catalano, P.J. (1997). Bivariate modelling of clustered continuous and ordered categorical outcomes. *Statistics in Medicine*, **16**, 883–900.

Catalano, P.J. and Ryan, L.M. (1992). Bivariate latent variable models for clustered discrete and continuous outcomes. *Journal of the American Statistical Association*, **87**, 651–658.

Catalano, P., Ryan, L., and Scharfstein, D. (1994). Modeling fetal death and malformation in developmental toxicity. *Risk Analysis*, **14**, 611–619.

Catalano, P.J., Scharfstein, D.O., Ryan, L.M., Kimmel, C.A., and Kimmel, G.L. (1993). Statistical model for fetal death, fetal weight, and malformation in developmental toxicity studies. *Teratology*, **47**, 281–290.

Chen, J.J. (1993). A malformation incidence dose–response model incorporating fetal weight and/or litter size as covariates. *Risk Analysis*, **13**, 559–564.

Chen, J.J., and Gaylor, D. W. (1992). Correlations of developmental end points observed after 2,4,5–trichlorophenoxyacetic acid exposure in mice. *Teratology*, **45**, 241–246.

Chen, J.J. and Kodell, R.L. (1989). Quantitative risk assessment for teratologic effects. *Journal of the American Statistical Association*, **84**, 966–971.

Chen, J.J., Kodell, R.L., Howe, R.B., and Gaylor, D.W. (1991). Analysis of trinomial responses from reproductive and developmental toxicity experiments. *Biometrics*, **47**, 1049–1058.

Claeskens, G. and Aerts, M. (2000a). Bootstrapping local polynomial estimators in likelihood-based models. *Journal of Statistical Planning and Inference*, **86**, 63–80.

Claeskens, G. and Aerts, M. (2000b). On local estimating equations in additive multiparameter models. *Statistics and Probability Letters*, **49**, 139–148.

Claeskens, G., Aerts, M., Molenberghs, G., and Ryan, L. (2002). Robust benchmark dose determination based on profile score methods. *Environmental and Ecological Statistics*, **00**, 000–000.

Claeskens, G. and Hjort, N.L. (2001). Goodness of fit via nonparametric likelihood ratios. *Submitted for publication*.

Clapp, D.E., Zaebst, D.D., and Herrick, R.F. (1984). Measuring exposures to glycol ethers. *Environmental Health Perspectives*, **57**, 91–95.

Cochran, W.G. (1977). *Sampling Techniques*. New York: John Wiley.

Collett, D. (1991). *Modelling Binary Data*. London: Chapman & Hall.

Conaway, M. (1989). Analysis of repeated categorical measurements with conditional likelihood methods. *Journal of the American Statistical Association*, **84**, 53–62.

Connolly, M.A. and Liang, K.Y. (1988). Conditional logistic regression models for correlated binary data. *Biometrika*, **75**, 501–506.

Copas, J.B. (1983). Plotting p against x. *Applied Statistics*, **32**, 25–31.

Corcoran, C.D., Mehta, C.R., and Senchaudhuri, P. (2000). Power comparisons for tests of trend in dose-response studies. *Statistics in Medicine*, **19**, 3037–3050.

Corcoran, C.D., Senchaudhuri, P., Ryan, L., Mehta, C., Patel, N., and Molenberghs, G. (2001). An exact trend test for correlated binary data. *Biometrics*, **57**, 93–101.

Cox, D.R. (1972). The analysis of multivariate binary data. *Applied Statistics*, **21**, 113–120.

Cox, N.R. (1974). Estimation of the correlation between a continuous and a discrete variable. *Biometrics*, **30**, 171–178.

Cox, D.R. and Wermuth, N. (1992). Response models for mixed binary and quantitative variables. *Biometrika*, **79**, 441–461.

Cox, D. R. and Wermuth, N. (1994). A note on the quadratic exponential binary distribution. *Biometrika*, **81**, 403–408.

Cressie, N.A.C. (1991). *Statistics for Spatial Data*. New York: John Wiley.

Crump, K. (1984). A new method for determining allowable daily intakes. *Fundamental and Applied Toxicology*, **4**, 854–871.

Crump, K.S. (1995). Calculation of benchmark doses from continuous data. *Risk Analysis*, **15**, 79–89.

Crump, K. S. and Howe, R.B. (1985). A review of methods for calculating statistical confidence limits in low dose extrapolation. In: Clayson, D. B., Krewski, D. and Munro, I. (Eds.), *Toxicological Risk Assessment. Volume I: Biological and Statistical Criteria*. CRC Press, Boca Raton, pp. 187–203.

Dale, J.R. (1986). Global cross-ratio models for bivariate, discrete, ordered responses. *Biometrics*, **42**, 909–917.

Davison, A.C. and Hinkley, D.V. (1997). *Bootstrap Methods and Their Application*. Cambridge: Cambridge University Press.

Declerck, L., Aerts, M., and Molenberghs, G. (1998). Behaviour of the likelihood ratio test statistic under a Bahadur model for exchangeable binary data. *Journal of Statistical Computations and Simulations*, **61**, 15–38.

Declerck, L., Molenberghs, G., Aerts, M., and Ryan, L. (2000). Litter-based methods in developmental toxicity risk assessment. *Environmental and Ecological Statistics*, **7**, 57–76.

Diggle, P.J. (1983). *Statistical Analysis of Spatial Point Patterns*. Mathematics in Biology. London: Academic Press.

Diggle, P.J., Liang, K-Y., and Zeger, S.L. (1994). *Analysis of Longitudinal Data*. New York: Oxford University Press.

Edwards, A.W.F. (1972). *Likelihood*. Cambridge: Cambridge University Press.

Efron, B. (1986). Double exponential families and their use in generalized linear regression. *Journal of the American Statistical Association*, **81**, 709–721.

Efron, B. and Hinkley, D.V. (1978). Assessing the accuracy of the maximum likelihood estimator: observed versus expected Fisher information. *Biometrika*, **65**, 457–487.

Eubank, R.L. and Hart, J.D. (1992). Testing goodness-of-fit in regression via order selection criteria. *The Annals of Statistics*, **20**, 1412–1425.

Fahrmeir, L. and Tutz, G. (1994). *Multivariate Statistical Modelling Based on Generalized Linear Models*. Heidelberg: Springer-Verlag.

Fan, J. and Gijbels, I. (1996). *Local Polynomial Modelling and Its Applications*. London: Chapman & Hall.

Fan, J., Heckman, N.E., and Wand, M.P. (1995). Local polynomial kernel regression for generalized linear models and quasi-likelihood functions. *Journal of the American Statistical Association*, **90**, 141–150.

Fitzmaurice, G.M. and Laird, N.M. (1993). A likelihood-based method for analysing longitudinal binary responses. *Biometrika*, **80**, 141–151.

Fitzmaurice, G.M. and Laird, N.M. (1995). Regression models for a bivariate discrete and continuous outcome with clustering. *Journal of the American Statistical Association*, **90**, 845–852.

Fitzmaurice, G.M., Laird, N.M., and Rotnitzky, A.G. (1993). Regression models for discrete longitudinal responses. *Statistical Science*, **8**, 284–309.

Fitzmaurice, G.M., Laird, N.M., and Tosteson, T.D. (1996). Polynomial exponential models for clustered binary outcomes. *Unpublished Manuscript*.

Fitzmaurice, G.M., Molenberghs, G., and Lipsitz, S.R. (1995). Regression models for longitudinal binary responses with informative dropouts. *Journal of the Royal Statistical Society, Series B*, **57**, 691–704.

Foreman, E. K. (1991). *Survey Sampling Principles*. New York: Marcel Dekker.

Gao, F., Wahba, G., Klein, R., and Klein, B. (2001). Smoothing spline ANOVA for multivariate Bernoulli observations, with application to opthalmology data (with discussion). *Journal of the American Statistical Association*, **96**, 127–158.

Gart, J.J., Krewski, D., Lee, P.N., Tarone, R.E. and Wahrendorf, J. (1986). *Statistical Methods in Cancer Research, Volume III: The Design and Analysis of Long-Term Animal Experiments*. Lyon: International Agency for Research on Cancer.

Gaylor, D.W. (1989). Quantitative risk analysis for quantal reproductive and developmental effects. *Environmental Health Perspectives*, **79**, 243–246.

Gaylor, D.W. and Slikker, W. (1990). Risk assessment for neurotoxic effects. *NeuroToxicology*, **11**, 211–218.

Gelman, A. and Speed, T.P. (1993). Characterizing a joint probability distribution by conditionals. *Journal of the Royal Statistical Society, Series B*, **70**, 185–188.

George, E.O. and Bowman, D. (1995). A full likelihood procedure for analysing exchangeable binary data. *Biometrics*, **51**, 512–523.

George, J.D., Price, C.J., Kimmel, C.A. and Marr, M.C. (1987). The developmental toxicity of triethylene glycol dimethyl ether in mice. *Fundamental and Applied Toxicology*, **9**, 173–181.

Geyer, C.J. and Thompson, E.A. (1992). Constrained Monte Carlo maximum likelihood for dependent data (with discussion). *Journal of the Royal Statistical Society, Series B*, **69**, 657–699.

Geys, H., Molenberghs, G., and Lipsitz, S.R. (1998). A note on the comparison of pseudo-likelihood and generalized estimating equations for marginal odds ratio models. *Journal of Statistical Computation and Simulation*, **62**, 45–72.

Geys, H., Molenberghs, G., and Ryan, L.M. (1997). Pseudo-likelihood inference for clustered binary data. *Communications in Statistics: Theory and Methods*, **26**, 2743–2767.

Geys, H., Molenberghs, G. and Ryan, L. (1999). Pseudo-likelihood modelling of multivariate outcomes in developmental toxicology. *Journal of the American Statistical Association*, **94**, 34–745.

Geys, H., Molenberghs, G., and Williams, P. (2001). Analysis of clustered binary data with covariates specific to each observation. *Journal of Agricultural, Biological, and Environmental Statistics*, **6**, 340–355.

Geys, H., Regan, M.M., Catalano, P.J., and Molenberghs, G. (2001). Two latent variable risk assessment approaches for mixed continuous and discrete outcomes from developmental toxicity data. *Journal of Agricultural, Biological and Environmental Statistics*, **6**, 340–355.

Ghosh, J.K. (1994). *Higher Order Asymptotics*. NSF-CBMW Regional Conference Series in Probability and Statistics, Vol. 4.

Glonek, G.F.V. and McCullagh, P. (1995). Multivariate logistic models. *Journal of the Royal Statistical Society, Series B*, **57**, 533–546.

Goldstein, H. (1986). Multilevel mixed linear model analysis using iterative generalized least squares. *Biometrika*, **73**, 43-56.

Goldstein, H. (1991). Nonlinear multilevel models, with an application to discrete response data. *Biometrika*, **73**, 43–56.

Goldstein, H. (1995). *Multilevel Statistical Models* (2nd edition). London: Edward Arnold.

Goldstein, H. and Rasbash, J. (1996). Improved approximations for multilevel models with binary responses. *Journal of the Royal Statistical Society A*, **159**, 505–513.

Goldstein, H., Rasbash, J., Plewis, I., Draper, D., Browne, W., Yang, M. *et al* (1998). *A User's Guide to MLwiN*. London: Institute of Education.

Graubard, B.I. and Korn, E.L. (1994). Regression analysis with clustered data. *Statistics in Medicine*, **13**, 509–522.

Gregoire, T., Brillinger, D.R., Diggle, P.J., Russek-Cohen, E., Warren, W.G., and Wolfinger, R.D. (1997). *Modelling Longitudinal and Spatially Correlated Data*. Lecture Notes in Statistics 122. New York: Springer-Verlag.

Haber, F. (1924). Zur Geschichte des Gaskrieges (On the History of Gas Warfare). In: *Funf Vortrage aus den Jahren 1920-1923 (five Lectures from the years 1920-1923)*. Berlin: Springer-Verlag, pp. 76–92.

Hannan, J.F. and Tate, R.F. (1965). Estimation of the parameters for a multivariate normal distribution when one variable is dichotomized. *Biometrika*, **52**, 664–668.

Hart, J.D. (1997). *Nonparametric Smoothing and Lack-of-Fit Tests*. New York: Springer-Verlag.

Haseman, J.K. and Kupper, L.L. (1979). Analysis of dichotomous response data from certain toxicological experiments. *Biometrics*, **35**, 281–293.

Hastie, T.J. and Tibshirani, R.J. (1990). *Generalized Additive Models*. London: Chapman & Hall.

Hauck, W.W. and Donner, A. (1977). Wald's test as applied to hypotheses in logit analysis. *Journal of the American Statistical Association*, **72**, 851–853.

Heagerty, P. and Zeger, S. (1996). Marginal regression models for clustered ordinal measurements, *Journal of the American Statistical Association*, **91**, 1024–1036.

Hedeker, D. and Gibbons, R.D. (1993). *MIXOR: a Computer Program for Mixed Effects Ordinal Probit and Logistic Regression*. Chicago: University of Illinois.

Hedeker, D. and Gibbons, R.D. (1994). A random-effects ordinal regression model for multilevel analysis. *Biometrics*, **50**, 933–944.

Hedeker, D. and Gibbons, R.D. (1996). MIXOR: a computer program for mixed-effects ordinal regression analysis. *Computer Methods and Programs in Biomedicine*, **49**, 157–176.

Herson, J. (1975). Fieller's theorem versus the delta method for significance intervals for ratios. *Journal of Statistical Computations and Simulations*, **3**, 265–274.

Hinde, J. and Demétrio, C.G.B. (1998). Overdispersion: models and estimation. *Computational Statistics and Data Analysis*, **27**, 151–170.

Hinkley, D. V., Reid, N., and Snell, E. J. (1991). *Statistical Theory and Modelling*. In Honour of Sir David Cox, FRS. London: Chapman & Hall.

Horton, N.J., Bebchuk, J.D., Jones, C.L., Lipsitz, S.R., Catalano, P.J., Zahner, G.E.P., and Fitzmaurice, G.M. (1999). Goodness-of-fit for GEE: an example with mental health service utilization. *Statistics in Medicine*, **18**, 213–222.

Hosmer, D.W. and Lemeshow, S. (1989). *Applied Logistic Regression*. New York: Wiley.

Hosmer, D.W., Hosmer, T., Lemeshow, S., and le Cessie, S. (1997). A comparison of goodness-of-fit tests for the logistic regression model. *Statistics in Medicine*, **16**, 965–980.

Huber, P.J. (1981). *Robust Statistics*. New York: John Wiley.

Johnson, R.A. and Wichern, D.W. (1992). *Applied Multivariate Statistical Analysis* (3rd ed.). Englewood Cliffs, NJ: Prentice-Hall.

Kavlock, R.J., Allen, B.C., Faustman, E.M., and Kimmel, C.A. (1995). Dose–response assessment for developmental toxicity. IV. Benchmark dose for fetal weight changes. *Fundamental and Applied Toxicology*, **26**, 211–222.

Kent, J.T. (1982). Robust properties of likelihood ratio tests. *Biometrika*, **69**, 19–27.

Kimmel, C.A. and Gaylor, D.W. (1988). Issues in qualitative and quantitative risk analysis for developmental toxicology. *Risk Analysis*, **8**, 15–19.

Kimmel, G.L., Williams, P.L., Kimmel, C.A., Claggett, T.W., and Tudor, N. (1994). The effects of temperature and duration of exposure on in vitro development and response-surface modelling of their interaction. *Teratology*, **49**, 366–367.

Kish, L. (1965). *Survey Sampling*. New York: John Wiley.

Klein, R., Klein, B.E.K., Moss, S.E., Davis, M.D., and DeMets, D.L. (1984). The Wisconsin epidemiologic study of diabetic retinopathy: II. Prevalence and risk of diabetic retinopathy when age at diagnosis is less than 30 years. *Archives of Ophthalmology*, **102**, 520–526.

Kleinman, J. (1973). Proportions with extraneous variance: single and independent samples. *Journal of the American Statistical Association*, **68**, 46–54.

Kodell, R.L. and West, R.W. (1993). Upper confidence limits on excess risk for quantitative responses. *Risk Analysis*, **13**, 177–182.

Kreft, I. and de Leeuw, J. (1998). *Introducing Multilevel Modeling*. London: Sage Publications.

Krewski, D. and Van Ryzin, J. (1981). Dose-response models for quantal response toxicity data. In: M. Csorgo, D. Dawson, J.N.K. Rao and E. Saleh (Eds.), *Statistics and Related Topics*. New York: North-Holland, pp. 201–231.

Krewski, D. and Zhu, Y. (1995). A simple data transformation for estimating benchmark doses in developmental toxicity experiments. *Risk Analysis*, **15**, 29–40.

Krzanowski, W.J. (1988). *Principles of Multivariate Analysis*. Oxford: Clarendon Press.

Kuk, A.Y.C. (1995). Asymptotically unbiased estimation in generalised linear models with random effects. *Journal of the Royal Statistical Society B*, **57**, 395–407.

Kupper, L.L. and Haseman, J.K. (1978). The use of a correlated binomial model for the analysis of certain toxicology experiments. *Biometrics*, **34**, 69-76.

Kupper, L.L., Portier, C., Hogan, M.D., and Yamamoto, E. (1986). The impact of litter effects on dose-response modeling in teratology. *Biometrics*, **42**, 85–98.

Lang, J.B. and Agresti, A. (1994). Simultaneously modeling joint and marginal distributions of multivariate categorical responses. *Journal of the American Statistical Association*, **89**, 625–632.

le Cessie, S. and van Houwelingen, J.C. (1991). A goodness-of-fit test for binary regression models, based on smoothing methods. *Biometrics*, **47**, 1267–1282.

le Cessie, S. and van Houwelingen, J.C. (1993). Building logistic models by means of a non parametric goodness of fit test: a case study. *Statistica Neerlandica*, **47**, 97–109.

le Cessie, S. and Van Houwelingen J.C. (1994). Logistic regression for correlated binary data. *Applied Statistics*, **43**, 95–108.

Ledwina, T. (1994). Data-driven version of Neyman's smooth test of fit. *Journal of the American Statistical Association*, **89**, 1000–1005.

Lefkopoulou, M., Moore, D., and Ryan, L. (1989). The analysis of multiple binary outcomes: application to rodent teratology experiments. *Journal of the American Statistical Association*, **84**, 810–815.

Lefkopoulou, M., Rotnitzky, A., and Ryan, L. (1996). Trend tests for clustered data. In: *Statistics in Toxicology*. B.J.T. Morgan (Ed). New York: Oxford University Press Inc., pp. 179–97

Lefkopoulou, M. and Ryan, L. (1993). Global tests for multiple binary outcomes. *Biometrics*, **49**, 975–988.

Lemeshow, S. and Levy, P. (1999). *Sampling of Populations: Methods and Applications* (3rd Edition). New York: John Wiley.

Liang, K.-Y. and Hanfelt, J. (1994). On the use of the quasi-likelihood method in teratological experiments. *Biometrics*, **50**, 872–880.

Liang, K.Y. and and Self, S. (1996). On the asymptotic behavior of the pseudolikelihood ratio test statistic. *Journal of the Royal Statistical Society, Series B*, **58**, 785–796.

Liang, K.-Y. and Zeger, S.L. (1986). Longitudinal data analysis using generalized linear models. *Biometrika*, **73**, 13–22.

Liang, K.-Y. and Zeger, S.L. (1989). A class of logistic regression models for multivariate binary time series. *Journal of the American Statistical Association*, **84**, 447–451.

Liang, K.Y., Zeger, S.L., and Qaqish, B. (1992). Multivariate regression analyses for categorical data. *Journal of the Royal Statistical Society, Series B*, **54**, 3–40.

Lin, X. and Breslow, N.E. (1996). Bias correction in generalized linear mixed models with multiple components of dispersion. *Journal of the American Statistical Association*, **91**, 1007–1016.

Lindsay, B. and Lesperance, M.L. (1995). A review of semiparametric mixture models. *Journal of Statistical Planning and Inference*, **47**, 29–39.

Lindström, P., Morrisey, R.E., George, J.D., Price, C.J., Marr, M.C., Kimmel, C.A., and Schwetz, B.A. (1990). The developmental toxicity of orally administered theophylline in rats and mice. *Fundamental and Applied Toxicology*, **14**, 167–178.

Lipsitz, S.R., Fitzmaurice, G.M., and Molenberghs, G. (1996). Goodness-of-fit tests for ordinal response regression models. *Applied Statistics*, **45**, 175–190.

Lipsitz, S.R., Laird, N.M., and Harrington, D.P. (1991). Generalized estimating equations for correlated binary data: using the odds ratio as a measure of association. *Biometrika*, **78**, 153–160.

Little, R.J.A. and Rubin, D.B. (1987). *Statistical Analysis with Missing Data*. New York: John Wiley.

Little, R.J.A. and Schluchter, M.D. (1985). Maximum likelihood estimation for mixed continuous and categorical data with missing values. *Biometrika*, **72**, 497–512.

Longford, N.T. (1993). *Random Coefficient Models*. London: Oxford University Press.

McCullagh, P. (1987). *Tensor Methods in Statistics.* London, New York: Chapman & Hall.

McCullagh, P. and Nelder, J.A. (1989). *Generalized Linear Models.* London: Chapman & Hall.

Meester, S.G. and MacKay, J. (1994). A parametric model for clustered correlated data. *Biometrics,* **50**, 954–963.

Mehta, C.R., Patel, N.R., and Senchaudhuri, P. (1992). Exact stratified linear rank tests for ordered categorical and binary data. *Journal of Computational and Graphical Statistics,* **1**, 21–40.

Mehta, C.R., Patel, N.R., and Senchaudhuri, P. (2000). Efficient Monte Carlo methods for conditional logistic regression. *Journal of the American Statistical Association,* **95**, 99–108.

Mendell, N.R. and Elston, R.C. (1974). Multifactorial qualitative traits: genetic analysis and prediction of recurrence risks. *Biometrics,* **30**, 41–57.

Mesbah, M., Cole, B.F., and Ting Lee, M.-L. (2001). *Statistical Methods for Quality of Life Studies: Design, Measurements, and Analysis.* New York: Kluwer Academic Publisher.

Molenberghs, G., Declerck, L., and Aerts, M. (1998). Misspecifying the likelihood for clustered binary data. *Computational Statistics and Data Analysis,* **26**, 327–349.

Molenberghs, G., Geys, H., and Buyse, M. (2001). Evaluation of surrogate endpoints in randomized experiments with mixed discrete and continuous outcomes. *Statistics in Medicine,* **20**, 3023–3038.

Molenberghs, G., Geys, H., Declerck, L., Claeskens, G., and Aerts, M. (1998). Analysis of clustered multivariate data from developmental toxicity studies. In: *Proceedings in Computational Statistics,* R. Payne and P. Green (eds), pp. 3–14.

Molenberghs, G. and Lesaffre, E. (1994). Marginal modelling of correlated ordinal data using a multivariate Plackett distribution. *Journal of the American Statistical Association,* **89**, 633–644.

Molenberghs, G. and Lesaffre, E. (1999). Marginal modelling of multivariate categorical data. *Statistics in Medicine,* **18**, 2237–2255.

Molenberghs, G. and Ritter, L. (1996). Methods for analyzing multivariate binary data, with association between outcomes of interest. *Biometrics,* **52**, 1121–1133.

Molenberghs, G. and Ryan, L.M. (1999). Likelihood inference for clustered multivariate binary data. *Environmetrics,* **10**, 279–300.

Moore, D.S. and Spruill, M.C. (1975). Unified large-sample theory of general chi-squared statistics for tests of fit. *Annals of Statistics,* **3**, 599–616.

Morgan, B.J.T. (1992) *Analysis of Quantal Response Data.* London: Chapman & Hall.

Müller, H.-G. and Schmitt, T. (1988). Kernel and probit estimates in quantal bioassay. *Journal of the American Statistical Association*, **83**, 750–759.

Nelder, J.A. and Wedderburn, R.W.M. (1972). Generalized linear models. *Journal of the Royal Statistical Society, Series B*, **135**, 370–384.

Neuhaus, J.M. (1992). Statistical methods for longitudinal and clustered designs with binary responses. *Statistical Methods in Medical Research*, **1**, 249-273.

Neuhaus, J.M. (1993). Estimation efficiency and tests of covariate effects with clustered binary data. *Biometrics*, **49**, 989–996.

Neuhaus, J.M., Kalbfleisch, J.D., and Hauck, W.W. (1991). A comparison of cluster-specific and population-averaged approaches for analyzing correlated binary data. *International Statistical Review*, **59**, 25–35.

Neuhaus, J.M. and Kalbfleisch, J.D. (1998). Between- and within-cluster covariate effects in the analysis of clustered data. *Biometrics*, **54**, 638–645.

Neuhaus, J.M. and Lesperance, M.L. (1996). Estimation efficiency in binary mixed-effects model setting. *Biometrika*, **83**, 441-446.

Neyman, J. (1937). 'Smooth' test for goodness of fit. *Skandinavisk Aktuarietidskrift*, **20**, 149–199.

NIOSH (1983). U.S. Department of Health and Human Services, Public Health Service, Center for Disease Control, National Institute for Occupational Safety and Health (1983). *Current Intelligence Bulletin 39: Glycol Ethers 2-Methoxyethanol and 2-Ethoxyethanol*.

Ochi, Y. and Prentice, R.L. (1984). Likelihood inference in a correlated probit regression model. *Biometrika*, **71**, 531–543.

Olkin, I. and Tate, R.F. (1961). Multivariate correlation models with mixed discrete and continuous variables. *Annals of Mathematical Statistics*, **32**, 448–465 (with correction in **36**, 343–344).

Pendergast, J., Gange, S.J., Newton, M.A., Lindstrom, M.., Palta, M., and Fisher, M. (1996). A survey of methods for analyzing clustered binary response data. *International Statistical Review*, **64**, 89–118.

Pfeffermann, D., Skinner, C.J., Holmes, D.J., Goldstein, H., and Rasbash, J. (1998). Weighting for unequal selection probabilities in multilevel models. *Journal of the Royal Statistical Society, Series B*, **60**, 23–40.

Phillips, P.C.B. and Park, J.Y. (1988). On the formulation of Wald tests of nonlinear restrictions, *Econometrica*, 56, 1065–1083.

Piantadosi, S. (1997) *Clinical Trials: A Methodologic Perspective*. New York: John Wiley.

Plackett, R.L. (1965). A class of bivariate distributions. *Journal of the American Statistical Association*, **60**, 516–522.

Prentice, R.L. (1988). Correlated binary regression with covariates specific to each binary observation. *Biometrics*, **44**, 1033-1048.

Prentice, R.L. and Zhao, L.P. (1991). Estimating equations for parameters in means and covariances of multivariate discrete and continuous responses. *Biometrics*, **47**, 825–839.

Price, C.J., Kimmel, C.A., George, J.D., and Marr, M.C. (1987). The developmental toxicity of diethylene glycol dimethyl ether in mice. *Fundamental and Applied Toxicology*, **8**, 115–126.

Price, C. J., Kimmel, C. A., Tyl, R. W., and Marr, M. C. (1985). The developmental toxicity of ethylene glycol in mice. *Toxicology and Applied Pharmacology*, **81**, 113–127.

Rai, K. and Van Ryzin, J. (1985). A dose-response model for teratological experiments involving quantal responses. *Biometrics*, **47**, 825–839.

Rao, C.R. (1973). *Linear Statistical Inference and Its Applications* (2nd ed.). New York: John Wiley.

Rao, J.N.K. and Scott, A.J. (1987). On simple adjustments to chi-square tests with sample survey data. *Annals of Statistics*, **15**, 385–397.

Regan, M.M. and Catalano, P.J. (1999a). Likelihood models for clustered binary and continuous outcomes: application to developmental toxicology. *Biometrics*, **55**, 760–768.

Regan, M.M. and Catalano, P.J. (1999b). Bivariate dose-response modeling and risk estimation in developmental toxicology. *Journal of Agricultural, Biological and Environmental Statistics*, **4**, 217–237.

Regan, M.M. and Catalano, P.J. (2000). Regression models for mixed discrete and continuous outcomes with clustering. *Risk Analysis*, **20**, 363–376.

Ripley, B.D. (1981). *Spatial Statistics*. New York: John Wiley.

Roberts, W.C. and Abernathy, C.O. (1996). Risk assessment: principles and methodologies. In: A. Fan and L.W. Chang (Eds.), *Toxicology and Risk Assessment, Principles, Methods and Applications*, New York: Marcel Dekker, pp. 245–270.

Roberts, J., Rao, J.N.K., and Kumar, S. (1987). Logistic regression analysis of sample survey data. *Biometrika*, **74**, 1–12.

Rodríguez, G. and Goldman, N. (1995). An assessment of estimation procedures for multilevel models with binary responses. *Journal of the Royal Statistical Society A*, **158**, 73–89.

Rosner, B. (1984). Multivariate methods in opthalmology with applications to other paired-data. *Biometrics*, **40**, 1025–1035.

Rotnitzky, A. and Jewell, N.P. (1990). Hypothesis testing of regression parameters in semiparametric generalized linear models for cluster correlated data. *Biometrika*, **77**, 485–497.

Rotnitzky, A. and Wypij, D. (1994). A note on the bias of estimators with missing data. *Biometrics*, **50**, 1163–1170.

Rowe, V.K., Glycols (1963). In: F.A. Patty (Ed.) *Industrial Hygiene and Toxicology,* New York: Wiley–Interscience, New York, Vol. 99, pp. 1497–1536.

Royston, P. and Altman, D.G. (1994). Regression using fractional polynomials of continuous covariates: parsimonious parametric modelling. *Applied Statistics,* **43**, 429–468.

Royston, P. and Wright, E. (1998). A method for estimating age-specific reference intervals ("normal ranges") based on fractional polynomials and exponential transformation. *Journal of the Royal Statistical Society, Series A,* **161**, 79–101.

Ryan, L. M. (1992). Quantitative risk assessment for developmental toxicity. *Biometrics,* **48**, 163–174.

Ryan, L.M. (1992). The use of generalized estimating equations for risk assessment in developmental toxicity. *Risk Analysis,* **12**, 439-447.

Ryan, L.M., Catalano, P.J., Kimmel, C.A., and Kimmel, G.L. (1991). Relationship between fetal weight and malformation in developmental toxicity studies. *Teratology,* **44**, 215-223.

Ryan, L. and Molenberghs G. (1999). Statistical methods for developmental toxicity: analysis of clustered multivariate binary data. In: *Uncertainty in the Risk Assessment of Environmental and Occupational Hazards: An International Workshop (Annals of the New York Academy of Sciences).* A.J. Bailer, C. Maltoni, and J.C. Bailar (Eds). New York: New York Academy of Sciences.

Salsburg, D. (1996). Estimating dose-response for toxic endpoints. In: Morgan, B.J.T. (Ed.) *Statistics in Toxicology.* Oxford: Oxford University Press.

Sammel, M.D., Ryan, L.M., and Legler, J.M. (1997). Latent variable models for mixed discrete and continuous outcomes. *Journal of the Royal Statistical Society, Series B,* **59**, 667–678.

SAS Institute Inc. (1997) *SAS/STAT Software: Changes and Enhancements through Release 6.12.* Cary, NC: SAS Institute Inc.

Sauerbrei, W. and Royston, P. (1999). Building multivariable prognostic and diagnostic models: transformation of the predictors by using fractional polynomials. *Journal of the Royal Statistical Society, Series A,* **162**, 71–94.

Schaumberg, D.A., Moyes, A.L., Gomes, J.A.P., and Dana, M.R. (1999). Corneal transplantation in young children with congenital hereditary endothelial dystrophy. *American Journal of Opthalmology,* **127**, 373–378.

Scheaffer, R.L., Mendenhall, W., and Ott, L. (1990). *Elementary Survey Sampling.* Boston: PWS-Kent.

Schwarz, G. (1978). Estimating the dimension of a model. *The Annals of Statistics,* **6**, 461–464.

Scientific Committee of the Food Safety Council, Proposed system for food safety assessment (1979, 1980). *Food and Cosmetic Toxicology*, **16**, Supplement 2, 1–136. Revised report published June 1980 by the Food Safety Council, Washington, DC.

Seber, G.A.F. (1984). *Multivariate Observations*. New York: John Wiley.

Serfling, R.J. (1980). *Approximation Theorems of Mathematical Statistics*. New York: John Wiley.

Shao, J. and Tu, D. (1995). *The Jacknife and the Bootstrap*. New York: Springer-Verlag.

Shepard, T.H., Mackler, B., and Finch, C.A. (1980). Reproductive studies in the iron-deficient rat. *Teratology*, **22**, 329–334.

Shiota, K., Chou, M.J., and Nishimura, H. (1980). Embryotoxic effects of di-2-ethylhexyl phthalate (DEHP) and di-*n*-butyl phthalate (DBP) in mice. *Environmental Research*, **22**, 245–253.

Simonoff, J.S. (1996). *Smoothing Methods in Statistics*. New York: Springer-Verlag.

Skellam, J.G. (1948). A probability distribution derived from the binomial distribution by regarding the probability of success as variable between the sets of trials. *Journal of the Royal Statistical Society, Series B*, **10**, 257–261.

Skinner, C.J., Holt, D., and Smith, T.M.F. (1989). *Analysis of Complex Surveys*. New York: John Wiley.

Smith, D.M., Robertson, B., and Diggle, P.J. (1996). *Object-oriented Software for the Analysis of Longitudinal Data in S*. Technical Report MA 96/192. Department of Mathematics and Statistics, University of Lancaster, LA1 4YF, United Kingdom.

Staniswalis, J.G. (1989). The kernel estimate of a regression function in likelihood-based models. *Journal of the American Statistical Association*, **84**, 276–283.

Staniswalis, J.G. and Cooper, V. (1988). Kernel estimates of dose response. *Biometrics*, **44**, 1103–1119.

StatXact, Version 5.0 (2001). Cytel Software Corporation, Cambridge, MA, USA.

Stiratelli, R., Laird, N., and Ware, J.H. (1984). Random-effects model for serial observations with binary response. *Biometrics*, **40**, 961–971.

Tanner, M.A. (1991). *Tools for Statistical Inference: Observed Data and Data Augmentation Methods*. Berlin: Springer-Verlag.

Tate, R.F. (1954). Correlation between a discrete and a continuous variable. *Annals of Mathematical Statistics*, **25**, 603–607.

Tate, R.F. (1955). The theory of correlation between two continuous variables when one is dichotomized. *Biometrika*, **42**, 205–216.

Ten Have, T.R., Landis, R., and Weaver, S.L. (1995). Association models for periodontal disease progression: a comparison of methods for clustered binary data. *Statistics in Medicine*, **14**, 413–429.

Thélot, C. (1985). Lois logistiques à deux dimensions. *Annales de l'Insée*, **58**, 123–149.

Tibshirani, R. and Hastie, T. (1987). Local likelihood estimation. *Journal of the American Statistical Association*, **82**, 559–568.

ToxTools, Version 1.0 (2001). Cytel Software Corporation, Cambridge, MA, USA.

Tsiatis, A.A. (1980). A note on a goodness-of-fit test for the logistic regression model. *Biometrika*, **67**, 250–251.

Tyl, R.W., Price, C.J., Marr, M.C., and Kimmel, C.A. (1988). Developmental toxicity evaluation of dietary di(2-ethylhexyl)phthalate in Fischer 344 rats and CD-1 mice. *Fundamental and Applied Toxicology*, **10**, 395–412.

U.S. Environmental Protection Agency (1991). *Guidelines for Developmental Toxicity Risk Assessment. Federal Register*, **56**, 63798–63826.

U.S. Environmental Protection Agency (1995). *The Use of the Benchmark Dose Approach in Health Risk Assessment*. Washington, D.C.: Office of Research and Development, EPA/630/R-94/007.

Verbeke, G. and Molenberghs, G. (2000). *Linear Mixed Models for Longitudinal Data*. New York: Springer-Verlag.

Verloove, S.P. and Verwey, R.Y. (1988). *Project on preterm and small-for-gestational age infants in the Netherlands, 1983 (Thesis, University of Leiden)*. University Microfilms International, Ann Arbor, Michigan, USA, no. 8807276.

Viraswami, K. and Reid, N. (1996). Higher-order asymptotics under model misspecification. *Canadian Journal of Statistics*, **24**, 263–278.

Wedderburn, R.W.M. (1974). Quasi-likelihood functions, generalized linear models, and the Gauss-Newton method. *Biometrika*, **61**, 439–447.

Weinberg, A., Betensky, R., Zhang, L., and Ray, G. (1998). Effect of shipment and storage, anticoagulant, and cell separation on lymphocyte proliferation assays in HIV-infected patients. *Clinical and Diagnostic Laboratory Immunology*, **5**, 804–807.

Weller, E.A., Catalano, P.J., and Williams, P.L. (1995). Implications of developmental toxicity study design for quantitative risk assessment. *Risk Analysis*, **15**, 567–574.

Welsh, A.H. (1997). *Aspects of Statistical Inference*. New York: John Wiley.

White, H. (1982). Maximum likelihood estimation of misspecified models. *Econometrica*, **50**, 1–26.

White, H. (1994). *Estimation, Inference and Specification Analysis*. Cambridge: Cambridge University Press.

Wild, C.J. and Yee, T.W. (1996). Additive extensions to generalized estimating equation models. *Journal of the Royal Statistical Society, Series B*, **58**, 711–725.

Williams, D.A. (1975). The analysis of binary responses from toxicological experiments involving reproduction and teratogenicity. *Biometrics*, **31**, 949–952.

Williams, P.L. and Ryan, L.M. (1996). Dose response models for developmental toxicity. In: Hood, R.D. (Ed.), *Handbook of Developmental Toxicology*. Boca Raton: CRC Press, pp. 609–640.

Williamson, J.M., Lipsitz, S.R., and Kim, K.M. (1997). GEECAT and GEEGOR: Computer programs for the analysis of correlated categorical response data. *Technical Report*.

Windholz, M. (1983). *The Merck Index: An Encyclopedia of Chemicals, Drugs, and Biologicals* (M. Windholz, Ed.), 10th ed., Merck and Co., Rahway, NJ.

Wolfinger, R. and O'Connell, M. (1993). Generalized linear mixed models: a pseudo-likelihood approach. *Journal of Statistical Computing and Simulation* **48**, 233–243.

Zeger, S.L. and Karim, M.R. (1991). Generalised linear models with random effects: a Gibbs sampling approach. *Journal of the American Statistical Association*, **86**, 79–102.

Zeger, S.L. and Liang, K.-Y. (1986). Longitudinal data analysis for discrete and continuous outcomes. *Biometrics*, **42**, 121–130.

Zeger, S.L., Liang, K-Y., and Albert, P.A. (1988). Models for longitudinal data: a generalized estimating equations approach. *Biometrics*, **44**, 1049–1060.

Zhao, L.P. and Prentice, R.L. (1990). Correlated binary regression using a quadratic exponential model. *Biometrika*, **77**, 642–648.

Zhao, L.P., Prentice, R.L., and Self, S.G. (1992). Multivariate mean parameter estimation by using a partly exponential model. *Journal of the Royal Statistical Society B*, **54**, 805–811.

Zhu and Fung (1996). Statistical methods in developmental toxicity risk assessment. In: A. Fan and L.W. Chang (Eds.), *Toxicology and Risk Assessment, Principles, Methods and Applications*, New York: Marcel Dekker, pp. 413–446.

Index